T0327467

Airline Network Planning and Scheduling

Wiley Series in
Operations Research and Management Science

Operations Research and Management Science (ORMS) is a broad, interdisciplinary branch of applied mathematics concerned with improving the quality of decisions and processes and is a major component of the global modern movement toward the use of advanced analytics in industry and scientific research. The *Wiley Series in Operations Research and Management Science* features a broad collection of books that meet the varied needs of researchers, practitioners, policy makers, and students who use or need to improve their use of analytics. Reflecting the wide range of current research within the ORMS community, the Series encompasses application, methodology, and theory and provides coverage of both classical and cutting-edge ORMS concepts and developments. Written by recognized international experts in the field, this collection is appropriate for students as well as professionals from private and public sectors including industry, government, and nonprofit organization who are interested in ORMS at a technical level. The Series is composed of four sections: Analytics; Decision and Risk Analysis; Optimization Models; and Stochastic Models.

Advisory Editors • Optimization Models
Lawrence V. Snyder, Lehigh University
Ya-xiang Yuan, Chinese Academy of Sciences

Founding Series Editor
James J. Cochran, University of Alabama

Analytics
Yang and Lee • *Healthcare Analytics: From Data to Knowledge to Healthcare Improvement*

Forthcoming Titles
Attoh-Okine • *Big Data and Differential Privacy: Analysis Strategies for Railway Track Engineering*
Kong and Zhang • *Decision Analytics and Optimization in Disease Prevention and Treatment*

Decision and Risk Analysis
Barron • *Game Theory: An Introduction,* Second Edition
Brailsford, Churilov, and Dangerfield • *Discrete-Event Simulation and System Dynamics for Management Decision Making*
Johnson, Keisler, Solak, Turcotte, Bayram, and Drew • *Decision Science for Housing and Community Development: Localized and Evidence-Based Responses to Distressed Housing and Blighted Communities*
Mislick and Nussbaum • *Cost Estimation: Methods and Tools*

Forthcoming Titles
Aleman and Carter • *Healthcare Engineering*

Optimization Models
Abdelghany • *Airline Network Planning and Scheduling*
Ghiani, Laporte, and Musmanno • *Introduction to Logistics Systems Management,* Second Edition

Forthcoming Titles
Smith • *Learning Operations Research Through Puzzles and Games*
Tone • *Advances in DEA Theory and Applications: With Examples in Forecasting Models*

Stochastic Models
Ibe • Random Walk and Diffusion Processes

Forthcoming Titles
Donohue, Katok, and Leider • *The Handbook of Behavioral Operations*
Matis • *Applied Markov Based Modelling of Random Processes*

Airline Network Planning and Scheduling

Ahmed Abdelghany
Khaled Abdelghany

This edition first published 2019
© 2019 John Wiley & Sons, Inc.

Edition History
John Wiley & Sons, Inc. (1e, 2009)

Registered Office(s)
John Wiley & Sons, Inc., 111 River Street, Hoboken, NJ 07030, USA

Editorial Office
111 River Street, Hoboken, NJ 07030, USA

For details of our global editorial offices, customer services, and more information about Wiley products visit us at www.wiley.com.

Wiley also publishes its books in a variety of electronic formats and by print-on-demand. Some content that appears in standard print versions of this book may not be available in other formats.

Library of Congress Cataloging-in-Publication Data

Names: Abdelghany, Ahmed, author. | Abdelghany, Khaled, author.
Title: Airline network planning and scheduling / Ahmed Faissal Said Abdelghany, Khaled Faissal Said Abdelghany.
Description: 1st edition. | Hoboken,NJ : John Wiley & Sons, Inc., 2019. | Series: Wiley series in operations research and management science | Includes bibliographical references and index. |
Identifiers: LCCN 2018021895 (print) | LCCN 2018029175 (ebook) | ISBN 9781119275879 (Adobe PDF) | ISBN 9781119275886 (ePub) | ISBN 9781119275862 (hardcover)
Subjects: LCSH: Airport slot allocation–Planning. | Airlines–Management. | Aeronautics, Commercial–Planning. | Operations research.
Classification: LCC HE9797.4.S56 (ebook) | LCC HE9797.4.S56 A23 2019 (print) | DDC 387.7/4042–dc23
LC record available at https://lccn.loc.gov/2018021895

Cover image: © iStock.com/vectorscore
Cover design by Wiley

Set in 10/12pt Warnock by SPi Global, Pondicherry, India

Printed in the United States of America

10 9 8 7 6 5 4 3 2 1

Contents

List of Figures

List of Tables

Preface

The task of network planning and scheduling is critical for the success of commercial airlines. It involves answering important questions on where to fly, when, and at what capacity. It is responsible for developing an operational flight schedule and determining resources required to operate this schedule. The network planning and scheduling task plays a significant role in determining the airline's brand, profitability, and operational performance. Regardless of the airline's size, the problem of network planning and scheduling is complex, which requires adopting high-level solution methodologies. The problem entails optimizing several interrelated decisions, handling conflicting objectives (profitability and operations performance), and responding timely to external factors such as change in economy, competition, and special events.

The book *Airline Network Planning and Scheduling* is a comprehensive textbook for students who seek career in the airline industry and in particular in airline network planning and scheduling. The book is primarily targeting graduate-level students. However, many chapters are also suitable to be taught at the undergraduate level. The book is targeting students who study degrees in business administration, operations management, transportation planning, operations research, management science, and logistics and is intended to be self-contained. Readers with basic background in spreadsheets, modeling, and mathematical programming will be able to fully follow all the content of the book. The book is also considered a valuable reference for practitioners to get deep understanding of the different aspects of airline network planning and scheduling process. It answers the why and how questions that practitioners usually face on a daily basis.

The book is organized in seven sections. Section 1 is an introductory section that provides the basic definitions and concepts that are used in airline network planning and scheduling. These concepts include the main brands of airlines, structure of airline networks, airline schedule planning decisions, measures of performance in the airline industry, freedom of airline services, and slot allocation. Section 2 gives an introduction to the concepts of competition in the airline industry, introducing new services/routes and estimating market share.

Section 3 describes the fleet assignment problem, which aims at determining the optimal fleet type for each flight in the schedule. The section presents the basic fleet assignment model (FAM). Section 4 extends the basic FAM to present the schedule adjustment problem, which involves scheduling decisions such as flight deletion, flight addition, and flight departure time adjustment. Section 5 provides another practical variation of the fleet assignment problem by introducing the itinerary-based fleet assignment model (IFAM). This model is more suitable for airlines that adopt hub-and-spoke network structure with significant number of connecting passengers. Section 6 presents the ISD-IFAM, which integrated schedule design problem using the IFAM. Finally, Section 7 provides a discussion on schedule robustness. It explains the different measures that airline network planners may consider to achieve an operationally robust schedule. The book provides several examples to further illustrate the applications of the presented models. These examples enable the reader to understand the details of each of the presented models and develop hands-on experience on applications.

Section 1

1

Brands of Airlines

Different types of airline brands find its way in the highly competitive airline market serving customers with diverse travel preferences. It is almost impossible to come across two identical airlines. There are several characteristics based on which airlines are defined and classified. These characteristics play a significant role in the undertaking of the airline's network planning and scheduling. Generally, airlines could be classified and branded based on schedule availability (chartered and scheduled airlines), size and domain of service (e.g. regional, national, and major), business model (legacy, low cost, and ultralow cost), ownership (publicly, privately, and mixed owned), network structure (hub-and-spoke and point-to-point), locality and network coverage (domestic and international), and transport service type (cargo only and passenger and cargo airlines). A combination of these characteristics defines the brand of the airline.

1.1 Schedule Availability

1.1.1 Charter Airlines

Charter airlines provide on-demand and as-needed service for a group of customers (or cargo) that share travel itinerary including origin, destination, and date of travel. Thus, these airlines do not provide a published flight schedule with pricing and seat availability. Any party that is interested in a customized travel service contacts the charter airline (or its agents) and agrees on initiating the flight. The agreement usually involves deciding on date of travel, origin, destination, pricing, and number of seats required. Charter airlines focus on serving customers with customized needs such as urgent or time-sensitive travel, privacy, and need of flexibility (French 1995; Buck and Lei 2004). Charter airlines are also used in peak travel periods or during special events, when demand temporarily increases above available capacity. Example of these events include pilgrimage seasons (e.g. pilgrimage in Saudi Arabia), major

Airline Network Planning and Scheduling, First Edition. Ahmed Abdelghany and Khaled Abdelghany.
© 2019 John Wiley & Sons, Inc. Published 2019 by John Wiley & Sons, Inc.

sports champions and games, tour operations for tourism, etc. (Karlaftis and Papastavrou 1998; Williams 2001). Some charter airlines may also follow a set schedule, with a certain number of flights per week to desirable tourist destinations. Even in this case, individual passengers do not book their tickets for the flight; they are sold in bulk to a tour company or a travel agency. The departure time of these flights are usually flexible based on the preferences of the service requesters.

1.1.2 Scheduled Airlines

The scheduled airlines plan and publish flight schedules on regular basis ahead of the scheduled dates of operation. The published schedule involves detailed information on served markets, departure and arrival times of each flight, available seat capacity, pricing and associated ticket restrictions, and fees. The flight schedule is usually published a few months before its implementation. During this period, which is also known as the booking horizon, prospective travelers can shop among the available itineraries offered by all airlines and select (book) the one that meets their travel preferences. The published schedule might be subject to several minor modifications and amendments made by the airline during the booking horizon. The seat availability is updated when a new booking is materialized on any itinerary. Pricing and ticket restriction information are subject to more frequent updates, as airlines respond to changes in demand and competition with other airlines. The published schedule is made available to customers and travel agents. Thus, customers can buy their tickets directly from the airline or through travel agents.

1.2 Size and Domain of Service

1.2.1 Major Airlines

The term major airlines is used in the United States to define large airlines that generate an annual operating revenue of one billion dollars or more. These airlines typically have a large network that covers most of the major destinations in the domestic market with reasonable service frequency. Major airlines usually connect its hubs and major destinations to other international major destinations. Most major airlines operate a mixed fleet of small-, medium-, and large-sized airplanes.

1.2.2 National Airlines

National airlines are scheduled airlines with annual operating revenues between $100 million and $1 billion. These airlines might serve certain regions of the country.

They may also provide long-distance routes and serve some international destinations. National airlines usually operate medium-sized airplanes.

1.2.3 Regional Airlines

As the name suggests, regional airlines focus their service on a region, where they carry traffic from small communities and connect them to a major nearby destination (Truitt and Haynes 1994; Forbes and Lederman 2007). Regional airlines typically use small aircraft to serve passengers originating from these small communities, as demand generated from these communities is not high enough to fill a large aircraft. Regional airlines could be operating as an independent airline under their own brand or as an affiliated airline operating under the brand of another airline. However, in most cases regional airlines operate as an affiliated airline, where they sign contracts with one or more major airlines to work under its brand. Under this contract, regional airline carries traffic from small communities and connect them to a hub served by the major airline. The major airline can then connect this traffic over its network to other destinations. In this case, the regional airline is known as a feeder airline, as it feeds traffic to the flights of the major airline. When the regional airline serves as an affiliated (feeder) airline, the regional airline only provides the aircraft and the crew to operate the flight. The schedule of flights as well as seat availability and pricing is determined by the major airline.

1.3 Business Model

1.3.1 Legacy Airlines (or Mainline)

Legacy airlines (or carriers) is a term used to describe major airlines in the United States. It originally refers to airlines that had established interstate service by the time of the route liberalization, which was permitted by the Airline Deregulation Act of 1978 (Reynolds-Feighan 1998; Goetz and Vowles 2009). Legacy airlines typically provide higher quality services, compared with other airlines. For example, a legacy airline typically offers first class and business class seating, convenient schedule in business markets, frequent-flyer program, and exclusive airport lounges. Many legacy airlines are also members of worldwide airline alliance. Being part of a worldwide alliance allows the airline to expand its service via its partner airlines. In addition, legacy airlines generally have better cabin services, where some of them offer free meal service on long-haul flights, free snacks, and in-flight entertainment. The term legacy airline is used in many cases in exchange of the terms major or mainline airline.

1.3.2 Low-cost Airlines

Low-cost airlines or low-cost carriers (LCC) refer to airlines that maintain low operating cost structure and typically can offer cheap fares to its customers. LCC usually focuses on attracting travelers that are price sensitive (e.g. leisure travelers). However, an airline that has a low-cost structure is not necessarily a low-fare airline. An airline can have a low-cost structure and still can charge high fares, when customers are willing to pay such high fares. There are different measures that low-cost airlines usually consider to reduce their cost of operation. For example, LCC usually fly single fleet type to reduce crew training and maintenance cost, increasing aircraft utilization by minimizing ground/turn time at connections, flying to less-congested secondary airports to avoid congestion at major airports, reducing marketing and sales cost by eliminating commissions to travel agents and sell tickets online via the airline's websites, and finally maintaining a less complex operation by focusing their services in a limited number of markets (Dobruszkes 2006; Vidović et al. 2013).

1.3.3 Ultralow-cost Airlines

Ultralow-cost carrier (ULCC) is a brand used to describe airlines that take additional measures to cut cost (and increase revenue) beyond those considered by LCCs. These airlines attract price-sensitive customers by offering very low fares. It is argued that these measures usually happen on the expenses of comfort, convenience, and amenities offered to customers. For example, aircrafts are configured to have all-couch cabin seats, with no cabin for first or business class seats. In addition, the seats are configured to add as many rows as possible by reducing the leg room between rows. This configuration allows the airline to have more seats per flight and increase their revenue and decrease the per-seat cost. Furthermore, to make up for revenue lost due to decreased ticket prices, the airline may charge for extras like food, priority boarding, seat allocating, baggage, etc. This nonticket revenue is known as ancillary revenue, and it became a very important stream of revenue for ULCC. As mentioned above, these airlines typically target customers who are sensitive to ticket price and less sensitive to quality of service or service frequency (i.e. not time sensitive). This demand is typically very elastic, as it significantly increases or decreases as the price of the ticket goes down or up. ULCC unbundles the service offered to the customers. Instead of having customers pay for a flight ticket that includes all service components such as free check-in bags, free carry-on, and flight meals/snacks, the service is unbundled where customers pay only for any ancillary services they want. Both ULCC and LCC are also commonly referred to as value airlines, budget airlines, or no-frills airlines (O'Connell and Williams 2005; Fageda et al. 2015; Vowles and Lück 2016).

1.4 Ownership

1.4.1 Public or State Ownership

Public- or state-owned airlines (also known as flag carriers or national carriers) refer to airlines that are under full control of a state government in terms of both ownership and management. States recognize the fact that airlines are critical to the country's economic development and national security, and thus they have to be fully under the control of the state. State-owned airlines are mostly considered as part of the national pride with the main goal of the management team is to maintain the airline and keep the business operating. Profitability might not be the first priority of a state-owned airline, as the government can always subsidize any reported losses. However, when government lacks funding, management will be under the pressure to minimize the losses and be self-sustainable. Besides the direct monetary funding, the state can offer other forms of subsidies to the airline including facilitating less expensive loans, less rate on using aviation infrastructure such as airports and terminals, subsidies on fuel price, tax breaks, etc. While it is inconclusive, many researchers argue that state-owned airlines are less efficient compared with privately owned airlines (Chang et al. 2004). State-owned airlines are usually characterized by excess employment, as politicians might force hiring to improve their image and reduce unemployment rates. It is suggested that politicians and bureaucrats may substitute their own goals and preferences, such as employment and image, over efficiency and productivity considerations.

1.4.2 Private Ownership

Private ownership airlines are completely owned by investors from the private sector. Thus, private ownership usually makes airlines more efficient, where owners pressure the management team to achieve higher returns for their investments (Backx et al. 2002). The possibility of losing investment funding forces private airlines to continuously find ways to reduce costs and improve revenues. There is no other objective of management of privately owned airlines other than improving efficiency of the business to maximize profitability. Private airlines are less subjected to bureaucratic practices compared to state-owned ones and are often more flexible. This flexibility gives these airlines better chances to take advantages of opportunities that arise to improve efficiency. From the consumer point of view, if a market is served only by privately owned airlines, the government usually interfere to ensure fair competition among the airlines and to prevent any form of monopoly practice by any of these privately owned airlines.

1.5 Network Structure

1.5.1 Hub and Spoke

Hub-and-spoke airlines adopt a network structure in which one or more major cities (airports) are selected to serve as hubs for the airline (Shaw 1993; Bryan and O'kelly 1999; Nero 1999; Brueckner and Zhang 2001; Burghouwt et al. 2003). Flights are scheduled to connect the hub to other destinations, known as spoke cities. Thus, any flight in the schedule is either starting or ending at a hub. Most airlines worldwide operate in a hub-and-spoke network structure. For these airlines, the hub is usually located at the capital city or at a major industrial or commercial city. The spokes represent other domestic and international cities. The flights of hub-and-spoke airlines carry local and connecting traffic from/to the hub(s). Local traffic is the traffic originating or ending at the hub. Connecting traffic includes passengers that travel between two spoke destinations through the hub.

1.5.2 Point-to-Point

Point-to-point airlines develop their networks by scheduling flights between city-pairs, as long as the demand between the two cities justifies scheduling these flights. In this case, the majority of the demand of the flight is local traffic, which is the demand between the origin and the destination of the flight (Alderighi et al. 2005; Alderighi et al. 2007; Marti et al. 2015).

1.5.3 Hybrid

In some cases, to gain the benefits of the hub-and-spoke structure, airlines that adopt the point-to-point structure tend to define some cities as concentration/focus points. Flights are scheduled to meet at these focus cities within a predefined time window to allow connection opportunities. Such structure is known as the hybrid network structure. The network structure of the hybrid airline includes both characteristics of the hub-and-spoke and point-to-point network structure. Thus, these airlines have some major destinations that operate as hubs, where they provide flights into and out of these hubs. They also connect spoke destinations directly with nonstop flights (Lordan et al. 2014). More discussion on the airline network structures is given Chapter 2.

1.6 Transport Service Type

1.6.1 Cargo Airlines

Cargo airlines (or airfreight carriers) are dedicated to the transport of cargo by air (Zhang and Zhang 2002). A cargo airline could be an independent airline or a subsidiary of larger passenger airlines. Cargo airlines typically carry perishable goods and other items that have high value and less volume including electronics and high-tech products, apparel, textile, footwear, medicine, mail, etc. Cargo airlines are usually part of a complicated intermodal supply chain that provides door-to-door pickup and delivery services. This supply chain includes customer, freight forwarders and consolidators, ground transportation, air flight, sorting, and distribution. Security checks are also a main concern within this supply chain. While cargo airlines share many characteristics of passenger airlines, there are several characteristics that are unique to cargo airlines. For example, cargo airlines suffer traffic imbalance as cargo usually travel in one-way direction. Unless there is cargo to be moved in the returning direction, aircraft might have to return empty. In addition, most cargo flights occur overnight such that the delivery of the cargo can happen during the business hours. Also, a cargo airline competes not only with other cargo airlines but also with other transportation modes including trucks, trains, and ships.

1.6.2 Passenger and Cargo Airlines

These airlines carry both passengers and cargo. As such, when selecting the type of aircraft for a flight, both the passenger and cargo demands are considered to determine the required capacity and appropriate aircraft size.

1.7 Network Coverage

1.7.1 Domestic

Domestic airlines refer to airlines that serve only domestic flights with the boundaries of the country.

1.7.2 International

International airlines are serving both domestic and international flights (or international flights only) (Doganis 2002; Heshmati and Kim 2016). An international flight is a flight where the departure and the arrival take place in different countries. Airports serving international flights are known as international airports.

2

Airline Network Structure

2.1 Introduction

As part of its strategic plan, each airline selects a flight network structure, which represents how it serves the different city-pairs and also how the scheduled flights are configured in the form of attractive itineraries. As discussed earlier, three major network types are commonly used:

- Hub-and-spoke network
- Point-to-point network
- Hybrid network structure

A city served by an airline could be a spoke, a hub, or a focus city. A spoke city for an airline is a city that is served by a small number of daily flights. A spoke city could be large or small in terms of population and economics activities. A spoke city could be served through a large- or small-sized airport. A hub for an airline is typically a major city that is served by a large number of flights by the airline. The hub city should have a major airport to support the operation of large number of flights. This large airport is selected by the airline as a base to offer service to the different spokes. The term spoke is sometimes used to refer to the route that connects the hub to the spoke city. For large-area countries, an airline may select more than one city to serve as hubs to form a multi-hub airline. Flights are also scheduled to connect between every pair of hubs. It should be emphasized that a hub for one airline could be a spoke of another competing airline and vice versa. Finally, a focus city is a destination that is served by considerable number of flights by the airline creating some opportunities for flight connections (Oum and Tretheway 1990).

The main logic behind the hub-and-spoke network structure is that the airline can serve passengers by nonstop flights between the hub and every served spoke city. In addition, it can serve traffic between any two spoke cities with one transfer by connecting passengers at the hub. Thus, passengers on the

Airline Network Planning and Scheduling, First Edition. Ahmed Abdelghany and Khaled Abdelghany.
© 2019 John Wiley & Sons, Inc. Published 2019 by John Wiley & Sons, Inc.

flights are a mix of local and connecting traffic. Local traffic represents passengers traveling from/to the hub, while connecting traffic represents passengers using the hub as a connecting airport. The ratio of local and connecting traffic on each flight could vary significantly depending on the hub location, markets demand, flight schedule, fares, competition, and seat availability.

As an example of a hub-and-spoke network with one hub, consider the hypothetical airline network in the US domestic market as shown in Figure 2.1. This airline has a hub at Hartsfield–Jackson Atlanta International Airport (ATL). It is assumed to provide nonstop service from ATL to 10 spoke destinations including O'Hare International Airport (ORD), Boston Logan International Airport (BOS), Denver International Airport (DEN), Detroit Metropolitan Airport (DTW), Philadelphia International Airport (PHL), Jacksonville International Airport (JAX), Orlando International Airport (MCO), Fort Lauderdale–Hollywood International Airport (FLL), Tampa International Airport (TPA), and Miami International Airport (MIA). It carries local traffic from/to ATL to the spoke cities served by these airports. The airline can also serve the demand between two spoke cities through its hub. For example, the airline can serve the demand of the Tampa-Boston market by connecting at ATL. As mentioned above, each flight in the hub-and-spoke network serves a mix of local and connecting traffic. For example, a flight between ATL and DEN could be serving local traffic traveling from ATL to

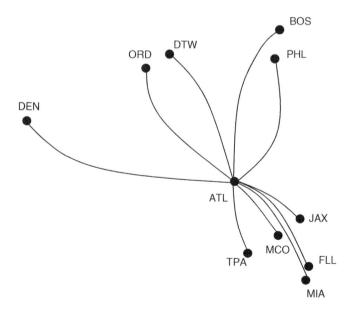

Figure 2.1 Example of a hub-and-spoke airline with a single hub at ATL.

DEN and connecting traffic traveling from TPA to DEN (through ATL), MCO to DEN, MIA to DEN, FLL to DEN, and JAX to DEN. Similarly, the flight FLL to ATL could be serving local traffic traveling from FLL to ATL and connecting traffic traveling from FLL to DEN (through ATL), FLL to ORD, FLL to DTW, FLL to BOS, and FLL to PHL.

In the point-to-point network, the airline connects between cities using nonstop flights. Thus, these airlines focus mainly on carrying nonstop local (i.e. nonconnecting) traffic, traveling between the origin and destination of the flight. Scheduling a flight between any two cities will be feasible only if there is sufficient local demand to justify the direct flight (Alderighi et al. 2005, 2007; Marti et al. 2015). Figure 2.2 gives an example of a hypothetical airline with a point-to-point network structure.

Finally, the hybrid network structure brings together some of the characteristics of the hub-and-spoke network and point-to-point networks. Airlines adopting this network structure can select some cities to serve as hubs, where passengers can connect through them. They can also schedule nonstop service between spoke cities with nonstop flights. Following such structure, some flights carry only local traffic, and some other flights carry a mix of local and connecting traffic.

It is important to differentiate between the topology and the function of an airline network. While an airline network might follow a hub-and-spoke network topology, if the flights are not well coordinated in terms of their arrival and departure times, opportunities for attractive flight connections could be

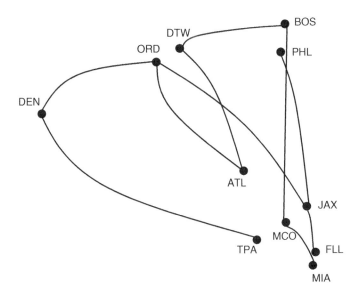

Figure 2.2 Example of a point-to-point network structure.

minimum. Thus, most flights will carry mainly local demand, and the network will functionally operate as a point-to-point network. The alignment of the flights at the hub to create attractive connecting itineraries requires developing the so-called time bank or flight complex, which is explained next.

2.2 Time Bank

Time bank is a core feature of the hub-and-spoke airline network (Dobson and Lederer 1993; Abdelghany and Abdelghany 2012). It is used to align flights scheduled at a hub to maximize the amount of connecting traffic. The process of designing the time banks at a hub starts by studying the origin-destination demand pattern for city-pairs connected through that hub. For example, consider the hypothetical hub-and-spoke network given in Figure 2.3. Assuming that there is a significant demand from MIA to ORD, the airline can schedule the arrival of its flight from MIA to ATL to arrive before the departure of the flight departing from ATL to ORD. The time between the arrival of the MIA-ATL flight and the departure of the ATL-ORD flight should be enough for passengers to connect between the two flights. Airlines plan the flight schedule such that the time window between the arrival of the MIA-ATL flight and the departure of the ATL-ORD flight is acceptable by passengers. For the time window to be acceptable, it should not be too short to allow enough time for passengers to connect between the two flights. Also, it should not be too long, as passengers do not prefer to wait a long period for flight connection.

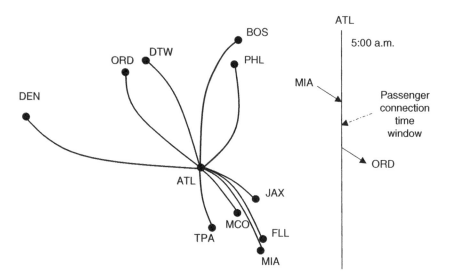

Figure 2.3 Example of connecting itinerary.

The coordination between the arrival time of the inbound flight and the departure time of the outbound flight creates a one-stop itinerary MIA-ATL-ORD that can serve the demand from MIA to ORD with a convenient connection at ATL.

In case there is another significant demand from FLL to ORD, a flight from FLL to ATL can also be scheduled before the departure of the ATL-ORD flight. Similar to the MIA-ATL flight, the FLL-ATL flight has to be scheduled to create attractive connecting itinerary from FLL to ORD, with acceptable connection time at ATL. Accordingly, the FLL-ATL flight is scheduled to arrive at ATL very close to the arrival of the MIA-ATL flight, as shown in Figure 2.4. In this case, passengers connecting from MIA and FLL, heading to ORD, connect at ATL on the scheduled ATL-ORD flight. Figure 2.4 shows the proposed schedule for these three flights. It should be noted that since the two flights MIA-ATL and FLL-ATL arrive at ATL around the same time, there should be enough resources to operate these two flights at ATL. Examples of these resources include gates, customer service, baggage and cargo handling, and other ground operation functions.

In case there is a significant demand from other spoke cities to ORD, inbound flights to ATL can be scheduled from these spoke cities, similar to the MIA-ATL and FLL-ATL flights. These inbound flights should maintain convenient connection time at ATL. For example, in case there is a significant demand from JAX, MCO, and TPA to ORD, inbound flights JAX-ATL, MCO-ATL, and TPA-ATL can be scheduled in a similar way. These flights are scheduled to allow passengers connectivity to the outbound flight ATL-ORD with convenient connection time at ATL. Figure 2.5 shows the proposed schedule of five ATL-inbound flights from JAX, MCO, MIA, TPA, and FLL. Passengers from these inbound flights can connect to the outbound flight ATL-ORD. Since the five inbound flights to ATL arrive around the same time, there should be enough resources to simultaneously handle these five flights.

Furthermore, if there is significant demand from MIA to DTW, an ATL-DTW outbound flight can be scheduled to allow connection from MIA to DTW through ATL. The ATL-DTW outbound flight is to be scheduled to allow convenient connection for passengers connecting from the MIA-ATL inbound flight to the ATL-ORD outbound flight. Accordingly, the ATL-DTW flight is scheduled close to the ATL-ORD outbound flight, as shown in Figure 2.6.

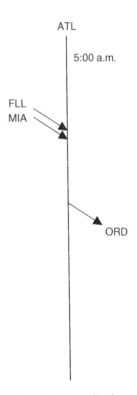

Figure 2.4 Example of two connecting itineraries.

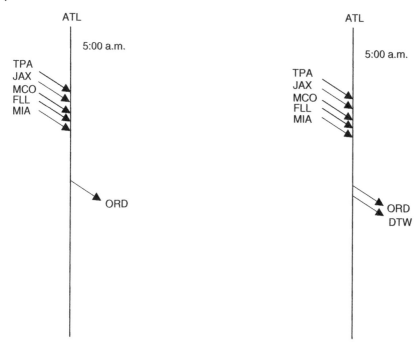

Figure 2.5 Example of several inbound flights connecting to ORD at ATL.

Figure 2.6 Example of several inbound flights connecting to ORD and DTW at ATL.

When the ATL-DTW outbound flight is added to the schedule, several additional city-pairs can now be served by connecting itineraries through ATL. These city-pairs include TPA-DTW, JAX-DTW, MCO-DTW, and FLL-DTW. This flight addition shows the significant advantage of the hub-and-spoke network structure. It allows serving several city-pairs by a few number of flights. In this example, adding the ATL-DTW flight to the schedule results in serving five city-pairs, which are MIA-DTW, TPA-DTW, JAX-DTW, MCO-DTW, and FLL-DTW, in addition to the ATL-DTW local market.

Other outbound flights can be resourcefully added at the hub to maximize passengers' connections from the inbound flights. For example, in Figure 2.7, three additional outbound flights are added out of ATL, near the departure time of the ATL-ORD and ATL-DTW outbound flights. These three flights are ATL-BOS, ATL-DEN, and ATL-PHL. Figure 2.7 shows a simplified example of a time bank at ATL, which is composed of five inbound flights and five outbound flights. This time bank includes 10 nonstop itineraries and 25 connecting itineraries serving a total of 35 city-pairs. These 35 city-pairs are given in Table 2.1.

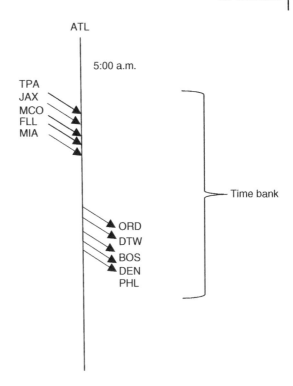

Figure 2.7 Example of a time bank at ATL.

Table 2.1 List of city-pairs served by the time bank.

City-pairs with nonstop service		City-pairs with connecting service				
TPA-ATL	ATL-ORD	TPA-ORD	JAX-ORD	MCO-ORD	FLL-ORD	MIA-ORD
JAX-ATL	ATL-DTW	TPA-DTW	JAX-DTW	MCO-DTW	FLL-DTW	MIA-DTW
MCO-ATL	ATL-BOS	TPA-BOS	JAX-BOS	MCO-BOS	FLL-BOS	MIA-BOS
FLL-ATL	ATL-DEN	TPA-DEN	JAX-DEN	MCO-DEN	FLL-DEN	MIA-DEN
MIA-ATL	ATL-PHL	TPA-PHL	JAX-PHL	MCO-PHL	FLL-PHL	MIA-PHL

Several important issues related to the design of time banks should be considered. First, there should be enough resources to handle multiple flights arriving and departing within a short time window. These resources include aircraft, cockpit crew, cabin crew, gates, customer service, and ground handling. Second, in most cases, resources (aircraft and crew) arriving on inbound flights are turned to operate outbound flights within the same time bank. For example, in the time bank shown in Figure 2.7, there are five aircraft arriving at the hub.

Typically, each of these five aircraft will be turned to operate one of the five outbound flights in the time bank. This is also the case for the cockpit crew and cabin crew unless the crew is terminating its duty at the hub and a new crew is assigned to the outbound flights. Third, the aircraft of the five inbound flights should be assigned to terminals and gates in a way that minimizes the walking distances for the connecting passengers between the inbound and the outbound flights. It also facilitates the handling baggage and cargo among the connecting flights. Fourth, demand is the main factor in selecting flights to be included as inbound or outbound flights in the time bank. This selection is typically related to the geographical locations of the spoke destinations relative to the hub. The time bank is usually connecting different geographical locations around the hub. In the example above, the time bank is connecting south destinations to north destinations and is usually referred to as south–north time bank. Another time bank is needed to connect the returning trips from north to south, as shown in Figure 2.8. This time bank has five inbound flights including DTW-ATL, PHL-ATL, ORD-ATL, BOS-ATL, and DEN-ATL and also five outbound flights including ATL-FLL, ATL-TPA, ATL-MIA, ATL-MCO, and ATL-JAX.

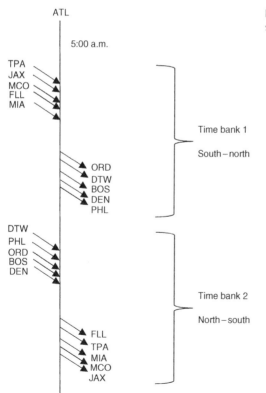

Figure 2.8 Example of two successive time banks.

Table 2.2 Information of spoke destinations included in the time bank.

Destination	Flight time to/from ATL (hh:mm)	Time zone	Time difference to ATL (hour)
ATL	N/A	Eastern Time Zone	N/A
MIA	01:55	Eastern Time Zone	0
FLL	01:50	Eastern Time Zone	0
MCO	01:25	Eastern Time Zone	0
TPA	01:30	Eastern Time Zone	0
JAX	01:10	Eastern Time Zone	0
DEN	02:50	Mountain Time Zone	−2
ORD	01:55	Central Time Zone	−1
BOS	03:10	Eastern Time Zone	0
PHL	02:20	Eastern Time Zone	0
DTW	02:10	Eastern Time Zone	0

The design of a time bank tries to achieve two main goals: (i) increasing the served demand/revenue through providing attractive nonstop and connecting itineraries and (ii) optimizing the utilization of resources (aircraft, crew, and ground services) required to operate the different flights (Abdelghany et al. 2017). To further explain these two factors, consider the information given in Table 2.2, which lists the different spoke cities/airports considered in the hypothetical airline network mentioned above. For each city, the flight time to/from ATL, the time zone, and the time difference from ATL are given.

Figure 2.9 gives four different examples that illustrate possible design of the schedule at ATL. In *case a*, the airline schedules a set of outbound flights departing from the hub airport between 9:00 and 10:00 a.m. These flights head to destinations in the southern part of the country. To maximize ridership on these flights, the airline schedules another set of inbound flights to arrive at the hub before these departures (roughly in the 8:00–9:00 a.m. time window). These inbound flights arrive from northern destinations and possibly carry connecting passengers to southern destinations. Accordingly, the schedule at ATL starts by a north–south time bank (i.e. time bank 1). In this time bank, flights are arriving at ATL from northern destinations and are followed by southbound flights, departing later from ATL. Considering this design, the inbound DEN-ATL flight has a flight duration of about 2 hours and 50 minutes. Thus, this flight has to depart from DEN around 5:10 a.m. EST (3:10 a.m. MST). Similarly, the BOS-ATL flight should depart from BOS around 4:50 a.m. EST, and the ORD-ATL flight should depart around 5:05 a.m. CST. The example in *case a* shows the need to consider flight time and time-zone difference, when

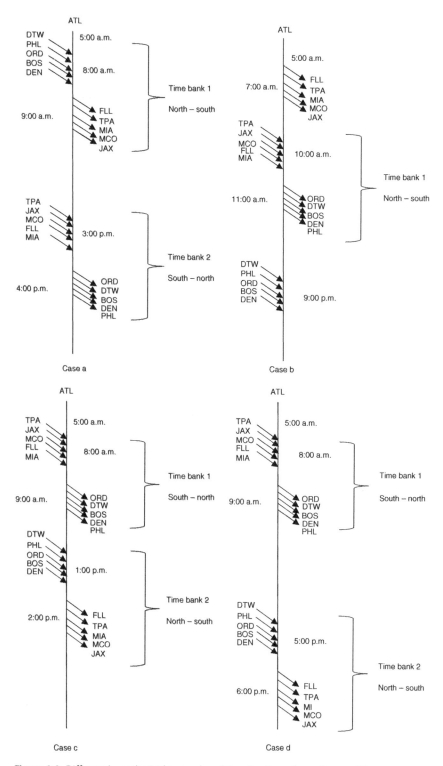

Figure 2.9 Different hypothetical examples of time bank configurations at ATL.

scheduling flights in the time bank. This consideration is important because some flights are scheduled to depart at inconvenient time for passengers. Typically, when airlines design an early-morning time bank at the hub, the inbound flights are typically selected to depart from time zones that allow convenient departure time (e.g. short-haul flights in the EST time zone in this example). Another consideration is to ensure that the scheduled departure times of the inbound flights do not violate any flying curfews at the origin cities. In *case a*, another south–north time bank is considered in the afternoon to balance the demand at the hub.

In *case b*, no early-morning inbound flights are scheduled into ATL. The schedule at ATL starts by a set of departures to southern destinations. The problem of this design is that these morning outbound flights will have no connecting traffic. Thus, the local demand originating from ATL heading to these southern destinations has to be large enough to fill these flights. To maintain balance of resources for these early-morning flights, there should be an equivalent set of arrivals in the previous night at the hub. Aircraft arriving on these last-night inbound flights will typically be prepared to operate early-morning outbound flights at the hub. For example, in *case b*, five inbound flights arrive to ATL during the 9:00–10:00 p.m. window. The resources on these five flights are prepared to fly the early-morning outbound flights on the next day. To balance demand at the hub, a set of inbound flights from the southern destinations and a set of outbound flights to the northern destinations are scheduled at the hub. These flights are arranged in a south–north time bank during the 10:00–11:00 a.m. time window to maximize connecting traffic on the flights.

In *case c*, the airline considers an early-morning time bank similar to the schedule in *case a*. However, the order of flights is reversed. The airline schedules a set of inbound flights from southern destinations to arrive at ATL during the 8:00–9:00 a.m. window. These flights are followed by a set of outbound flights heading to northern destinations during the 9:00–10:00 a.m. window. This arrangement forms a south–north morning time bank at the hub. This time bank arrangement is more acceptable over the one presented in *case a*. In *case c*, early-morning inbound flights are relatively short-haul flights with short flight times. In addition, these inbound flights are originating from destinations that are located in the same time zone of ATL (i.e. Eastern Time Zone). For example, the flight time of the MCO-ATL flight is one hour and 25 minutes, and both MCO and ATL are located in the same time zone. Thus, for example, for this flight to arrive ATL at 8:00 a.m., this flight has to depart MCO at 6:35 a.m. Similarly, the TPA-ATL flight has to depart at 6:30 a.m., and the JAX-ATL flight has to depart at 6:50 a.m. All these departure times are relatively acceptable as early-morning departures. In *case c*, another time bank is considered in the afternoon at 1:00–2:00 p.m., which is a north–south time bank to balance the demand at the hub.

Another factor that should be considered when scheduling flights is the rotation of aircraft needed to operate this schedule. In *case c*, five aircraft are needed to operate the five inbound flights at the hub. Upon arrival at the hub during the 8:00–9:00 a.m. window, these five aircraft are turned to operate the five outbound flights to the northern destinations in the first time bank (i.e. the south–north time bank) during the 9:00–10:00 a.m. window. Each of these outbound flights needs about two to three hours of flight time to reach at their corresponding destinations. Thus, each aircraft will need twice this amount of time to make a round trip to return back to the hub, in addition to the ground time (e.g. one hour) to turn the aircraft and to get it ready at the spoke destination. For example, the aircraft operating the ATL-BOS flight requires at least 7 hours and 20 minutes to return back to ATL. Thus, if this aircraft departs at 9:00 a.m. from ATL, the earliest time it can return back to the hub is 4:10 p.m. Thus, if the airline adopts the schedule in *case c*, where the second time bank is scheduled at 1:00–2:00 p.m. (the north–south time bank), another five aircraft are needed to operate these other 10 flights.

In *case d*, the schedule is modified such that the second time bank is scheduled later in the day at 5:00 p.m. This schedule allows enough time to have the aircraft operating the first time bank to make a round trip and return back to the hub. Thus, in *case d*, all the 20 flights in the two time banks are operated by only five aircraft, which represent an efficient utilization of the resources. The only drawback is that there could be loss of passenger revenue as the schedule focuses more on optimizing aircraft rotations rather than satisfying passengers preferred departure/arrival times.

As mentioned earlier, a key consideration in determining the optimal time bank configuration is the availability of resources that are needed to simultaneously handle several flights at the same time. For example, as explained in the example above, each of the time banks given in Figure 2.9 (*case d*) requires at least five aircraft along with their crew to operate the scheduled flights. It also requires at least five gates (or parking spaces) along with ground staff to handle all inbound and outbound flights at the hub. For large airlines, the air traffic control agency at the airport might also raise a concern due to scheduling a large number of arriving and departing flights within a short period. Clearly, availability of limited resources at the hub could significantly impact the configuration of the time banks. In any case, the schedule should be optimized to maximize revenue from local and connecting passengers from all city-pairs. In addition, the schedule should optimize the schedule of all the resources needed to operate the different flights including aircraft, cockpit crew, and cabin crew. Another factor to consider is that for airlines that have more than one hub, the design of a time bank at one hub will influence the design of time banks at other hubs. For flights that are connecting between hubs, it might be difficult to align them within the time banks at the origin and destination.

2.3   Advantages of the Hub-and-spoke Network

2.3.1   Better Network Coverage

One main advantage of incorporating time banks in the hub-and-spoke network structure is the coverage of many city-pairs using minimum number of flights (Burghouwt 2007; Derudder et al. 2007). For example, Figure 2.10 (*case a*) shows a hypothetical time bank similar to the one discussed above. As shown in the figure, this south–north time bank includes five inbound flights and five outbound flights. These 10 flights serve 10 nonstop itineraries and 25 connecting itineraries. In total, these itineraries serve 35 city-pairs. If the airline expands its network to serve Tallahassee International Airport (TLH) and Port Columbus International Airport (CMH), two new flights are added to this time bank at ATL, which are TLH-ATL and ATL-CMH. These two flights are represented by dotted lines in Figure 2.10 (*case b*). Adding these two flights expanded the service to 12 nonstop itineraries and 36 connecting itineraries serving 48 city-pairs. Compared with the point-to-point structure, the airline has to offer 48 flights to serve the same 48 city-pairs, which requires considerable amount of operation resources.

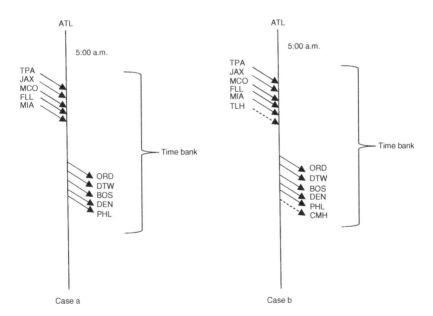

Figure 2.10 Adding more flights in the time bank.

2.3.2 Mixed Portfolio of Passenger Demand

The flights of hub-and-spoke airlines serve a mix of local and connecting traffic, compared with point-to-point airlines that mostly serve local traffic (Glover et al. 1982). Figure 2.11 and Figure 2.12 give an example of a hub-and-spoke airline and a point-to-point airline. Both airlines operate an ATL-ORD flight. For the hub-and-spoke airline, the ATL-ORD flight is part of a time bank with the following inbound flights to ATL: TPA-ATL, MCO-ATL, JAX-ATL, MIA-ATL, and FLL-ATL. As a result, for the hub-and-spoke airline, the ATL-ORD flight is serving the local market ATL-ORD. In addition, it serves the connecting demand from TPA, MCO, JAX, MIA, and FLL heading to ORD. Accordingly, the demand on the ATL-ORD flight is a diversified demand representing a mixed portfolio of local and connecting demand. On the other hand, the ATL-ORD flight served by the point-to-point airline is only serving the local traffic from ATL to ORD with no connecting traffic.

Assume that there is another competing airline that decided to serve the city-pair ATL-ORD and compete with the two airlines in the example above. This new flight is shown by dotted lines in Figures 2.13 and 2.14. For the hub-and-spoke airline, only the local demand (i.e. demand traveling from ATL to ORD) on the ATL-ORD flight could be affected as a result of the new competition. Meanwhile, the connecting traffic on the ATL-ORD flight is expected to

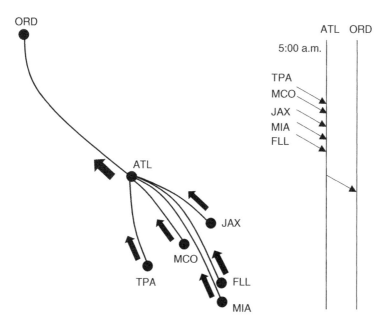

Figure 2.11 Demand mix on flight ATL-ORD in a hub-and-spoke network.

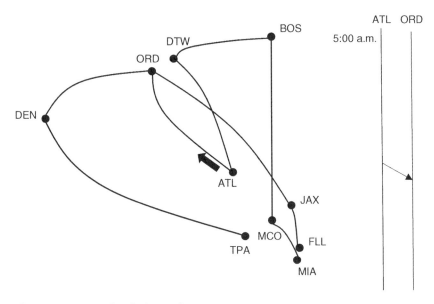

Figure 2.12 Demand on flight ATL-ORD in a point-to-point network.

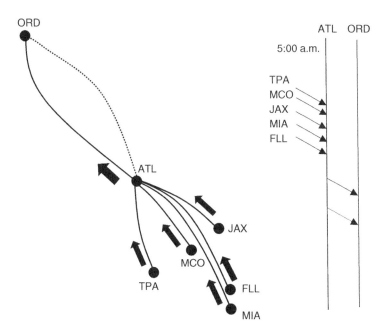

Figure 2.13 Competition in a hub-and-spoke network.

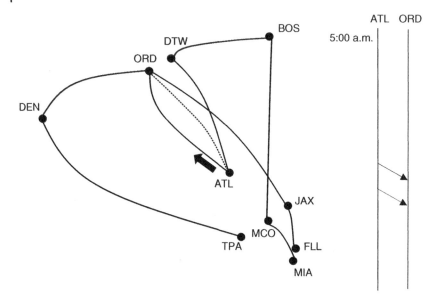

Figure 2.14 Competition in a point-to-point network.

remain unchanged, since there is no change in competition in the connecting city-pairs. Accordingly, due to demand diversity on the ATL-ORD flight, this flight is likely to sustain the competition with the new airline. On the other hand, for the ATL-ORD flight operated by the point-to-point airline, the demand on this flight is only local demand. This local demand is expected to be significantly affected by the competition. Accordingly, this flight might not be able to sustain the competition. This example shows that hub-and-spoke airlines are generally more likely to sustain the competition due to demand diversity of their flights. On the other hand, point-to-point airlines need to find a way to make their flights less affected by the competition. They usually offer cheap fare (e.g. low-cost point-to-point airlines) to make it difficult for other competing airlines to enter the market.

2.3.3 Dominance at the Hub

By definition, airlines that adopt the hub-and-spoke network structure preserve a form of dominance at the hub, where a large number of flights are scheduled from/to the hub. This dominance provides the airline a competition advantage in cities served through the hub (Toh and Higgins 1985; Borenstein 1989; Brueckner et al. 1992; Lee and Luengo-Prado 2005). At the hub, the airline will have more flexibility to respond to competition by shifting capacity, adding/removing flight frequency, and adjusting pricing. The high dominance and associated schedule adjustment flexibility make it difficult for other airlines to compete at the hub city.

2.3.4   Economy of Scale Operations at the Hub

The term economy of scale refers to the ability to reduce the production cost per unit as the production volume increases. One can think of an airline as a flight production factory. Thus, since airlines are producing more flights at the hub, the average flight operation cost is expected to fall. Resources required to operate the flight include aircraft, cockpit crew, cabin crew, customer service agents, gate agents, baggage and cargo handlers, ground staff, catering, aircraft cleaning, mechanics, and fueling. These resources can be scheduled and also can be shifted between jobs as needed to generate efficient operation plan with minimum cost. When an airline has small number of flights at any stations, these resources might not be fully utilized. In addition, in many cases, the airline depends on a third-party company to support the different operation functions, which increases its overall operation cost (Caves et al. 1984; Horner and O'Kelly 2001).

2.4   Limitations of the Hub-and-spoke Network

2.4.1   Congestion at the Hub

As mentioned above, time banks are used to coordinate inbound and outbound flights at hubs to construct itineraries that maximize connecting demand. Time banks represent a set of flight arrivals (inbound) followed by a set of flight departures (outbound) at the hub airport, which are scheduled within a short time window. Scheduling many flights to arrive at and depart from the airport within a short time window usually results in congestion at the airport (Slack 1999; Elhedhli and Hu 2005; Daniel and Harback 2008). This congestion is experienced at all levels of operation including the terminal services, ground operations, and flight operations. For example, at the terminal, time banks result in large number of simultaneous passengers' check-ins and connections that need customer service attention. These passengers are associated with large number of baggage to be handled in the handling facilities. In addition, efficient ground operations are needed to handle simultaneous aircraft arrivals and prepare these aircraft for departure within short time intervals. Several ground services are performed simultaneously including aircraft parking assistance, cleaning, refueling, cargo/baggage handling, and catering. Furthermore, time banks result in busy taxi ways, runways, and congested airspace around the airport.

This congestion at the hub is usually accounted for in the flight time planning by increasing the time needed to taxi in and taxi out at the hub. In addition, aircraft airborne time is extended to account for possible airspace delays around the hub. Accounting for extended taxi in, taxi out, and flight time at the hub typically results in a poor aircraft utilization. Each flight in the schedule of a hub-and-spoke airline has to visit a hub. Thus, airlines with hub-and-spoke network structure would suffer congestion, which results in poor utilization of their resources. Point-to-point airlines typically avoid flying to busy hubs. When flying to large cities, the airline usually targets secondary less congested airports instead of primary busy airports.

2.4.2   Schedule Vulnerability to Disruption at the Hub

Adverse weather conditions at the hub typically cause significant disruption to the schedule of hub-and-spoke airlines (Abdelghany et al. 2004b). When the hub is subjected to adverse weather conditions, the aviation authority typically issues programs that control flight arrivals (e.g. ground delay programs). These control programs usually result in delays and cancellations for the inbound flights. Since outbound flights depend on resources of inbound flights, the outbound flights also suffer disruption. Furthermore, inbound flights feed connecting traffic to outbound flights in the different time banks. Thus, disruptions occurring to inbound flights of any time bank directly affect most connections. The impact of this disruption propagates to other destinations causing what is known as snowball effect.

2.4.3   Extended Ground Time for Resources

One major drawback of the hub-and-spoke airlines is related to excessive ground time of resources used to build the schedule. To explain this problem, consider the inbound and outbound flights at ATL airport presented in

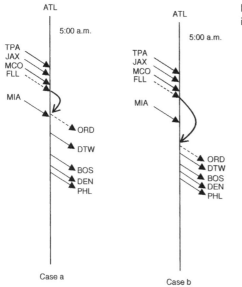

Figure 2.15 Ground time of resources in a time bank.

Figure 2.15a. Assume that the aircraft of the MCO-ATL inbound flight is scheduled to operate the ATL-ORD outbound flight. This aircraft is scheduled such that it spends minimum required ground time at ATL. This minimum ground time is the time needed to turn the aircraft including passenger and baggage handling, aircraft cleaning, refueling, catering, cargo handling, crew preparation, and aircraft safety checks. However, if the minimum aircraft turn time is used, considering the scheduled arrival time of the MIA-ATL inbound flight, the itinerary serving MIA and ORD is no longer feasible. Thus, the demand between MIA and ORD shifts to other airlines, which represents a potential loss of revenue. To overcome this problem, the airline might decide to adjust the schedule of the ATL-ORD flight by pushing its departure later in the day. This change creates a connecting itinerary between MIA and ORD by one stop at ATL, as shown in Figure 2.15b. Nonetheless, this schedule results in an unnecessary ground time for the aircraft scheduled to operate the MCO-ATL and ATL-ORD flights. While the airline might gain additional revenue by serving connecting traffic between MIA and ORD, the less aircraft utilization is expected to increase the overall operation cost. Thus, the airline should determine the optimal flight schedule that considers the trade-off between the additional revenue gained from connecting itineraries and the loss due to the extended ground time of resources (e.g. aircraft) (Wu and Caves 2000).

3

Airline Schedule Planning Decisions

The airline schedule planning involves developing the flight schedule for a future operation horizon. It also involves developing the schedule of all resources (i.e. aircraft, gates, crew, and airport personnel) used to support the flight operation. To accomplish this task, several subtasks are considered in the airline schedule planning process. These subtasks include demand forecasting, competition analysis, determining flight frequencies, determining flight departure times, and identifying flight seat capacities. The schedule of resources entails determining the resources required to operate each flight including optimal schedule of these resources. The main objective is to maximize utilization of the resources while maintaining regulations and other required work rules (Barnhart et al. 2003a; Abdelghany and Abdelghany 2012; Belobaba et al. 2015; Bazargan 2016; Teodorovic 2017). In this chapter, we define the different subtasks involved in airline schedule planning. Then, the interdependence among these decisions is discussed.

3.1 Definitions

3.1.1 Demand Forecasting and Competition Analysis

Airlines spend considerable effort to estimate current and future number of passengers in the different city-pairs, which they currently serve or aim to serve in the future. They also try to estimate their market share in each of these markets, given existing or potential competition (Proussaloglou and Koppelman 1995; Witt and Witt 1995; Coldren et al. 2003). For this purpose, airlines compile historical data on demand capturing seasonality, economic growth, special holidays, and schedule of competitors (Armstrong 2001). They also collect information on events that might cause significant change in demand including sport events, large conventions, and religious gatherings. Airlines also collect as much information as possible on the schedule of

Airline Network Planning and Scheduling, First Edition. Ahmed Abdelghany and Khaled Abdelghany.
© 2019 John Wiley & Sons, Inc. Published 2019 by John Wiley & Sons, Inc.

competitors. This information includes flight schedule, available seat capacity, and fares. They also collect information on other transportation modes including rail and bus services, especially in short-haul markets. These data are compiled and analyzed using different tools to estimate the airline's market share in the different city-pairs.

3.1.2 Served Markets

A key decision by airlines is to determine which city-pairs (routes or markets) to serve. The main factors behind these decisions are cost and revenue (i.e. profitability), availability of resources, and service sustainability. Airlines typically decide on routes that maximize their total profitability (Reiss and Spiller 1989; Barrett 2001). A route could be profitable on its own or contributing to the profitability of the entire network. Airlines increase their resources to add profitable routes. They also shift resources from nonprofitable or less profitable markets to ones that are more profitable. When an airline provides service in a city-pair, they aim at strengthening their existence and sustaining service. Airlines would avoid markets that have low demand or significant competition with high risk of not continuing their service in these markets.

3.1.3 Flight Frequency

Given the estimated demand level, competition, and available resources, airlines decide on number of flights to provide in each city-pair. Customers typically prefer airlines to provide more frequent service, as it gives them more flexibility, when booking their tickets (Hassan et al. 2009; Pai 2010; Brueckner and Luo 2014). The flight frequency may be counted on a daily or a weekly basis. A flight schedule could be designed to repeat itself every day with some flights added/removed during certain days of the week. Also, airlines tend to provide frequent services in business market because business travelers usually demand flexibility in travel options.

3.1.4 Flight Departure/Arrival Time

The airline selects an optimal departure time for each flight such that it creates attractive itineraries that maximize ridership for both local and connecting traffic (Borenstein and Netz 1999; Abdelghany et al. 2017). The goal is to schedule the flights around the passengers' preferred departure/arrival times. The flight departure time is also selected to optimize the schedule of all operation resources of the airline including aircraft, crew, landing slots, and gates. In slot-constrained airports, airline must schedule their flights considering the timing of their available slots. When a set of inbound flights at an airport are scheduled, the departure times of these flights should be coordinated to prevent congestion built-up at the host airport. The host airport might have limited number of gates, runways, and ground services. Airports with such limited resources might not be able to handle large number of flights arriving within a short-time window.

3.1.5 Fleet Assignment

For airlines that operate more than one aircraft type, the fleet assignment task involves determining the most suitable aircraft type for each flight. Several factors determine the fleet assignment solution (Gu et al. 1994; Subramanian et al. 1994; Hane et al. 1995). These factors include, for example, number of aircraft in every fleet type, aircraft seat and cargo capacity, aircraft flying range, and runway and infrastructure suitability for the aircraft at the origin and destination airports. Airlines select the optimal fleet assignment to maximize revenue from passengers and cargo and minimize operation cost including fuel cost, crew cost, and overflight and landing fees. Figure 3.1 shows a representation of the fleet assignment solution for a hypothetical airline that operates three fleet types. As shown in the figure, the daily flight schedule of this airline is assigned to three fleet types.

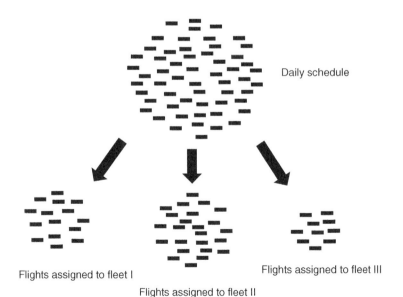

Figure 3.1 Fleet assignment among three different fleet types.

3.1.6   Aircraft Schedule

Aircraft schedule (also known as aircraft rotation or line of flying (LOF)) specifies the sequence of flights that are to be flown by each aircraft. The goal is to make sure that each flight in the schedule is covered by an aircraft (Clarke et al. 1997; Desaulniers et al. 1997; Barnhart et al. 1998). The schedule of the aircraft, which usually extends over a few days (two to seven days), is designed to maximize aircraft utilization. The schedule should also satisfy the aircraft maintenance requirements by ensuring that each aircraft is sent to a designated maintenance facility at the appropriate time. In addition, the aircraft schedule has to satisfy all operation constraints including minimum ground time at the different stations. This ground time is needed to turn the aircraft between two consecutive flights.

Solving for aircraft rotation is performed after the fleet assignment, where the flights assigned to each fleet type are identified. Figure 3.2 illustrates an example of a 6-day aircraft rotation for an aircraft of fleet type I that includes 17 flights. As shown in the figure, since the aircraft's LOF extends over several days, the flights assigned to fleet type I on each day are identified. Then, for a given aircraft, a sequence of flights is selected on each day to be flown by this aircraft. When selecting this sequence, the airline ensures optimal aircraft utilization with minimum turn times between flights and aircraft maintenance. Figure 3.3 shows the rotations of the aircraft over time and space for the same example. As shown in the figure, three flights are assigned to this aircraft on day 1, day 2, and day 3. Two flights are assigned to this aircraft on day 4, which are followed by a maintenance activity. The destination of the second flight on day 4 has to be one of the airline's maintenance stations. Typically, the maintenance activity is performed overnight, as the number of flights significantly drops.

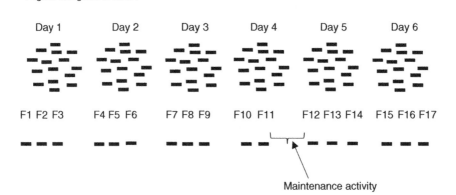

Figure 3.2   An example of an aircraft rotation.

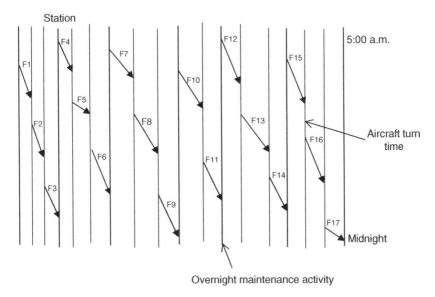

Figure 3.3 The rotation of the aircraft over time and space.

3.1.7 Crew Schedule

Each crew member (cabin crew or cockpit crew) is assigned a monthly schedule typically known as a line of flying (LOF) (Kohl and Karisch 2004). The line consists of several trippairs (also known as trips or pairings). A trippair is a sequence of flights that extend over a few days, where the first flight starts at the crew member's home base (domicile) and the last flight ends at her/his home base (Barnhart et al. 2003b). Each working day in the trippair represents a duty period. The duty period starts one hour before the departure time of the first flight in the duty period and ends within fifteen minutes after the arrival of the last flight. Crew members get enough ground time to connect between flights (crew connection). Between every two duty periods, the crew is given a layover (rest) period. The length of the duty period, crew connection, and rest period should be according to the regulations that are set by the aviation authority and negotiated agreements with unions representing the crew. These regulations ensure prudent and safe workloads for the crew.

3.1.8 Gate Assignment

For airports with gateways, flights are typically assigned to departure and arrival gates. At busy airports, where gates are scarce, the gates have to be scheduled efficiently to accommodate all flights (Bihr 1990; Ding et al. 2005). Flights are assigned to gates in a way that efficiently facilitate the movement of local and connecting passengers and the handling of baggage and cargo.

3.1.9   Other Resources

The airline schedule planning involves scheduling all other resources, facilities, and personnel needed to support the operation, including baggage handling facilities and personnel (de Neufville 1994; Abdelghany et al. 2006), customer service agents, catering, maintenance, aircraft cleaning, fueling and dispatching, cargo agent, and ground handling staff (Holloran and Byrn 1986; Clausen and Pisinger 2010; Stolletz 2010).

3.2   Relationships Among Scheduling Decisions

The main challenge of the airline schedule planning process lies in understanding the interdependence and relationships among the different decisions involved in this process. As sketched in Figure 3.4, a change in one of these planning decisions could result in a significant impact on other decisions. This dependability complicates the schedule planning process and usually requires examining the effect of these decisions iteratively. In this section, we provide several examples to illustrate the relationships among these decisions.

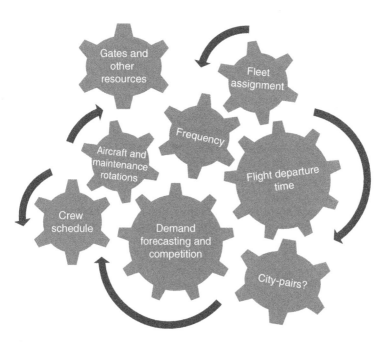

Figure 3.4 Interaction among the different airline's schedule planning decisions.

3.2.1   Flight Frequency and Fleet Assignment

For an airline that operates multiple fleet types, the airline's planning team should decide on seat capacity provided in the different city-pairs. This seat capacity is a function of the flight frequency and fleet type assigned for each flight. For instance, for any given city-pair, the airline might select to provide service with small aircraft (i.e. less seat capacity per flight) with high service frequency. Alternatively, it could serve this city-pair with a fewer flights with large aircraft. For example, Figure 3.5 shows three different service scenarios (*case a*, *case b*, and *case c*) for a hypothetical airline serving the city-pair ATL-ORD. The size of the arrow represents the seat capacity of the used aircraft. As shown in the figure, in *case a*, the airline provides four flights using small aircraft. In *case b*, the airline serves the market with large aircraft by scheduling two flights only. Finally, in *case c*, the airline provides three flights: Two of them are operated by a small aircraft, and one flight is operated by a large aircraft. Selecting the best option requires a detailed comparison among all options taking into consideration service options in other markets to optimally assign the available resources and to select the most optimal service plan across the entire network (de Pitfield et al. 2010). City-pairs that are served by the same fleet type are known as pure stations.

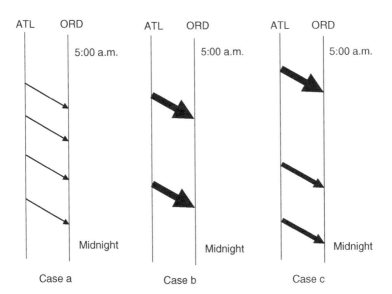

Figure 3.5 Interaction between flight frequency and fleet assignment.

3.2.2 Departure Time and City-pairs

Consider the example of the airline network shown in Figure 3.6, with a hub at ATL. This airline is assumed to provide nonstop services from ATL to ORD, JAX, MCO, FLL, TPA, and MIA. This airline is considered to provide connecting service through its hub at ATL. For example, this airline may serve the MIA-ORD city-pair by connecting at ATL. However, this is true only if the departure time of a flight from ATL to ORD is aligned with the arrival time of a flight from MIA to ATL. This alignment is to generate an attractive connecting itinerary between MIA and ORD. Figure 3.6 *case a* and *case b* shows the timeline at ATL. In *case a*, the ATL-ORD flight departs (at 3:00 p.m.) before the arrival time of the flight MIA-ATL (at 4:00 p.m.), creating no itinerary from MIA to ORD. In *case b*, the ATL-ORD flight is scheduled to depart (4:30 p.m.) after the arrival of the MIA-ATL flight (3:30 p.m.), resulting in an attractive itinerary between MIA and ORD. Accordingly, adjusting the flight departure times is crucial to provide attractive service in markets served with connecting itineraries.

3.2.3 Departure Time and Demand

The passenger demand for a given flight depends not only on the schedule of this flight but also on the schedule of other flights. For example, in Figure 3.7, the demand of the ATL-ORD flight depends on the schedule of this flight and the schedule of the TPA-ATL, MCO-ATL, JAX-ATL, MIA-ATL, and FLL-ATL

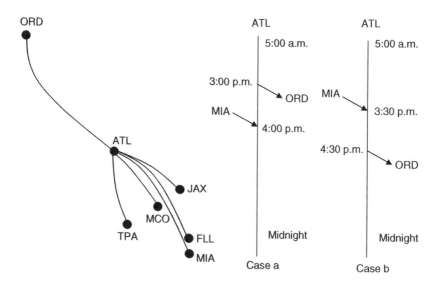

Figure 3.6 Interaction between flight departure time decision and city-pair service.

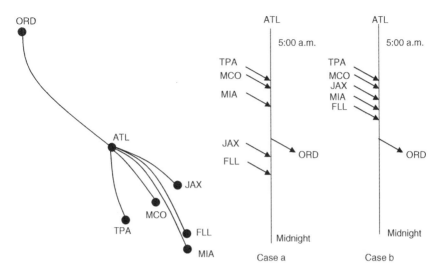

Figure 3.7 Interaction between flight departure time and demand.

flights. These five flights could be feeding connecting traffic destined to ORD, if these flights are scheduled to arrive before the departure of the ATL-ORD flight. For instance, in the schedule presented in *case a* in Figure 3.7, only flights TPA-ATL, MCO-ATL, and MIA-ATL arrive before the departure of the ATL-ORD flight, creating attractive itineraries between from TPA, MCO, and MIA to ORD. These three connecting markets will feed traffic on the ATL-ORD flight, in addition to the local ATL-ORD demand. In the schedule presented in *case b* in Figure 3.7, inbound flights from TPA, MCO, JAX, MIA, and FLL are aligned such that they connect to ATL-ORD flight and carry traffic to ORD. Thus, all these five flights will feed traffic to the ATL-ORD flight in addition to the local traffic between ATL and ORD.

3.2.4 Fleet Assignment and Flight Arrival Time

The flight gate-to-gate time (also known as the flight block time) is composed of the flight taxi-out time (time from the gate of departure to the runway) at the origin airport, airborne time, and taxi-in time (time from runway to gate of arrival) at the destination airport. The flight airborne time depends on the aircraft type and its speed. In addition, the taxi-out and taxi-in times depend on the aircraft type, varying on where each aircraft type is gated at the airport. Accordingly, the flight arrival time cannot be estimated without knowledge of the aircraft type assigned to this flight (i.e. fleet assignment). For example, Figure 3.8 shows a hypothetical flight between ATL and ORD. This flight is scheduled to depart at time D1. The arrival time of the flight depends on the aircraft assigned to the flight. The flight is estimated to arrive at A1 if assigned to a fast aircraft and arrive at A2 if assigned to a slower aircraft.

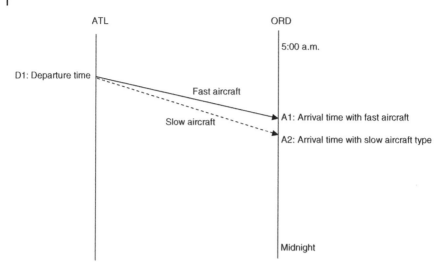

Figure 3.8 Interaction between flight arrival time and fleet assignment.

3.2.5   Fleet Assignment and Flight Departure Time

When an aircraft parks at the arrival gate, it requires service to get it ready for its next scheduled flight. This service includes cleaning, fueling, supply of catering, cargo and baggage handling, and passenger boarding. The time needed for these activities (also known as minimum required aircraft turn time) depends on the aircraft type, as large aircraft typically requires longer turn time compared with that of small aircraft (Lohatepanont and Barnhart 2004). Figure 3.9 *case a* and *case b* presents examples of the minimum required turn time for a small and large aircraft, respectively. In *case a*, both flights F1 and F2 are operated by small aircraft, which requires a minimum turn time of T1. Thus, flight F2 can be scheduled to depart at D2. In *case b*, when a large aircraft is used to operate flights F1 and F2, this aircraft typically requires longer turn time T2, which requires flight F2 to be scheduled later at D4.

3.2.6   Flight Departure Time, Arrival Time, and Block Time

Each flight in the schedule is defined by its origin, destination, scheduled departure and arrival times, and aircraft type. The scheduled departure and arrival times of a flight are linked together by the block time planned for this flight. A flight block time is defined as the period between the aircraft push-back time at the departure gate and the arrival time at the destination gate. As the scheduled departure time of a flight is determined, the flight's scheduled arrival time is determined by adding the block time of the flight to its scheduled departure time. Alternatively, if the scheduled arrival time of the flight is

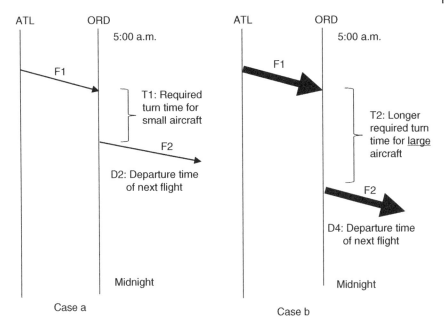

Figure 3.9 Interaction between fleet assignment and departure time decisions.

determined, its scheduled departure time is determined by subtracting the block time of the flight from its scheduled arrival time. To link the scheduled departure time and arrival times of the flight, the block time of each flight needs to be accurately estimated.

As mentioned above, the flight block time consists of three main components, which are the taxi-out time, the flight (airborne) time, and the taxi-in time. The taxi-out time is the period between the aircraft pushback time at the departure gate and the aircraft wheels-off time at the departure runway (Rosenberger et al. 2002; Sohoni et al. 2011). The flight time is the period between the aircraft wheels-off time at the departure runway and the aircraft wheels-on time at the destination runway. The taxi-in time is the period between the aircraft wheels-on time at the arrival runway and the aircraft arrival time at the destination gate. Block times for all flights in the schedule are typically estimated using historical data of similar flights operated in the past. It is critical for the airline to precisely estimate the block time for each flight in the schedule.

Estimating the flight block time differs by city-pair, as it significantly depends on the departure and arrival airports of the flight. Airlines account for uncertainty in block time, which typically increases when a flight departs from and/or arrives at a congested airport. When estimating the block time, the airline should select a value that captures the trade-off between the possibility of late arrivals and the idling of the resources.

3.2.7 Flight Departure Time and Aircraft Rotation

The departure time of a flight is selected to generate attractive itineraries with respect to the preferences of travelers in the city-pair. However, it is critical to minimize the idle (ground) time of aircraft and crew between flights. To further explain this concept, consider six different schedules for three hypothetical flights F1, F2, and F3 as given in Figure 3.10 *case a* through *case f*. For instance, when flights F1 and F2 are scheduled as shown in *case a*, an aircraft has to stay on the ground most of the day and overnight at ATL to operate the next flight F2. Similarly, in *case b*, when one aircraft is operating flights F2 and F3, this aircraft has to stay overnight and also spend long ground time during the next day at ORD. In *case c*, one aircraft can operate the three flights F1, F2, and F3, with minimum ground idle time between flights. Flights F1, F2, and F3 can also be scheduled to connect between a late-night arrival and early-morning departure. For example, in *case d*, flights F2 and F3 are scheduled such that flight F2 is a late-night arrival at ORD and flight F3 is an early-morning departure at ORD. Similarly, in *case e*, flights F1 and F2 are scheduled such that flight F1 is a late-night arrival at ATL and flight F2 is an early-morning departure at ATL. In *case f*, the schedule of the three flights F1, F2, and F3 is not coordinated efficiently, as an aircraft has to stay overnight and has long ground time during the next day at ATL to operate both F1 and F2. Similarly, an aircraft has to stay overnight and spend long ground time during the next day at ORD to operate flights F2 and F3.

3.2.8 Flight Schedule and Fleet Assignment Balance

To optimize aircraft rotations, flights that are assigned to the same fleet type have to be scheduled in a way that creates optimal turns for the aircraft and the crew at each station (Hane et al. 1995). For example, as shown in *case a* of Figure 3.11, flights F1 and F4 are assigned to a large aircraft, while flights F2 and F3 are assigned to a small aircraft. It is expected that one large aircraft operates flights F1 and F4 with long ground time at ORD. Meanwhile, one small aircraft operates flights F3 and F2 with overnight turn at ORD. This combination of flight scheduling and fleet assignment does not produce efficient aircraft rotations. Accordingly, one might recommend changing the schedules of flights F2 and F4, as shown in Figure 3.11 *case b*. Modifying the schedules of flights F2 and F4, more efficient rotations are obtained at ORD. A large aircraft can be used to operate flights F1 and F4, and a small aircraft can be used to operate flights F3 and F2 with efficient turn times.

3.2.9 Maintenance Rotations and Fleet Assignment

As part of scheduling the aircraft rotations, maintenance activities are required to be scheduled along the aircraft route (Clarke et al. 1996). Accordingly, the aircraft has to be positioned at an appropriate maintenance station at the right

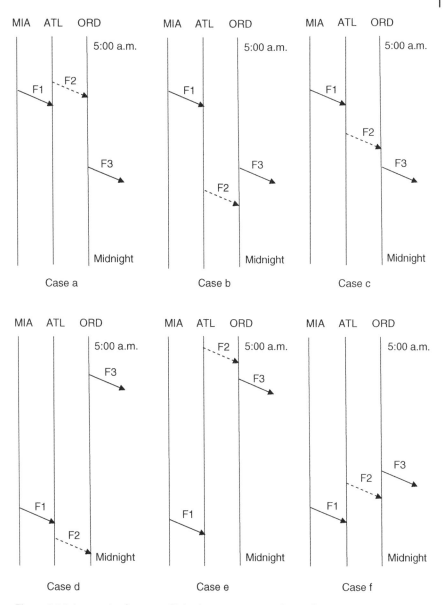

Figure 3.10 Interaction between flight departure time and aircraft rotation.

time. For example, Figure 3.12 shows an example of an aircraft rotation for an aircraft of fleet type I. As shown in the figure, there is a scheduled maintenance activity after day 4 at ORD. Thus, the destination of the last flight of day 4 should be ORD. In this example, a flight from ATL-ORD is used to position the

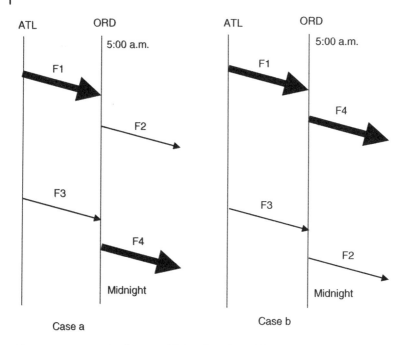

Figure 3.11 Interaction between flight schedule and fleet assignment balance.

aircraft at ORD. In some situations, schedulers may not be able find a suitable flight to position the aircraft at ORD toward the end of the day. A possible reason could be that all flights between ATL and ORD are assigned to other fleet types (e.g. fleet type II or fleet type III). To address this issue, the fleet assignment solution is modified, and a flight between ATL and ORD is reassigned to fleet type I to allow sending the aircraft to ORD. This fleet reassignment is shown in the lower part of Figure 3.12. As shown in the figure, a flight between ATL and ORD is assigned to fleet type I to rotate the aircraft to satisfy the required maintenance activity at ORD.

3.2.10 Seat Capacity/Frequency and Demand

It is generally observed that when airlines add more flights in a certain city-pair, the frequent presence in the market attracts more demand to the airline. Having frequent service is expected to generate attractive itineraries that are more likely to be selected by customers. Higher (lower) flight frequency results in disproportionately higher (lower) market shares. A similar phenomenon is represented by the S-curve that gives the relationship between the airline seat capacity in a given market and the percentage of passengers selected to fly on

Flights assigned to fleet I

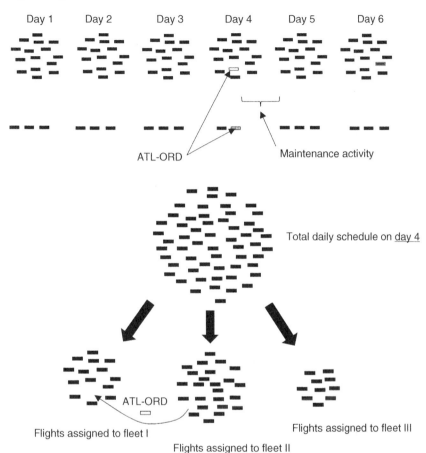

Figure 3.12 Interaction between aircraft rotation and fleet assignment.

this airline. There is much empirical evidence that higher (lower) seat capacity by one airline in the market is associated with disproportionately higher (less) market share. Figure 3.13 shows a representation of the S-curve. As shown in the figure, the dotted line provides the relationship between the provided seat capacity by one airline and its proportional market share. As this airline provides additional seat capacity, empirical evidence shows that this airline gains additional market share beyond the proportional level, and vice versa, as shown by the solid line (i.e. the S-curve) (Wei and Hansen 2005).

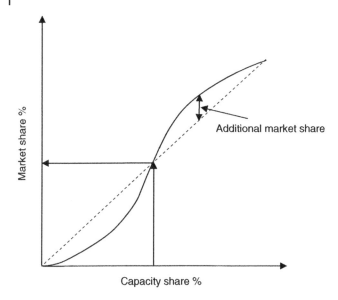

Figure 3.13 The S-curve showing the relationship between seat capacity and market share.

3.2.11 Feet Assignment and Flight Demand

As mentioned above, the objective of the fleet assignment problem is to assign the most suitable fleet type to each flight in the schedule. A key information required to obtain the optimal fleet assignment solution is to estimate the expected demand (number of passengers) on each flight. This demand depends on several factors including the airline's schedule attractiveness, competition, and pricing. When solving the fleet assignment problem, it should be clear that the demand of a flight depends on the aircraft type assigned to other flights. For example, in Figure 3.14, an airline is scheduling two flights, F1 and F2, in the ATL-ORD market. The demand on flight F2 depends on what fleet type is assigned to flight 1 and vice versa. For instance, as shown in *case a*, when a small aircraft is assigned to flight F1, some demand might be spilled from flight F1 to flight F2, due to the limited seat capacity of flight F1. On the contrary, in *case b*, when a large aircraft is assigned to flight F1, there will be no demand spilled to flight F2.

3.2.12 Frequency and Departure Time

For an airline that operates a hybrid network structure (i.e. hub and spoke and also point-to-point), deciding on serving a given city-pair could be through a nonstop service with a certain frequency or through connecting service or combination of both. For example, Figure 3.15 shows an airline that can serve the

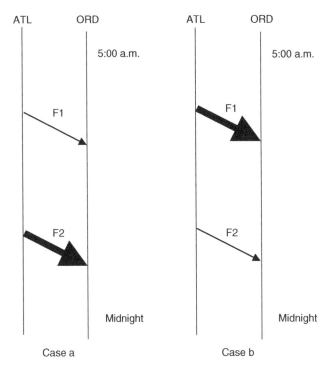

Figure 3.14 Interaction between fleet assignment and flight demand.

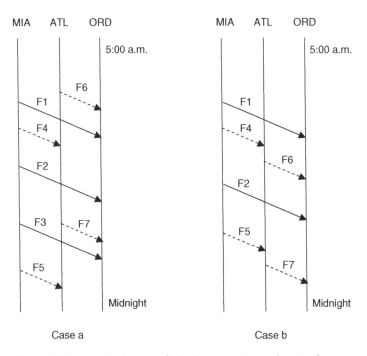

Figure 3.15 Interaction between flight departure time and service frequency.

MIA-ORD market through three nonstop flights F1, F2, and F3, as shown in *case a*. Alternatively, this airline can change the schedule of F4, F5, F6, and F7 to generate connecting itineraries between MIA and ORD through ATL. As shown in *case b*, two connecting itineraries, F4–F6 and F5–F7, are serving the MIA-ORD market. This airline might decide to reduce the nonstop service between MIA and ORD from three to two flights only. In this case, two nonstop itineraries and two connecting itineraries are serving the city-pair MIA-ORD.

3.2.13   Departure/Arrival Time and Gate Availability

Airline might have limited number of gates at some airports. Accordingly, a certain number of flights, equal to the number of gates, can be scheduled to simultaneously exist at these gates (Dorndorf et al. 2007). For example, *case a* of Figure 3.16 shows the schedule of an airline with eight flights. Assume that this airline has access to only four gates at ORD. Consequently, this airline can schedule a maximum of four flights that arrive at ORD at the same time window. As shown in *case a* of Figure 3.16, flights F1, F2, F3, and F4 are scheduled to arrive at ORD at the same time window. Then, later in the day, another four flights (F5, F6, F7, and F8) are scheduled to depart simultaneously at these four gates.

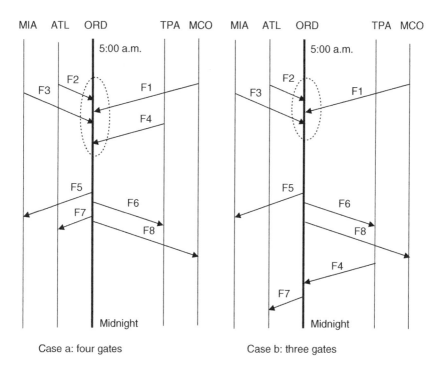

Figure 3.16 Interaction between departure/arrival times of flights and gates availability.

Now, assume that the airline has access to only three gates at ORD. The airline cannot schedule the four flights to arrive at the same time window anymore. In this case, the airline can schedule one of the four flights to arrive at ORD at a different time, when a gate becomes available. For example, in *case b* of Figure 3.16, flight F4 is scheduled later in the day. The flights F1, F2, and F3 are arriving simultaneously occupying the three available gates. Similarly, one of the outbound flights F4, F5, F6, and F7 has to be rescheduled. As shown in the figure, flight F7 is scheduled later in the day, allowing flights F5, F6, and F8 to depart simultaneously at the three gates.

This example shows that the departure and arrival times of the different flights are interrelated decisions, as they are controlled by available resources at the different airports (e.g. gates). The same example can be extended to represent the case where the departure/arrival time decisions of flights are controlled by the availability of aircraft, crew, and other resources.

3.2.14 Departure Time and Crew Schedule

The flight departure times of the flights could have significant impact on the crew schedule and associated operation cost. Airline crews are typically positioned at the airline's hub(s), where each crew member has a home base or a

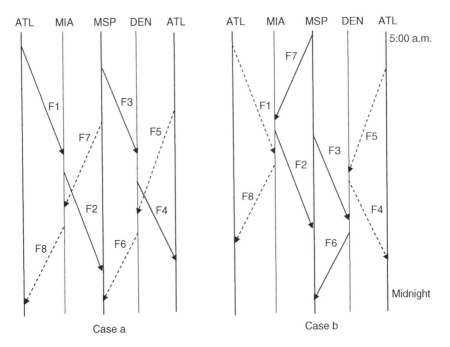

Figure 3.17 Interaction between departure/arrival times of flights and crew schedule.

domicile. When the crew makes a layover night at his/her domicile, there is no hotel or accommodation cost required. However, if the layover is away from the domicile, the airline pays for the crew accommodation.

To explain this issue, consider the schedule of the hypothetical airline given in Figure 3.17, where the airline has two hubs at ATL and MSP. The airline also serves two spoke destinations, which are MIA and DEN. This airline operates a schedule of eight flights. Two flight schedule alternatives are shown in *case a* and *case b* of Figure 3.17. Two crews are needed to operate these eight flights. The crew scheduling typically avoids having long connection periods for the crew between two successive flights on the same day to increase crew utilization and efficiency.

In *case a*, the two crews start and end their schedule at their ATL domicile. The first crew operates the flight sequence F1-F2-F3-F4 with a layover night at MSP. The second crew operates the flight sequence F5-F6-F7-F8 with a layover night at MSP. In both cases, hotel and accommodation cost are needed for the two crews, since the layover is not at their domicile. In *case b*, the departure time of some of the flights is adjusted. Thus, one crew positioned at ATL, which operates the flight sequence F1-F8-F5-F4 with a layover night at ATL. The other crew positioned at MSP, which operates the flight sequence F7-F2-F3-F6 with a layover night at MSP. In this case, there is no hotel or accommodation cost needed for the two crews, since both crews are making the layover at their domiciles. In either case, the airline has to examine the trade-off between saving the crew accommodation cost and possible revenue loss due to changing the flight departure times.

4

Measures of Performance

Several measures of performance are used by the airline network planning practitioners. Some of these measures are undisclosed information and kept confidential by the airline, while other measures are reported to governments, airports, and aviation agencies. Airlines use their own performance measures to evaluate the impact of their decisions. They also observe the performance of their competitors to extract useful information on their strategies and possible moves. It is very crucial to understand these measures and interpret their meaning. These performance measures are easy to understand when compared across airlines or when compared against historical values. Thus, these measures are typically presented in the form of a time series data that illustrates their values over a defined past horizon.

4.1 Operating Cost

Operating costs are the expenses associated with the operation of the airline. They include the cost of resources used by an airline to maintain the operations (Tsoukalas et al. 2008; Zou and Hansen 2012). Operating cost is usually classified as direct operating cost and indirect operating cost. Direct operating costs are the cost elements that depend on the aircraft type(s) used by the airline. Indirect operating costs are cost elements that do not depend on the aircraft type. Example of direct operating cost includes cockpit crew cost, maintenance cost, landing fees, aircraft insurance, aircraft rents and ownership, fuel consumption, and food and beverages. Example of indirect operating cost includes salaries of management personnel; utilities and office supplies; employee business expenses such as lodging, meals, and transportation; passenger insurance; commissions to sales agents; advertising and promotions; etc.

Airline Network Planning and Scheduling, First Edition. Ahmed Abdelghany and Khaled Abdelghany.
© 2019 John Wiley & Sons, Inc. Published 2019 by John Wiley & Sons, Inc.

About two thirds of the airline's operation cost can be considered as fixed cost irrespective of how many passengers it carries. The marginal cost of carrying an additional customer is usually very small to the airline. In the United States, airline reports their cost in a designated form known as form 41.

4.2 Revenue or Income

Revenue (or income) represents the amount of money collected by an airline during a specific period from selling tickets for seats, cargo transportation services, and other ancillary revenue sources including baggage fees, ticket change fees, meals sale, etc. Most of monetary values like revenue are reported quarterly. Figure 4.1 shows examples of the revenues reported by American Airlines (AA) and Delta Air Lines (DL) for the period 2000–2016. Delta Air Lines merged with Northwest Airlines in 2009, while American merged with US Airways in 2015, which explains the sudden increase in revenue for both airlines in 2012 and 2015, respectively. An increase in the revenue does not necessarily mean that the airline is profitable. Operating cost is subtracted from the revenue to determine the airline's net income or profitability. In some cases, an increase in the revenue could also be associated with a significant cost increase that results in net losses.

Operating revenue for American Airlines (AA) and Delta Air Lines (DL) (quarterly)

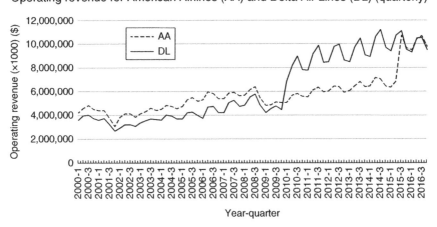

Figure 4.1 Operating revenue for American Airlines and Delta Air Lines for the period 2000–2016.

4.3 Net Income (Net Profit) and Operating Profit

Net income is the main indicator of airline's profitability and is calculated as the total revenue minus all expenses for a defined accounting period (quarterly). Net income (profit or loss) and operating profit or loss are two different measures of airline financial performance. Net profit or loss includes nonoperating income and expenses and nonrecurring items or income taxes. Operating profit or loss is calculated from operating revenues and expenses before taxes and other nonrecurring items. Figure 4.2 shows examples of the net income for American Airlines (AA) and Delta Air Lines (DL) for the period 2000–2016. As shown in the figure, American Airlines were experiencing significant losses until early 2014. Net income improved in 2015 after the merge with US Airways. Similarly, Delta Air Lines reported significant loss until the airline merged with Northwest Airlines in 2012 after which positive net income is reported.

4.4 Flights

Number of flights (scheduled or actually operated by the airline) for a given time period is another important measure of performance. The scheduled and operated number of flights could be different in case of flight cancellations due to irregular operations conditions (e.g. adverse weather). Scheduled flight data are available for past periods and could also be available for few months in the

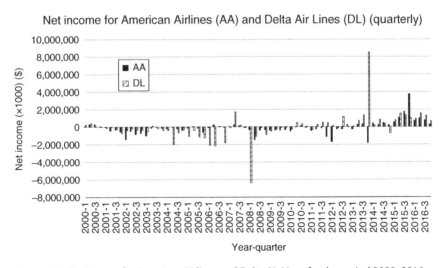

Figure 4.2 Net income for American Airlines and Delta Air Lines for the period 2000–2016.

Flights scheduled by American Airlines (AA) and Delta Air Lines (DL) (monthly)

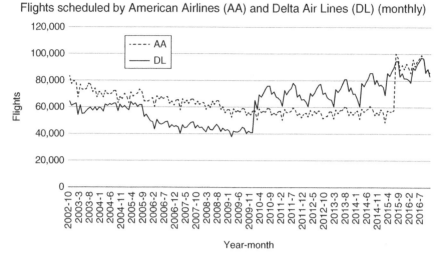

Figure 4.3 Number of flights scheduled by American Airlines and Delta Air Lines for the period 2002–2016.

future after airlines publish their future schedules. Operated flights data are available for past periods only. They are recorded for the schedule-based airlines as well as for other services including charter, business jets, and general aviation. This data is usually available through airports records and aviation authorities. Flights are usually classified based on whether they are scheduled or nonscheduled; domestic or international flights; regional or mainline; short haul, medium haul, or long haul; or per aircraft type.

Figure 4.3 shows the number of flights scheduled by American Airlines (AA) and Delta Air Lines (DL) for the period 2002–2016. The numbers do not include the flights by their regional airlines partners. As explained above, the sudden increase in the number of flights for AA and DL is due to the merge with US Airways and Northwest Airlines, respectively. Before merging, both airlines were continuously and gradually cutting number of scheduled flights to eliminate nonprofitable flights. In addition, the figure shows that the number of flights is changing in a systematic pattern from year to year. This change reflects how airlines adjust their schedule to respond to seasonal changes in demand.

When the number of scheduled flights increases for a given airline, it means that this airline is expanding their flight frequency in one or more of the city-pairs they serve. Some analysts might depend on number of flights as an indication of the airline size. In fact, number of scheduled flights is not an accurate measure of the airline size. Number of flights cannot solely explain the increase/decrease in airline size. To elaborate, consider two airlines with 100 scheduled daily flights each. Assume the average travel distance per flight is 1000 miles and 2000 miles for the two airlines, respectively. While these two airlines are equivalent in terms of number of scheduled flights, the second airline will require approximately double the resources to run its schedule compared with the first airline including aircraft, fuel consumption, and crew duty periods.

Consider another example in which there are two airlines with 100 scheduled daily flights each. The average travel distance per flight for both airlines is 1000 miles. The main difference between the two airlines is that the first airline operates small aircraft with average seat capacity of 50 seats, while the second airline operates larger aircraft with average seat capacity of 250 seats. While these two airlines are equivalent in terms of the number of scheduled flights and average trip length, the two airlines cannot be considered equal in size due to the difference in the number of passengers they could possibly serve.

4.5 Available Seat Miles

Available seat miles (ASM) (or available seat kilometers (ASK) in countries that adopt the standard metric system) is a performance measure that simultaneously accounts for number of flights, flight distance, and number of seats. It is calculated as the product of the seat capacity of the flight and its distance and is summed for all flights. To illustrate the calculations of the ASM, consider a hypothetical airline that operates six daily flights, given in Table 4.1. The distance and seat capacity of each flight are as given in the second and third columns, respectively. The fourth column gives the product of the distance and seat capacity for each of the six flights. The bottom cell of the column gives the total ASM for the airline.

Table 4.1 Example of ASM calculation.

Flight	Distance (miles)	Aircraft seat capacity (seats)	Distance × seats
1	1,500	220	330,000
2	1,500	220	330,000
3	1,800	230	414,000
4	1,800	230	414,000
5	950	120	115,900
6	950	120	115,900
Total available seat miles			1,719,800

Available seat miles for American Airlines (AA) and Delta Air Lines (DL) (monthly)

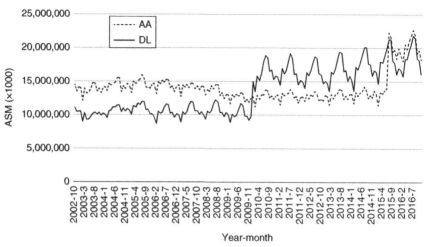

Year-month

Figure 4.4 Available seat miles for American Airlines and Delta Air Lines for the period 2002–2016.

The unit of airline productivity is one seat moved a unit distance (e.g. one mile). It does not matter whether this seat is full or empty. The more an airline moves seats, it is considered to have more production. In the above example, this hypothetical airline has 1,719,800 units of production (i.e. seat miles). Since ASM is a combined measure, the change in its value should be carefully interpreted. For example, an increase in ASM for a given airline could be due to the increase in number of flights, increase in the size of the aircraft (i.e. seat capacity or seat density), and/or increase in length of haul of their trips. Some ultralow-cost airlines increase their ASM by changing the configuration of their aircraft to increase the number of seat rows (i.e. increase seat density). Figure 4.4 shows the ASM by American Airlines (AA) and Delta Air Lines (DL) for the period 2002–2016. As explained above, the sudden increase in the ASM for AA and DL is due to the merge with US Airways and Northwest Airlines, respectively. Similar to the number of flights, the figure shows that the ASM is changing in a systematic pattern from year to year. This change reflects how airlines adjust their ASM to respond to seasonal demand changes.

4.6 Cost per Available Seat Miles (CASM)

The CASM is defined as the total operating cost of an airline divided by its ASM. The CASM measures how much on average it costs an airline to move one seat for a unit distance. When metric system is used, this measure is

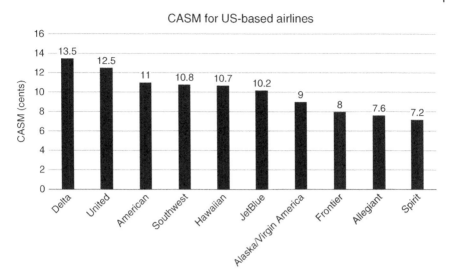

Figure 4.5 CASM for US-based airlines for Q2 of 2016.

typically known as cost per available seat kilometer (CASK). The airline's operating cost includes all cost elements involved in the operations. CASM is considered as a normalization of the operating cost, and hence it is used to compare the operating cost of different size airlines. Figure 4.5 shows the CASM values for the US-based airlines for Q2 of 2016. As shown in the figure, CASM ranges from 7.2 cents to about 13.5 cents.

It should be noted that the cost per mile for a short-haul flight is higher than that of a long-haul flight because many cost elements are distributed over more miles. In other words, CASM depends on the average flight length (also known as average stage length) of the airline. For instance, the CASM for a 500 miles flight is likely greater than that for a 1500 miles flight. Thus, in some cases, an airline whose average stage length is 500 miles and has a CASM of 12 cents is likely more efficient than an airline with a CASM of 10 cents but with a 1500-mile average stage length. Accordingly, when comparing CASM for different airlines, a stage-length-adjusted (SLA) CASM is used.

4.7 CASM-EX or CASM-EX Fuel

CASM-EX is calculated similar to CASM, except that fuel cost is excluded from the total operating cost. This measure is more suitable to compare cost of airlines by excluding the fuel cost that is more volatile and instable due to

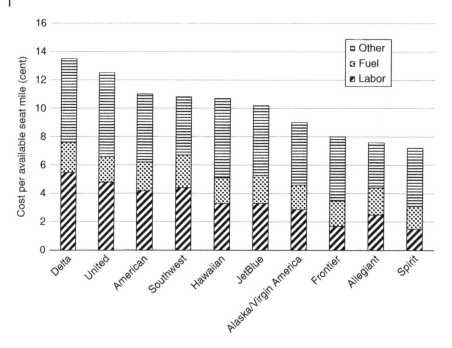

Figure 4.6 CASM components for Q2 2016.

market conditions. In addition, fuel cost could vary among airlines by region. Some airlines might also have fuel hedging contracts or fuel subsidies that allow them to have lower fuel cost. Figure 4.6 shows the main components of CASM including labor and fuel cost for the US-based airlines in Q2 of 2016. As shown in the figure, the fuel cost varies among airlines significantly, and when excluded, a more meaningful comparison among airline operations cost can be performed.

4.8 Passengers

Number of passengers carried by an airline is another performance measure. The data could be reported for a given flight, for the city-pair, or for the entire schedule. Passengers are usually classified based on whether their travel is domestic or international. They are also classified based on whether they are traveling in the local or connecting market. They are also classified as revenue and nonrevenue passengers. Nonrevenue passengers include employees of the airline and their families that usually get standby travel benefits. The time

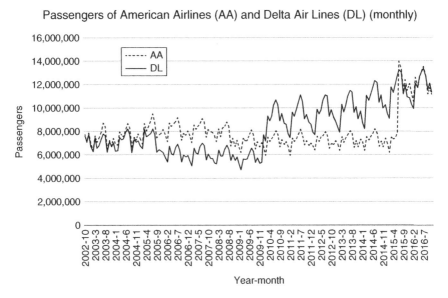

Figure 4.7 Number of passengers for American Airlines and Delta Air Lines for the period 2002–2016.

period over which the data is aggregated and reported varies depending on whether it is for internal use by the airline or for external reporting. The passenger data is usually reported monthly at the origin-destination (OD) level. Treated as a proprietorial information, passenger data at the flight level are rarely disclosed. Figure 4.7 shows the number of passengers for American Airlines (AA) and Delta Air Lines (DL) for the period 2002–2016. As explained above, the sudden increase in the number of passengers for AA and DL is due to the merge with US Airways and Northwest Airlines, respectively. The figure shows that the number of passengers is changing in a systematic pattern. The change reflects demand seasonality, as more passengers travel in the summer months.

It is important to correctly interpret number of passengers as a measure of performance. For example, what does it mean when number of passengers goes up? An increase in number of passengers for a given airline does not mean that the airline is necessarily more profitable, and vice versa. Airline could be attracting more passengers by selling cheap tickets. The main limitation of using number of passengers as a performance measure is that they do not reveal any information on revenue. In addition, when these numbers are aggregated and compared among airlines, they might result in misleading conclusions, as they do not reveal the distance of travel for these passengers. To

Table 4.2 Example of RPM calculation.

Flight	Distance (miles)	Revenue passengers	Distance passengers
1	1,500	208	312,000
2	1,500	212	318,000
3	1,800	217	390,600
4	1,800	225	405,000
5	950	111	105,450
6	950	109	103,550
Total revenue passenger miles			1,634,600

elaborate, consider two airlines that carried 15,000 passengers on a given day. The average distance per flight for the first airline is 1000 miles, and the average distance per flight for the second airline is 2000 miles. While these two airlines have the same number of passengers, the second airline will require approximately double the time/resources compared with the first airline to carry those passengers.

4.9   Revenue Passenger Miles (RPM)

RPM accounts for both number of passengers and also the distance traveled by each revenue passenger in miles. It is calculated by multiplying the number of revenue passengers of the flight by the flight distance, and the product is summed for all flights. To illustrate the calculations of the RPM, consider the hypothetical airline that operates six daily flights given in Table 4.2. The distance and number of revenue passengers of each flight are as given in the second and third columns, respectively. The fourth column shows the multiplication of the distance and number of revenue passengers for each of the six flights. The bottom cell of the column gives the total RPM for the whole airline, which is equal to 1,634,600. As discussed hereafter, RPM is typically less than ASM. RPM is equal to ASM only if all seats are sold to revenue passengers. Similar to ASM, since RPM is a combined metric, the change in its value should be carefully interpreted. For example, an increase in RPM for a given airline could be due to an increase in number of flights, number of revenue passengers, and/or increase in length of haul of their trips, and vice versa. Figure 4.8 presents the revenue passenger miles for American Airlines (AA) and Delta Air Lines (DL) for the period 2002–2016.

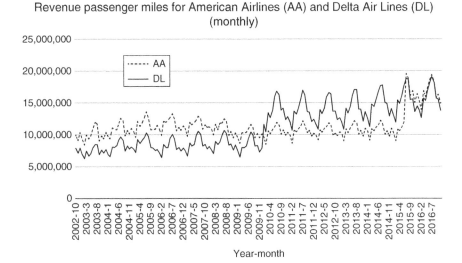

Figure 4.8 Revenue passenger miles for American Airlines and Delta Air Lines for the period 2002–2016.

4.10 Total Revenue per Available Seat Mile (TRASM or Simply RASM)

RASM is a measure used to compare the revenue of airlines of different sizes. Similar to CASM, RASM normalizes the revenue by dividing revenue over the number of produced units (ASM) for each airline. It includes all revenue sources including seat tickets, cargo, ancillary revenue, reservation change fee, and other revenue sources. Figure 4.9 shows an example of RASM values for several US-based airlines in Q2 of 2016. As shown in the figure, RASM values range from 8.5 cents to 16.5 cents. Higher RASM does not necessarily mean the airline is profitable. Instead, it indicates that the airline generates more revenue. The difference between RASM and CASM gives indication on profitability. Similar to CASM, revenue per mile for a short-haul flight is higher than that of a long-haul flight because revenue elements are distributed over more miles. Accordingly, when comparing RASM for different airlines, an SLA RASM is used.

4.11 Passenger Revenue per Available Seat Mile (PRASM)

It is another derivative of RASM, when only passenger revenue and not total revenue is divided by ASM. Passenger revenue is a revenue from selling tickets and other ancillary revenue such as baggage fees, meals, etc. Some versions of

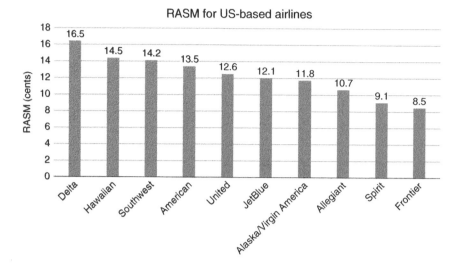

Figure 4.9 RASM for several US-based airlines for Q2 of 2016.

PRASM calculations might exclude ancillary revenue and account for ticket sales only. It is often referred to as a measure of passenger "unit revenue."

4.12   Passenger Yield

Yield is calculated by dividing passenger revenue by revenue passenger miles. Yield is a useful measure in assessing changes in fares and ancillary fees over time. Both yield and PRASM can be calculated per route, per market, and for the entire schedule. The values of yield and PRASM usually vary by stage length. Thus, in many cases, they are not useful for comparisons across markets with significant difference in stage length. The change of these two measures is usually reported over time for a given route or market.

4.13   Average Load Factor (LF)

The average load factor (LF) indicates an average ratio (percentage) of full seats on the flights by the airline, calculated over the whole distance flown by the airline. It is calculated by dividing the airline's RPM over its ASM. For example, for the hypothetical airline given above, the RPM and the ASM are 1,634,600 and 1,719,800, respectively. Accordingly, the average LF is equal to 1,634,600/1,719,800 = 0.95 (or 95%). This means that on average for every 100 seats offered by this airline, 95 seats are full by revenue passengers and

Revenue passenger miles and available seat miles for American Airlines (monthly)

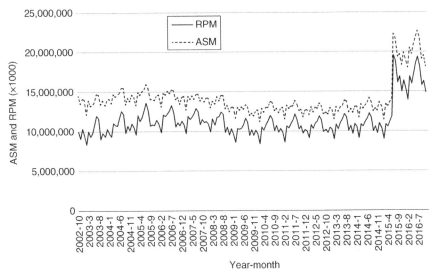

Figure 4.10 Revenue passenger miles and available seat miles for American Airlines for the period 2002–2016.

5 seats are either empty or filled by nonrevenue passengers. Figure 4.10 gives a comparison of RPM and ASM for American Airlines over the period 2002–2016. As shown in the figure, RPM is always less than ASM. It is a good sign that there is no large gap between RPM and ASM. It indicates that the airline is filling its available seats with revenue passengers. Figure 4.11 shows the average LF for American Airlines (AA) and Delta Air Lines (DL) for the period 2002–2016. The figure shows that LF is changing in a systematic pattern from year to year. This change reflects demand seasonality, where flights gets busier in summer months. Airlines might need to further cut schedule in winter months to avoid having empty flights, especially when these flights have low revenue.

Another inaccurate (possibly wrong) method for calculating average LF is through dividing number of revenue passengers over number of seats averaged for all the flights. For example, in the example given above, the parentage of full seats is calculated for each flight, as shown in Table 4.3. Then, the average over the six flights is calculated to be 94.4%. The limitation of this calculation is that it does not reflect the flight size or distance over which the seat was full. To further explain the limitation of this method, consider an airline that has four flights as given in Table 4.4. Flights 1 and 3 are short-haul flights (500 miles) operated with small aircraft (50 seats) and fly empty. Flights 2 and 4 are long-haul flights (3000 miles) operated with large aircraft (300 seats) and fly

Average load factor of American Airlines (AA) and Delta Air Lines (DL) (monthly)

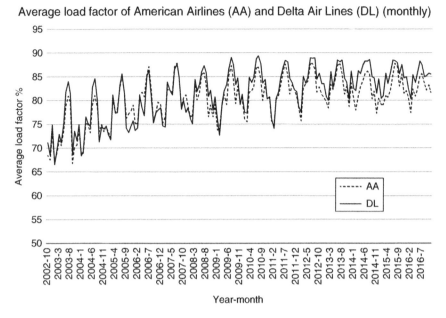

Figure 4.11 Average load factor for American Airlines and Delta Air Lines for the period 2002–2016.

Table 4.3 Calculations for LF for individual flights.

Flight	Revenue passengers	Aircraft seat capacity (seats)	% of full seats
1	208	220	94.5
2	212	220	96.4
3	217	230	94.3
4	225	230	97.8
5	111	120	92.5
6	109	120	90.8
Average percentage of full seats			94.4%

completely full. For this hypothetical airline, the RPM and the ASM are calculated as 1,800,000 and 1,850,000, as shown in Tables 4.5 and 4.6, respectively. Thus, the average LF is equal to 1,800,000/1,850,000 = 0.973 (or 97.3%). When calculating the percentage of full seats on each flight independently and then calculating the average over the four flights, it results in an LF of 50%, as given in Table 4.7. The difference in the LF obtained from both methods is significant.

Table 4.4 Information of a hypothetical airline with four flights.

Flight	Distance (miles)	Revenue passengers	Aircraft seat capacity (seats)
1	500	0	50
2	3000	300	300
3	500	0	50
4	3000	300	300

Table 4.5 Calculations of ASM for a hypothetical airline with four flights.

Flight	Distance (miles)	Aircraft seat capacity (seats)	Distance × seats
1	500	50	25,000
2	3000	300	900,000
1	500	50	25,000
2	3000	300	900,000
Total available seat miles			1,850,000

Table 4.6 Calculations of RPM for a hypothetical airline with four flights.

Flight	Distance (miles)	Revenue passengers	Distance × passengers
1	500	0	0
2	3000	300	900,000
1	500	0	0
2	3000	300	900,000
Total revenue passenger miles			1,800,000

Table 4.7 Calculations of LF for individual flights for a hypothetical airline with four flights.

Flight	Revenue passengers	Aircraft seat capacity (seats)	% of full seats
1	0	50	0
2	300	300	100
3	0	50	0
4	300	300	100
Average percentage of full seats			50%

This difference highlights the inaccuracy of the second method as it treats flights equally regardless of their sizes or travel distances.

The change in the value of the average LF has to be carefully interpreted. An increase in the LF does not necessarily indicate an increase in airline's revenue or profitability. Increase in the average LF could be due to increase in demand and/or reduction in supply (e.g. flight frequency and/or aircraft size). The LF is a good measure that assists the airline in matching its capacity to the demand observed in the different markets.

4.14   Block Hours

The block hour for a given flight is the time from which the aircraft pushes back from the departure gate until the moment the aircraft arrives at the arrival gate after landing. The block hour of the flight includes its taxi-out time, flight airborne time, and taxi-in time. Taxi-out time is the time from pushing back at the departure gate to the departure runway. Taxi-in time is the time from the landing runway to the arrival gate. The number of block hours for an airline for a given period of time (e.g. month, quarter, or year) measures the total time that its aircraft were utilized during that period.

4.15   Aircraft Utilization

Aircraft utilization is used to normalize block hours. Presented in terms of block hours per day, it is calculated by dividing aircraft block hours by number of aircraft days assigned to service. Aircraft utilization is usually calculated for every fleet type. It is a measure of aircraft productivity that is used to compare performance across different airlines and for the same airline over time (Wu and Caves 2004).

4.16   Stage Length

Stage length represents the average distance flown by the airline. It is calculated by dividing the total aircraft miles (statute miles) flown by the number of flights (aircraft departures performed). Stage length is used to differentiate between airlines serving short-haul and long-haul markets, where they typically have different characteristics. When all other factors remain equal, the CASM, RASM, and yield decrease as the stage length increases, as cost and revenue are distributed over more miles.

4.17   On-time Performance Measures

As discussed in more details later in Chapter 21, there is a strong relationship between the schedule design and service punctuality (Ageeva 2000; Jiang and Barnhart 2013). Therefore, it is important to understand the major measures that are used to evaluate service punctuality. Examples of these measures include:

Completion rate: The percentage of scheduled flights completed as a ratio of total number of scheduled flights.

Cancellation rate: The percentage of scheduled flights that are canceled (100% minus completion rate). This cancellation could be due to aircraft mechanical problems, adverse weather conditions, computer systems problems, etc.

D + zero (D+0): Percentage of flight departures that push back from the gate on time as scheduled.

D + five (D+5): Percentage of flight departures that push back from the gate within five minutes of the scheduled time.

A + 15: Percentage of flight arrivals within 15 minutes of the published arrival time.

4.18   Aircraft Life Cycle

Aircraft is the main asset of the airline industry. Airlines make sure that their aircraft are utilized efficiently to guarantee that the highest return on investment is achieved. Typically, each aircraft is designed to perform a certain number of cycles. A cycle is defined as a takeoff and a landing (i.e. one flight). For example, the Airbus aircraft A320 is designed for a lifespan of 48,000 cycles. The Boeing aircraft B737 is designed for a lifespan of 75,000 cycles. The Boeing aircraft B777 is designed for a lifespan of 40,000 cycles. The lifespan of the aircraft can easily be exceeded with the proper maintenance. As the aircraft reaches its maximum lifetime, it usually goes through a major maintenance to keep it in operation. The viability of the decision to extend the lifespan of the aircraft depends primarily on its operating cost.

It is important to understand the cost–revenue ratio for an aircraft. Consider a Boeing 737-800 aircraft with a seat capacity of 162 seats, which has a lifespan of 75,000 cycles. In other words, this aircraft is designed to fly about 75,000 one-way flights. Assume that the airline that operates this aircraft can schedule 5 daily flights using this aircraft, where each flight is about two hours. This means that theoretically the aircraft can operate for more than 40 years. We assume that the airline operates the aircraft for only 20 years (some aircraft go much beyond that) with an average LF of 85% and an average one-way fare of $150. Accordingly, assuming zero interest, the revenue generated using this aircraft is revenue = seats × LF × ticket price × number of daily flights × years × 365 = 162 × 0.85 × $150 × 5 × 20 × 365 = $753,907,500.

This means that an aircraft can generate about $0.75 billion during its lifespan. When comparing this number with an average price of the aircraft, which is about $50–100 million (some aircraft might be cheaper than this range), it translates to the fact that the ratio of the aircraft price to its revenue over its lifespan is in the range of 6–12%. In other words, the cost of buying the aircraft is very small compared with the revenue it generates. Other operation cost elements such as fuel cost, labor cost, landing and overflying fees, and maintenance cost play more significant impact on airline profitability.

4.19   Aircraft Number and Diversification

In some cases, the size of an airline is measured in terms of the number of aircraft that the airline operates. The measure is more useful when the number of aircraft is defined per each fleet type. Airlines select the type of aircraft that is most suitable for its service and network structure. Seat capacity, aircraft range, purchase price, operations cost, age, and delivery date are among the variables that determine the airline's decision on aircraft acquisition. Figure 4.12 gives the seat capacity and maximum range for most common commercial aircraft available in the market. The seat capacity is shown on the horizontal axis, and the aircraft maximum range is shown on the vertical axis. Aircraft are usually divided based on their range as short-haul, medium-haul, and long-haul aircraft. Short-haul aircraft are typically used by regional/shuttle airlines, while long-range aircraft are used for long international markets by international airlines. Airlines usually use aircraft with high seat capacity at constrained airports that have limited landing slots. In each city-pair, when selecting aircraft seat capacity, the airline might decide on operating less number of flights with large aircraft or frequent service with small aircraft.

Some airlines adopt a strategy of diversifying its fleet types. This diversification allows these airlines to compete in most city-pairs with variant demand and haul length. On the other end, other airlines use a single fleet type. Typically, low-cost airlines fall under this category. Operating a single fleet reduces cost associated with aircraft maintenance and crew training. It also simplifies recovery during irregular operations, where there are more opportunities to swap aircraft and crew at the different stations. However, on the other hand, these airlines only operate in the

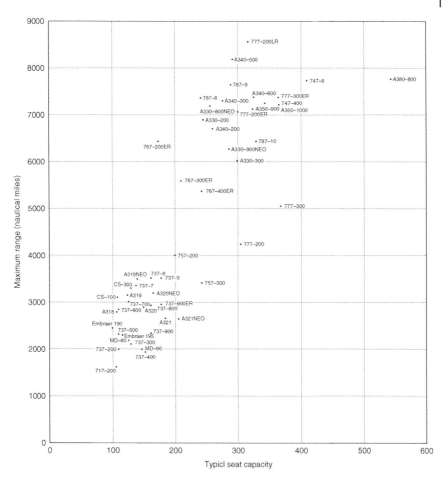

Figure 4.12 Aircraft seat capacity and maximum range.

city-pairs that can be served by the available fleet type. Tables 4.8 and 4.9 give the aircraft portfolio for major US and European carriers, respectively. The tables give the aircraft that are currently in operation and those that are in order from the manufacturer. The tables also show the average aircraft age for each fleet type. As shown in Table 4.8, Southwest Airlines operate only the Boeing 737 family, while American Airlines, Delta Air Lines, and United Airlines have more diversified fleet. Similarly, as shown in Table 4.9, Ryanair also operates only the Boeing 737 family. easyJet operates primarily the Airbus 318 and 320 families. Lufthansa and KLM operate more diversified fleet.

Table 4.8 Aircraft portfolio for major US carriers.

	Delta Air Lines			American Airlines			United Airlines			Southwest Airlines		
	In service	Average age	In order	In service	Average age	In order	In service	Average age	In order	In service	Average age	In order
717-200	91	16										
737 CL												
737NG	173	9	34	301	8	3	329	11	5	679	11	21
737 MAX				2	0	98			160	12	0	188
747-400	6	26										
747-8												
757-200	110	21		38	19		56	22				
757-300	16	15					21	15				
767-300	60	22		25	19		34	22				
767-400	20	17					16	16				
777-200	18	13		47	17		74	19				
777-300				20	4		14	1	4			
777-8X												
777-9x												
787-8				20	2		12	5				
787-9				12	1	10	21	2	4			
787-10									14			
A319-100	57	16		125	14		63	17				
A320-200	64	22		48	17		97	20				

A321-100												
A321-200	28	1	94	218	5							
A320neo						100						
A330-200	11	13		15	6							
A330-300	31	9		9	17							
A330neo			25									
A340-300												
A340-600												
A350-900	3	0	22			22			45			
A350-1000												
A380-800												
MD-82				10	29							
MD-83				36	20							
MD-88	111	27										
MD-90-30	62	21				14						
CS100			75									
CRJ900												
E175									2			
E190				20	10							
MRJ90												
72-500												
72-600												
Total	861	17	250	946	10	247	737	14	234	691	11	209

Table 4.9 Aircraft portfolio for major European carriers.

	KLM			Lufthansa			Ryanair			easyJet		
	In service	Average age	In order	In service	Average age	In order	In service	Average age	In order	In service	Average age	In order
717-200												
737 CL												
737NG	50	11					411	7	52			
737 MAX									110			
747-400	15	24		13	19							
747-8				19	4							
757-200												
757-300												
767-300												
767-400												
777-200	15	13										
777-300	14	5										
777-8X												
777-9x						20						
787-8												
787-9	10	2	1									
787-10												
A319-100				30	16					129	11	

A320-200	8	11		67	11					114	4	15
A321-100		11		20	22							
A321-200				43	8					3	0	127
A320neo				9	1	92						
A330-200	5											
A330-300		5		19	11							
A330neo												
A340-300				18	18							
A340-600				16	12							
A350-900			7	5	1	20						
A350-1000												
A380-800				14	6							
MD-82												
MD-83												
MD-88												
MD-90-30												
CS100												
CRJ900				2	7							
E175												
E190												
MRJ90												
72-500												
72-600												
Total	117	11	8	275	12	132	411	7	162	246	7	142

5

Freedoms of Air Service

The air service freedoms (articles) are a set of commercial aviation rules that define the rights of an airline belonging to a country to fly over another country's airspace and also to land in that country (Button and Taylor 2000; Button 2009; Bowen 2010; Fu et al. 2010; Williams 2017). They also regulate the rights of airlines to carry people, cargo, and mail internationally. These freedoms are formulated as a result of disagreements among countries over the extent of aviation liberalization in the Convention on International Civil Aviation of 1944, known as the Chicago Convention. There are typically nine major articles that define the rights of an airline to provide international service. The first two articles are related to the right of a commercial aircraft to pass through foreign airspace and airports (known as the transit rights). The other articles are related to the right to carry passengers, mail, and cargo in the international markets (known as the traffic rights). Specifically, the last four articles are not officially recognized because they are not mentioned by the Chicago Convention. However, they could be made possible by some air services agreements. Table 5.1 gives a description of the nine air service articles. The table also gives a hypothetical example to explain their application. In addition, Figure 5.1 shows a graphical representation of these nine air service articles.

The first article gives an international airline the right to fly over or cross the airspace of a foreign country without landing. It gives the right to fly over the territory of a treaty country without landing. This freedom got more publicity recently when Qatar Airways is prevented from flying over some of Qatar's neighbor countries. The second article allows technical stops without the embarking or disembarking of passengers or cargo. This article allows an airline to stop in a foreign country for refueling and/or other maintenance activities. The first and second freedoms are known as the transit rights. Some states are members of what is known as International Air Services Transit Agreement (IASTA). These states grant these two freedoms to other member states. Non-IASTA-member states prefer to maintain control over foreign airlines flying over their airspace. They typically negotiate transit agreements

Airline Network Planning and Scheduling, First Edition. Ahmed Abdelghany and Khaled Abdelghany.

Table 5.1 Description of the air service freedoms.

Freedom	Description	Classification					Hypothetical illustrative example
First	To fly over a foreign country without landing	Recognized freedoms by the Chicago Convention	Transit freedoms				A flight from Canada to Mexico, flown by a Canadian airline, flying over the United States
Second	To refuel or carry out maintenance in a foreign country without embarking or disembarking passengers or cargo						A flight from Canada to Mexico, flown by a Canadian airline, with a fuel/maintenance stop in the United States
Third	To fly from one's own country to another			Traffic freedoms			A flight from Canada to Mexico, flown by a Canadian airline, which carries traffic from Canada to Mexico
Fourth	To fly from another country to one's own						A flight from Mexico to Canada, flown by a Canadian airline, which carries traffic from Mexico to Canada
Fifth	To fly between two foreign countries on a flight originating or ending in one's own country				Beyond freedoms		A flight from the Chile to Canada, flown by a Canadian airline, with a full stop in Peru. Passengers and cargo originating from Chile may disembark the flight in Peru, with no intention to continue the flight to Canada. Of course, other passengers can originate from Peru and board the aircraft to travel to Canada, which is the fourth right
Sixth	To fly from a foreign country to another while stopping in one's own country for nontechnical reasons						A flight flown from Canada to Mexico, flown by an airline based in the United States, with a full stop in the United States
Seventh	To fly between two foreign countries while not offering flights to one's own country						A flight flown between Canada and Mexico, flown by an airline based in the United States

Table 5.1 (Continued)

Freedom	Description	Classification					Hypothetical illustrative example
Eighth	To fly inside a foreign country, continuing to one's own country					Cabotage	A flight by a US airline flown between Vancouver (Canada) and Boston (USA), with a full stop in Toronto (Canada). Passengers and cargo may board or disembark the flight in Toronto, with no intention to continue the flight to Boston
Ninth	To fly inside a foreign country without continuing to one's own country						A flight between Vancouver (Canada) and Toronto (Canada), flown by a Mexican airline

with other countries on a case-by-case basis. In all cases, the granting county designates special routes for overflight, requires prior notification before an overflight, and may charge substantial fees for the transit privilege. As an example of the transit fees, as of January 2018, the US Department of Transportation (USDOT), through the Federal Aviation Administration (FAA), charges $60.07 per 100 nautical miles (nm) from point of entry to point of exit from US-controlled airspace. It also charges $24.77 for flying over the international waters, where air traffic is controlled by the United States.

The third and fourth articles grant fundamental international service between two countries. The third article gives the privilege to carry passengers or cargo from one's own country to another. The fourth article gives the privilege to carry passengers or cargo from another country to one's own. Third and fourth articles are always granted concurrently in bilateral agreements between countries. However, the agreement may still restrict several aspects including the airlines permitted to fly, the airports permitted to be served, the frequency of flights, and the seat capacity.

The fifth article gives the privilege for an airline to carry traffic between two or more foreign countries as part of a service from/to the airline's own country. Tickets can be sold for any segment(s) of the route. The fifth article is also known as the "beyond" right, which allows an airline to carry traffic between or within other foreign countries. The main purpose of allowing the fifth article is to promote the airlines' long-haul route and enhance their economic viability.

The fifth article requires negotiation between three countries. The fifth freedom article is usually viewed by local airlines and flag carriers (airlines owned by governments) as unfair competition. The percentage of traffic (market share) carried between any two countries by an airline that owned by a third foreign country is always watched by local airlines that belong to these two countries (Weber and Dinwoodie 2000).

The sixth article, which is unofficial, combines the third and fourth articles. It defines the right to carry passengers or cargo from a second country to a third country by stopping in one's own country. For most airlines, the

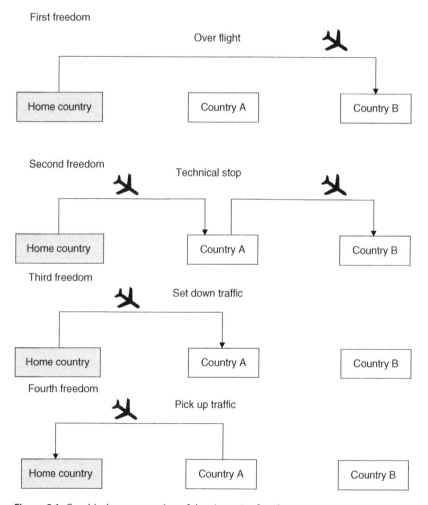

Figure 5.1 Graphical representation of the air service freedoms.

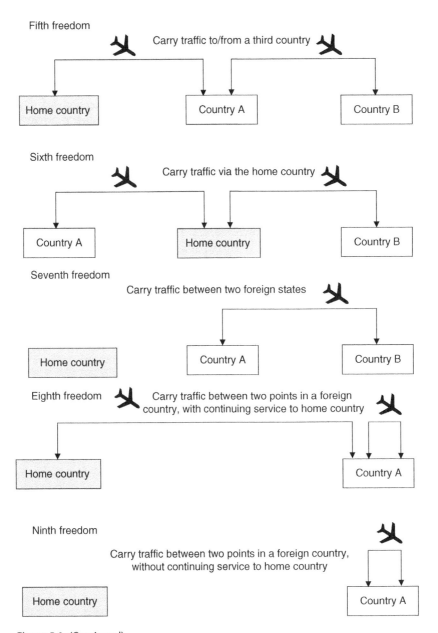

Figure 5.1 (Continued)

connected traffic is secondary compared with the local traffic (i.e. traffic carried based on the third and fourth articles). However, some international airlines rely in their business model on carrying large amount of connecting traffic. These airlines typically belong to small states that do not generate much local traffic. Examples of these airlines including Emirates, Qatar Airways, Etihad Airways, and historically KLM (now Air France–KLM). The percentage of connecting traffic on these airlines may reach up to 70–80%.

The seventh article is the right to carry passengers or cargo between two foreign countries without any continuing service to one's own country. The eighth and the ninth articles are known as the cabotage freedoms. Cabotage is the privilege to carry traffic between two airports in the same country by an airline registered in another country. The eighth freedom is the right to carry passengers or cargo between two or more points in a foreign country, with continuing service to/from one's own country. The ninth article (also known as stand-alone cabotage) is the right to carry passengers or cargo between two or more points in one foreign country, without continuing service to/from one's own country. The main example of these freedoms exists in the European Union, where airlines that belong to all state members are granted these freedoms. It is worth mentioning that two states that have no restrictions on the fifth to the ninth freedom articles are called to have an open sky agreement (OSA). Again, articles six to nine are unofficial as these articles are not part of the Chicago Convention.

6

Slot Availability

A landing/takeoff slot is a right granted by an airport authority to airlines interested in serving this airport (De Boe 2005; Condorelli 2007). The right allows the slot holder to schedule a flight landing or a departure during a specific time period. The International Air Transport Association (IATA), through its Worldwide Airport Slots Group, develops the procedure according to which the landing slots are allocated to airlines and other operators using or plan to use the airport. Slot allocation becomes critical in case the airport has limited capacity. In this situation, some coordination is required between the airport authority and the different airlines that are using or plan to use the airport. The prime objective of this coordination is to ensure the most efficient use of airport infrastructure to maximize benefits to the airlines (De Wit and Burghouwt 2008; Madas and Zografos 2008; Santos and Robin 2010).

The IATA's Worldwide Airport Slots Group periodically publishes what is known as the Worldwide Slot Guidelines (WSG), which gives detailed description of the slot allocation procedure at the different airports (IATA 2017). We encourage the readers of this chapter to refer to the most recent IATA's WSG to get the most recent information on slot allocation practice (IATA 2017). This chapter highlights the main relevant aspects of slot allocation to airline network planning. Given the legal and commercial nature of the slot allocation practice, this section follows in most cases the text given in the IATA's WSG report (IATA 2017). Additional explanation is given as needed.

The slot allocation practice varies based on the congestion level at the airport to be served by the airlines. In that regard, airports are categorized based on their congestion level. Airports are categorized by the responsible authorities according to the following levels of congestion:

Level 1: At all times, the capacity of the airport infrastructure is generally adequate to meet the demands of airport users.

Level 2: During some periods (daily, weekly, or seasonally), there is a potential for congestion. This congestion can be resolved by schedule adjustments

Airline Network Planning and Scheduling, First Edition. Ahmed Abdelghany and Khaled Abdelghany.
© 2019 John Wiley & Sons, Inc. Published 2019 by John Wiley & Sons, Inc.

made by the different airlines. The schedule adjustments are mutually agreed between the airlines and an appointed facilitator.

Level 3: Airports that have demand exceeding their capacity due to lack of adequate infrastructure. For these airports, a coordinator is appointed by the authority to allocate available slots to airlines and other aircraft operators. According to the IATA's WSG report (IATA 2017), a Level 3 airport is one where

> "a) Demand for airport infrastructure significantly exceeds the airport's capacity during the relevant period;
> b) Expansion of airport infrastructure to meet demand is not possible in the short term;
> c) Attempts to resolve the problem through voluntary schedule adjustments have failed or are ineffective; and
> d) As a result, a process of slot allocation is required whereby it is necessary for all airlines and other aircraft operators to have a slot allocated by a coordinator in order to arrive or depart at the airport during the periods when slot allocation occurs."

6.1 Level 1 Airports

Level 1 airports typically have no congestion or capacity problems at all times. Thus, there is no major coordination needed between the airports and the airlines. It is only required that airlines operating or planning to operate at a Level 1 airport provide adequate notice of their planned operations to the airport managing authority. They should also provide adequate notice to their appointed handling agent and any appointed data collection agent, if any. The airport managing authority should monitor the demand. The airport should also consider developing additional capacity when required to meet that demand. It is also responsible of facilitating the work of handling agents and other agencies to avoid any disruptions to airline operations. The airport managing authority may request information from airlines on planned operations in specified formats. In some cases, it may appoint a data collection agent to perform this task. The flight handling agents are responsible of arranging and coordinating with the airport managing authority to handle planned operations. Handling agents are responsible for planning and operating in a way that avoids the creation of any unnecessary constraints that affect airport operations.

6.2 Level 2 Airports

For Level 2 airports, a facilitator must be appointed by the airport authority to facilitate the coordination of the available capacity among the airlines using or plan to use the airport. This appointment is done in consultations with the airlines using the airport and their representative organizations (e.g. IATA). The facilitator must be qualified to perform the job, where they must have previous airline scheduling knowledge and/or coordination experience as a prerequisite for appointment. The selected facilitator must be independent, neutral, transparent, and nondiscriminatory. In addition, it must have the adequate resources to provide facilitation and coordination services. The key principles of schedule facilitation at a Level 2 airport are as follows (IATA 2017):

"a) Schedule facilitation is based on a process of schedule adjustments mutually agreed between the airlines and facilitator to avoid exceeding the airport's coordination parameters.

b) No slots are allocated at a Level 2 airport. The concepts of historic precedence and series of slots do not apply at Level 2 airports. These concepts are used with Level 3 airports.

c) The facilitator should adjust the smallest number of operations by the least amount of time necessary to avoid exceeding the airport's coordination parameters.

d) Facilitators must be independent and act in a neutral, transparent and non-discriminatory way.

e) An airline or other aircraft operator must advise the facilitator of all planned operations prior to operating at a Level 2 airport and of all changes to planned operations.

f) Airlines and other aircraft operators must not intentionally operate services at a significantly different time or in a significantly different way than agreed with the facilitator.

g) Planned times of operation are based on the planned on-block (arrival) and off-block (departure) times. Actual times of arrival and departure may vary due to operational factors."

When identifying the necessary schedule adjustments at a Level 2 airport, the following priorities are considered and applied by the facilitators to avoid exceeding the airport's coordination parameters (e.g. airport capacity) (IATA 2017):

"a) Services from the Previous Equivalent Season: Services operated as approved during the previous equivalent season should have priority over new demand for the same timings. Services that plan to operate unchanged from the previous equivalent season should have priority over services that plan to change time or other capacity relevant parameter, for example, operations with a larger aircraft where terminal capacity is a coordination parameter.

b) Introduction of Year-Round Operations: New operations that extend an existing operation into a year-round operation should have priority over other new operations. In evaluating whether the year-round priority applies, facilitators should allow flexibility on timings to allow for the differing requirements of short and long-haul services.

c) Effective Period of Operation: The schedule that will be effective for a longer period of operation in the same season should have priority.

d) Ad Hoc operations: Regularly planned operations should have priority over ad hoc operations.

e) Operational Factors: Operations that are constrained by slots or a curfew period at the other end of the route, or other relevant operational factors, should have priority over other demand where the airline may have timing flexibility."

All airlines operating or planning to operate at a Level 2 airport must submit details of their planned operations to the facilitator before operating at that airport. However, airlines should be prepared to accept an alternative time that may be suggested by the facilitator to avoid exceeding the coordination parameters. The roles of the facilitator at Level 2 airports include ensuring the feasibility of the plans submitted by the airlines, having details of the coordination parameters and utilization of the declared capacity available to all interested parties, advising airlines if planned operations will exceed coordination parameters, and facilitating a process of mutually agreed schedule adjustments to avoid exceeding these parameters. The role of the airport managing authority includes providing support to the facilitator in seeking full airline cooperation, providing the infrastructure necessary to handle planned airline operations within agreed levels of service, and keeping all stakeholders properly informed about any capacity limitations or changes.

6.3 Level 3 Airports

For Level 3 airports, the responsible authority must ensure the appointment of a coordinator. This appointment is made following consultation with the different stakeholders including the airport managing authority, the airlines using the airport, and their representative organizations (e.g. IATA). The appointed coordinator should have previous experience in airline scheduling and/or coordination experience. They also must have sufficient time and resources to provide coordination services to airlines and other operators currently using or plan to use the airport. Coordinators must be completely independent of any interested party. They should perform the job in a neutral, transparent, and nondiscriminatory way.

It is required that before operating at a Level 3 airport, all airlines must be allocated a slot by the coordinator. Airlines should have adequate resources and expertise to effectively participate in the coordination process. At Level 3 airports, airlines should be willing to develop alternative scheduling plans in case it is not possible to obtain the slots they initially require. In some busy Level 3 airports, airlines might not be able to get a slot to fly to these airports due to capacity limitations. In these cases, airlines should consider alternative airports (possibly adjacent airports, if any) that could accommodate their planned operations.

The primary role of the coordinator at a Level 3 airport is to allocate available slots to airlines and other aircraft operators in a neutral, transparent, and non-discriminatory way. The allocation should be in accordance with the priority criteria of the IATA's WSG and any other local guidelines and regulations (IATA 2017). The coordinator should get updates on the capacity parameters at the airport (which is typically updated twice a year). In addition, to ensure transparency, the coordinator should provide interested parties with details on the applicable coordination parameters, local guidelines and regulations, and any other criteria used for slot allocation. The coordinator should make available to the airlines all information about allocated slots and other unassigned/open slots. The coordinator should be able to inform the airlines on the reasons why slots were not allocated as requested. The coordinator should also monitor cancellations and any nonutilization of slots for the purpose of applying what is known as the Use it or Lose it rule. In addition, the coordinator should monitor planned and actual use of slots to identify any possible instances of intentional misuse of slots. When any misuse is observed, the coordinator initiates a dialogue with the airline to resolve this issue. The coordinator should be willing to offer advice to stakeholders on all actions that might be taken to improve airport performance including capacity or slot allocation flexibility. They should also be able to offer advice on any actions that could help the airport return to Level 2 or Level 1. Finally, the coordinator should address problems arising from conflicting requirements to avoid any need for external intervention.

The key principles of slot allocation at a Level 3 airport are (IATA 2017):

> "a) Slots are only allocated for planning purposes by a duly appointed coordinator at a Level 3 airport.
> b) Slots are only allocated to airlines and other aircraft operators.
> c) An airline or other aircraft operator must have a slot allocated to it before operating at a Level 3 airport. Certain types of flights (for example, humanitarian or state flights) may be exempt or subject to special local procedures.
> d) Airlines and other aircraft operators must not intentionally operate services at a significantly different time or use slots in a significantly different way than allocated by the coordinator.
> e) A series of slots is at least 5 slots requested for the same time on the same day-of-the-week, distributed regularly in the same season, and allocated in that way or, if that is not possible, allocated at approximately the same time.

f) An airline is entitled to retain a series of slots on the basis of historic precedence.

g) Historic precedence applies to a series of slots that was operated at least 80% of the time during the period allocated in the previous equivalent season.

h) Historic slots may not be withdrawn from an airline to accommodate new entrants or any other category of aircraft operator. Confiscation of slots for any reason other than proven, intentional slot misuse is not permitted.

i) Slots may be transferred or swapped between airlines, or used as part of a shared operation, subject to the provisions of these guidelines and applicable regulations.

j) Coordinators must be functionally and financially independent of any single interested party and act in a neutral, transparent and non-discriminatory way.

k) The allocation of slots is independent from the assignment of traffic rights under bilateral air service agreements.

l) Airlines and coordinators must use the SSIM message formats for communications at Level 3 airports.

m) Slot times are based on the planned on-block (arrival) and off-block (departure) times. Actual times of arrival and departure may vary due to operational factors."

When slots cannot be allocated using the primary criteria as set above, consideration should be given to the following factors (IATA 2017):

"a) Effective Period of Operation: The schedule that will be effective for a longer period of operation in the same season should have priority.

b) Type of Service and Market: The balance of the different types of services (scheduled, charter and cargo) and markets (domestic, regional and long haul), and the development of the airport route network should be considered.

c) Competition: Coordinators should try to ensure that due account is taken of competitive factors in the allocation of available slots.

d) Curfews: When a curfew at one airport creates a slot problem elsewhere, priority should be given to the airline whose schedule is constrained by the curfew.

e) Requirements of the Travelling Public and Other Users: Coordinators should try to ensure that the needs of the travelling public and shippers are met as far as possible.

f) Frequency of Operation: Higher frequency such as more flights per week should not in itself imply higher priority for slot allocation.

g) Local Guidelines: The coordinator must take local guidelines into account should they exist. Such guidelines should be approved by the Coordination Committee or its equivalent."

At Level 3 airports, airlines may only hold slots that they intend to operate. They can also swap, transfer, or use slots in a shared operation. Airlines must immediately return any slots they know they will not use to avoid capacity waste. Airlines are given a designated due date (slot return deadline), before which they must return to operators any slots that they do not intend to use. Even at short notice, it may be possible to reallocate returned slots to other airlines. Airlines that intentionally return slots after the slot return deadline usually receive a lower priority by the coordinator during the initial coordination of the next equivalent season. A list of airlines that return slots after the slot return deadline is maintained and published by the coordinator.

Airlines are given priority for historic precedence. Historic precedence is only granted for slots if the airline can demonstrate to the satisfaction of the coordinator that the slots were operated at least 80% of the time during the period allocated in the previous equivalent season. Coordinators should provide timely feedback to the airlines about flights at risk of failing to meet the minimum 80% usage requirement during the season to allow the airlines to take appropriate action. The following guidelines are used to determine which slots are eligible for historic precedence and number of operations required to achieve 80% usage (IATA 2017):

"a) The series of slots held on the Historics Baseline Date of 23:59 UTC 31 January (summer) and 23:59 UTC 31 August (winter) is used as the basis for determining eligibility for historic precedence.

b) For a series of slots newly allocated after the Historics Baseline Date, the number of slots in the series on the date of first allocation forms the basis of the 80% usage calculation.

c) If the period of operation of a series of slots is extended after the Historics Baseline Date, then the airline is eligible for historic precedence for the extended period of operation, subject to the 80% usage of the extended series.

d) Slots allocated on an ad hoc basis are not eligible for historic precedence. However, slots requested as a series but initially allocated on an ad hoc basis, which form a series by the end of the season, may be eligible for historic precedence.

e) If an airline holds more than one series of slots at the same time with identical or overlapping periods of operation, then the usage of each series is calculated separately.

f) If a flight operates on more than one day-of-week, then each day-of-week is considered as a separate series of slots.

g) Time changes allocated by the coordinator for part of a series of slots (for example, daylight saving time) do not affect eligibility for historic precedence, provided the 80% usage requirement is met over the full period of operation of the service.

h) Historic precedence applies to the latest times approved by coordinators for a series of slots, unless otherwise agreed between the coordinator and airline.

i) Ad hoc non-time related changes to a series of slots (for example, aircraft type, flight number, route or service type) do not affect eligibility for historic precedence. The 80% usage is calculated over the full period of operation of the service. Historic precedence normally applies to the series of slots as operated the majority of the time, unless otherwise agreed between the coordinator and airline."

As mentioned above, before operating at a Level 3 airport, an airline must have a slot allocated to it. Coordinators should request any airline that operates in a Level 3 airport without a slot to stop service immediately. In addition, airlines must not intentionally operate services at a significantly different time from the time of the slots allocated to them by the coordinator. Airlines that do so on a regular basis might not be entitled to historic precedence for either the times they operated or for the allocated times. Continued slot misuse may result in a lower priority for future slot requests. Additionally, the coordinator may seek to have sanctions applied under local regulations and/or national law. The following actions also constitute slot misuse (IATA 2017):

"a) Holding slots that the airline does not intend to operate, transfer, swap, or use in a shared operation;

b) Holding slots for an operation other than that planned for the purpose of denying capacity to another aircraft operator;

c) Requesting new slots that the airline does not intend to operate; or

d) Requesting slots for an operation other than that indicated, with the intention of gaining improved priority."

It should be noted that when an airline is given a slot at a Level 3 airport, they can use this slot for any route, aircraft, or flight number and may be changed by the airline from one route or type of service to another. Such changes are subject to final confirmation and approval by the coordinator.

At Level 3 airports, airlines are allowed to swap historic slots with other airlines on a one-for-one basis. Airlines engaging in slot swaps must notify the coordinator of every swap. The role of the coordinator is to confirm the feasibility of each swap and update the slot database. Similarly, airlines are allowed to transfer slots to other airlines. Slot transfer may involve compensation between airlines. Slot transfers between airlines, whether for compensation or not, may only take place if such practice is not prohibited by the laws of the relevant country. The transfer of newly allocated slots, which are slots other than historic slots or changed historic slots, is not allowed until such slots have been operated for two equivalent seasons. This is to prevent airlines taking advantage of an enhanced priority, such as new entrant status, to obtain slots simply to transfer them to another airline. Airlines engaging in a transfer of slots must notify the coordinator of every transfer. The coordinator will confirm the feasibility of the transfer and amend its database.

At Level 3 airports, shared operation may be allowed, given that laws at the relevant country do not prohibit such practice. Shared operation of the slots is when slots held by one airline can be shared by another airline. The coordinator must be notified by airlines engaged in a shared operation in advance to confirm the nature and feasibility of the operation. Main issues of shared operations include (IATA 2017):

> "a) Under a shared operation, the original slot holder retains historic precedence, not the operator of the slots.
> b) The slot holder is responsible for Initial Submissions and typically retains control of the slots until the Slot Return Deadline.
> c) Day-to-day management of the slots (the authority to amend and cancel slots on an ad hoc basis) after the Slot Return Deadline should be agreed between the airlines concerned and the coordinator, but will typically transfer to the operating airline.
> d) The operating airline is responsible for all usage and performance requirements.
> e) At the end of the shared operation or if the operating airline loses its operating license, the slots involved in a shared operation remain allocated to the original slot holder."

When an airline ends operations at an airport of Level 3, it must immediately return all slots for the remainder of the season and for the next season (if already allocated). The airline also must advise the coordinator whether or not it will use the slots in the future. The coordinator may decide to withdraw the slots from the airline and return them to the slot pool, if this airline fails to provide any needed information before a predetermined deadline. In the case of bankruptcy (or similar proceedings), the representatives of the airline should enter into dialogue with the coordinators to discuss their future intentions for the slots. Finally, slots can only be held by an airline with a valid operating license. If an airline fails to hold a valid operating license, its slots return to the slot pool.

Section 2

7

Feasibility of a New Route

Several factors promote the airlines' schedule updates. For example, in order to respond to demand seasonal fluctuations, airlines typically publish a different flight schedule for every month or quarter of the year (Lohatepanont and Barnhart 2004; Abdelghany et al. 2017). In addition, airlines continuously keep eye on changes in strategies or actions taken by their competitors. A competitor airline might add or reduce capacity in one or more city-pairs, which might require the airline to respond and update its schedule. Moreover, airlines might update their schedule to respond to changes in the schedule of its alliance and/ or code-share partners. Furthermore, airlines might update their schedule to meet long-term strategies related to service expansion or contraction in a market. Finally, airlines might also update their schedule to respond to changes in available resources such as number of aircraft and airport slots.

A newly published schedule could include flights or routes operated previously by the airline and new routes that are to be flown for the first time. For these new routes, airlines usually perform preliminary studies to investigate the feasibility of these new routes and estimate their potential profitability. The main objective is to understand how this route could sustain competition and how it fits in the airline network (i.e. relationship with other flights). This chapter presents an example that illustrates how airlines perform these preliminary studies to evaluate the feasibility of new routes.

One can think of an airline initiating a new route as entrepreneurs starting a new business or capital ventures evaluating a new investment opportunity. Accordingly, the techniques used by airlines to evaluate the feasibility of new routes are generally similar to those adopted by entrepreneurs and capital ventures to evaluate the profitability of a new business. Thus, we take advantage of this similarity and provide an example of a business feasibility study that the reader can easy grasp. Next, we illustrate the application of the steps of this study to evaluate the feasibility of a new route considered by a hypothetical airline.

Airline Network Planning and Scheduling, First Edition. Ahmed Abdelghany and Khaled Abdelghany.
© 2019 John Wiley & Sons, Inc. Published 2019 by John Wiley & Sons, Inc.

7.1 Business Plan

Consider an investor who is evaluating a business opportunity in the form of a new fast-food restaurant. This investor performs a feasibility study to evaluate the feasibility of this new business. The feasibility study can be summarized in six main steps as given below.

7.1.1 Proposed Property

In case there is no property available to host the restaurant, the first step of the plan is to select a suitable property for the new restaurant. The selection of the property depends on several factors including, for example, location, price/rent, available space, and expected cost of operations. In most cases, real-estate or property development agents would recommend a property along with its basic attributes to the investor. Figure 7.1 shows an example of a proposed property for the new fast-food restaurant that is located at the intersection of two major streets.

7.1.2 Identifying Demand Feeders

Upon selecting a location for the restaurant, the next step is to determine the potential customer demand of the restaurant. This step is usually performed by studying the traffic pattern in the restaurant's vicinity and learning about major businesses and attractions near the restaurant location. High traffic is expected

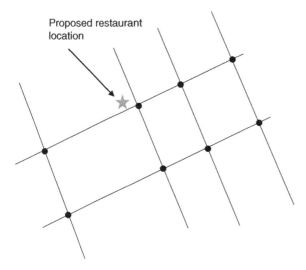

Figure 7.1 Location of the proposed property.

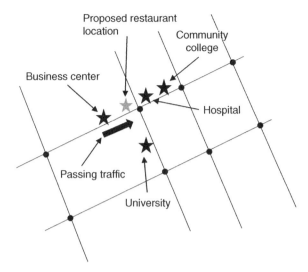

Figure 7.2 Locations of possible demand feeders.

to yield more customers. Also, the restaurant is expected to attract customers from nearby major attractions. For example, considering the restaurant location given above, one can identify four major businesses or attractions (demand feeders). These attractions include a hospital, a community college, a university, and a business center. The locations of these attractions are as given in Figure 7.2. While there might be other small attractions near the proposed restaurant location, their demand is minor and can be ignored in this high-level feasibility study.

7.1.3   Identifying the Size of the Demand Feeders

As the size of the demand feeders increases, the number of potential customers who might visit the restaurant is expected to increase. For example, as the number of traffic passing the streets around the restaurant location increases, there is higher chance that some of this traffic stop for a meal. Also, as the size of the hospital increases (i.e. number of beds), the number of people working and visiting the hospital is expected to increase and they are likely to stop by the restaurant. There is always a mechanism that can be used to reliably estimate the size of demand feeders. For example, the traffic passing in the streets around the restaurant can be obtained from traffic counts. Also, the number of people working and visiting the hospital can be obtained from the hospital records or estimated as a function of its number of beds. Table 7.1 gives an estimate of the size of the demand feeders for this example. There are 1,900 vehicles that pass by the restaurant during the lunch hour. Also, there

Table 7.1 Demand from each feeder.

Passing traffic (vehicles/hour)	Hospital (persons)	University (persons)	Community college (persons)	Business center (persons)
1,900	2,500	3,000	2,000	1,000

are 2,500 persons associated with the hospital (employees and visitors). The number of persons associated with the university, the community college, and the business center are 3000, 2000, and 1000, respectively.

7.1.4 Analyzing Competition

It cannot be assumed that the entire demand originating from the demand feeders visits the new fast-food restaurant under consideration. Typically, customers have several restaurant options. Thus, it is important to determine competing restaurants considered by the demand originating from each feeder. For example, if a passing driver decided to stop for a meal in that area, she/he may select one restaurant from several existing restaurants in that area (i.e. Restaurant 1, Restaurant 2, Restaurant 3, etc.). Similarly, for the people associated with the hospital, they could have the option to bring lunch from home, have lunch in the hospital's cafeteria, or visit any other restaurant including the new one. The set of restaurants available for the demand originating from each feeder is known as the choice set of this demand. As an example, Figure 7.3 shows the choice set of lunch places of the people associated with the hospital. The choice set of the demand is a feeder specific. In other words, the choice set associated with any feeder could be different from the choice set available for the demand originating from other feeders. For example, the passing traffic is not likely to consider the cafeteria of the hospital or the cafeteria of the university. Hence, this step identifies the set of places that a customer from each feeder might be considering (i.e. the customer's choice set). Table 7.2 gives the choice set of places for demand from each feeder.

7.1.5 Estimating Market Share

The objective of this step is to estimate the percentage of demand that might select the proposed new restaurant out of each feeder. This percentage is also referred to as the market share. For example, out of the 2,500 individuals associated with the hospital, what is the percentage of individuals who might visit the new restaurant? This percentage is equivalent to the "probability" that an average individual associated with the hospital visits the new restaurant. Several questions need to be answered in order to accurately estimate the market share of the new restaurant. First, what factors affect the choice of the

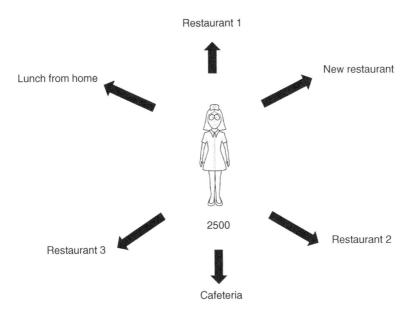

Figure 7.3 The different competing lunch options for demand generated out of the hospital.

Table 7.2 The choice set of lunch options for each demand group.

Demand feeder	Passing traffic (vehicles/hour)	Hospital	University	Community college	Business center
Demand	1900	2500	3000	2000	1000
Choice set	New restaurant	New restaurant	New restaurant	New restaurant	New restaurant
	Restaurant 1	Restaurant 1	Restaurant 1	Restaurant 1	Restaurant 1
	Restaurant 2	Restaurant 2	Restaurant 2	Restaurant 2	Restaurant 2
	Restaurant 3	Restaurant 3	Restaurant 3	Restaurant 3	Restaurant 3
	Eat at home	Hospital cafeteria	University cafeteria	College cafeteria	Lunch from home
		Lunch from home	Eat at home	Eat at home	
			Lunch from home	Lunch from home	

individuals? In other words, what makes a restaurant most attractive to an individual? Second, what is the sensitivity of the market share to these factors? Third, can this market share be estimated? Fourth, how can it be estimated? Fifth, how accurate is this estimation? Sixth, is the estimation accurate enough to build on for studying the feasibility of the new restaurant?

Many factors might affect restaurant selection by an individual. These factors depend on the individual's characteristics, the characteristics of the restaurants, and the meal type or its setting. Examples of the individual's factors include income, age, gender, car ownership, etc. For instance, one could expect that high-income individuals would prefer expensive restaurants and vice versa. In addition, females might be more willing to bring lunch from home. Some age groups might prefer restaurants that offer healthy meals. Examples of factors related to the characteristics of the restaurants include price, menu options, customer service, waiting time, travel time to the restaurant, parking availability, etc. For instance, it is expected that a restaurant that is less expensive is likely to be selected. Also, a restaurant that has a variety of food options, good customer service, and plenty of parking spaces is more likely to attract customers. Finally, the type or the setting of the meal could be business, casual, or a routine meal. Customers select a restaurant that is suitable for the meal setting.

Historical data can be used to estimate the percentage of people who visit each restaurant in the choice set of each demand feeder. For example, historical data might indicate that 40% of individuals originating from the hospital are eating lunch in the cafeteria, 25% are bringing lunch from home, and the remaining 35% eat lunch in the restaurants around the hospital. As mentioned above, these percentages are equivalent to the market share of each lunch place options or the probability that an average individual selects each of these places.

The problem of using historical data to estimate market share is that the new restaurant was not in the individual's choice set historically. Hence, another technique is needed to predict these market shares considering the existence of the new restaurant as part of the choice set. Typically, using historical data, an analytical model can be developed to predict the probability that an average individual selects one of the places in her/his choice set as a function of the characteristics listed above (i.e. individual characteristics, restaurant characteristics, and lunch type). This model can be in the following form:

$$p_i^j = f\left(x, y_j, z\right)$$

where

p_i^j: The probability that individual j selects lunch place option i
x: Set of variables describing the characteristics of lunch place options available in the choice set of individual j
y_j: Set of variables describing the characteristics of individual j
z: Variables related to the lunch type

Assume that such a model is already developed and available to estimate the market shares. Assume that the model estimates the market shares as given in

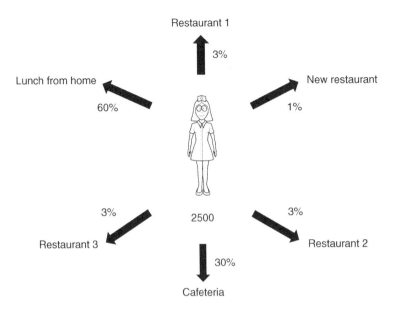

Figure 7.4 Market share of each option in the choice set for the hospital demand.

Figure 7.4. These market shares represent the probabilities that an average individual associated with the hospital selects each of the available lunch places. For example, the model estimates that 60% of the individuals bring lunch from home. Another 30% eat lunch in the cafeteria. The market share of each of the three existing restaurants is 3%, while the share of the new restaurant is 1%. Of course, the sum of these shares should add to 100%. Considering these percentages (probabilities), out of the 2500 individuals of the hospital, 25 individuals are predicted to visit the new restaurant (2500 × 1%).

The same probabilities can be estimated for all potential demand feeders (i.e. the passing traffic, the university, the community college, and the business center). For example, Figure 7.5 gives the shares for the passing traffic, which indicates that 90% of the traffic passes through without stopping at any of the restaurants in the area. The probability that a traveler visits the new restaurant is estimated at 2%. Accordingly, out of the 1900 travelers, 38 individuals are expected to stop at the new restaurant (1900 × 2%). It should be noted that the probability that a restaurant is selected may differ based on the demand feeders. For instance, based on the given example, the probability that the new restaurant is selected is equal to 1% for the demand originating from the hospital and is equal to 2% for the passing traffic. This is logical because the demand of each feeder has different choice sets and that the options in each set are perceived differently. Table 7.3 gives the estimated probabilities (in brackets) for each option considered in the choice set of the demand originating from the different feeders.

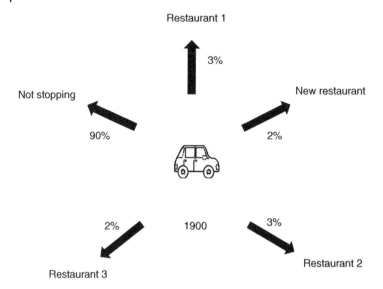

Figure 7.5 Market share of each option in the choice set for the passing traffic.

Table 7.3 The estimated probabilities (in brackets) for each lunch place option for each of the demand feeders.

Passing traffic (vehicles/hour)	Hospital	University	Community college	Business center
New restaurant (2%)	New restaurant (1%)	New restaurant (3%)	New restaurant (2%)	New restaurant (10%)
Restaurant 1 (3%)	Restaurant 1 (3%)	Restaurant 1 (4%)	Restaurant 1 (3%)	Restaurant 1 (20%)
Restaurant 2 (3%)	Restaurant 2 (3%)	Restaurant 2 (1%)	Restaurant 2 (8%)	Restaurant 2 (20%)
Restaurant 3 (2%)	Restaurant 3 (3%)	Restaurant 3 (2%)	Restaurant 3 (2%)	Restaurant 3 (10%)
Not stopping (90%)	Hospital cafeteria (30%)	University cafeteria (35%)	College cafeteria (32%)	Lunch from home (40%)
	Lunch from home (60%)	Eat at home (5%)	Eat at home (8%)	
		Lunch from home (50%)	Lunch from home (45%)	

7.1.6 Estimating Total Demand and Unconstrained Market Share

As explained above, the number of individuals from each demand feeder who visits the proposed new restaurant can be estimated by multiplying the total demand of the feeder by the estimated market share (i.e. the probability that this restaurant is selected). Adding the number of customers from each feeder gives the total number of customers forecasted to visit the new restaurant. The bottom part of Table 7.4 gives the number of customers forecasted to visit the new restaurant from each demand feeder.

Based on these calculations, the new restaurant is expected to attract about 243 customers. These 243 customers represent the market share of the new restaurant. More correctly, it actually represents the unconstrained market share as no information is provided yet on the capacity of the restaurant. In other words, the market share of the restaurant is 234 customers, only if the restaurant has the capacity to handle all this demand. However, in case the capacity of the restaurant is less than the demand, the market share will be constrained by the capacity. For example, if the restaurant can handle only 200 customers, then the market share will be only 200 customers. Thus, the estimated demand of 243 customers is referred to as the unconstrained market share, which is the customer demand assuming no capacity limitations. The market share is the least of the customer demand and the capacity of the

Table 7.4 The estimated demand of the new restaurant.

Passing traffic (vehicles/hour)	Hospital	University	Community college	Business center
1900	2500	3000	2000	1000
New restaurant (2%)	New restaurant (1%)	New restaurant (3%)	New restaurant (2%)	New restaurant (5%)
Restaurant 1 (3%) Restaurant 2 (3%)	Restaurant 1 (3%)	Restaurant 1 (4%)	Restaurant 1 (3%)	Restaurant 1 (22%)
Restaurant 3 (2%) Not stopping (90%)	Restaurant 2 (3%)	Restaurant 2 (1%)	Restaurant 2 (8%)	Restaurant 2 (23%)
	Restaurant 3 (3%)	Restaurant 3 (2%)	Restaurant 3 (2%)	Restaurant 3 (10%)
	Hospital cafeteria (30%)	University cafeteria (35%)	College cafeteria (32%)	Lunch from home (40%)
	Lunch from home (60%)	Eat at home (5%)	Eat at home (8%)	
		Lunch from home (50%)	Lunch from home (45%)	
$1900 \times 0.02 = 38$	$2500 \times 0.01 = 25$	$3000 \times 0.03 = 90$	$2000 \times 0.02 = 40$	$1000 \times 0.05 = 50$

restaurant. Given this demand estimate, the investor may be able to decide on the feasibility of investing in this new restaurant. In reality, there are many other factors that need to be considered in the process including operation cost, availability of trained labor, competition with new restaurants to open in the future, and meal pricing. The decision of investing in the new restaurant that can be made requires careful studying of these factors.

7.2 Application of Feasibility Study on a New Airline Route

In this section, the methodology presented above to evaluate the feasibility of a new restaurant is applied to evaluate the feasibility of a new airline route in a given city-pairs. For this purpose, consider the hypothetical two-hub airline given in Figure 7.6. This airline has two hubs: the first is at Hartsfield–Jackson Atlanta International Airport (ATL) and the second is at Salt Lake City International Airport (SLC). This airline is serving several spoke destinations through at least one of these two hubs. These spoke destinations include O'Hare International Airport (ORD), Dallas/Fort Worth International Airport (DFW), Newark Liberty International Airport (EWR), John F. Kennedy International Airport (JFK), Denver International Airport (DEN), Detroit Metropolitan Airport (DTW), Orlando International Airport (MCO), Tampa International Airport (TPA), Los Angeles International Airport (LAX), Phoenix Sky Harbor International Airport (PHX),

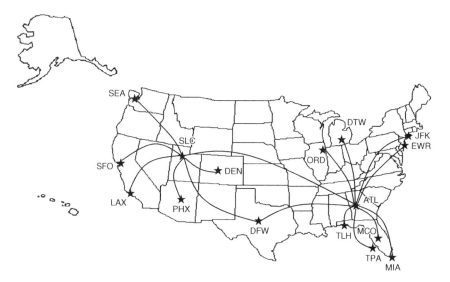

Figure 7.6 A hypothetical airline with two hubs.

Seattle–Tacoma International Airport (SEA), Tallahassee International Airport (TLH), San Francisco International Airport (SFO), and Miami International Airport (MIA).

7.2.1 The Proposed Route

There are different methods by which an airline can decide on the feasibility of adding a new route to expand its network. With the help of specialized analysts, airlines continuously evaluate potential new markets to expand their services. In some situations, the airline is approached by different agencies requesting the airline to serve a certain city-pair. For example, airlines may be approached by airports, local governments, and city commissioners to serve their city. In any case, the proposed new route has to undergo a feasibility study to examine its sustainability. The selection of a new route to be added to the airline network depends on several factors including, for example, available demand, integration with the existing network, competition with other airlines, cost of operations, and expected revenue. For the hypothetical airline network given above, assume that hypothetical airline (Airline 1) is proposing to add a new route between ATL and SEA, as shown by the dash-dotted line in Figure 7.7.

7.2.2 Identifying Demand Feeders

Once a route is identified, the next step is to determine the potential customers of this route. This new route (flight) is expected to serve leisure and business demand

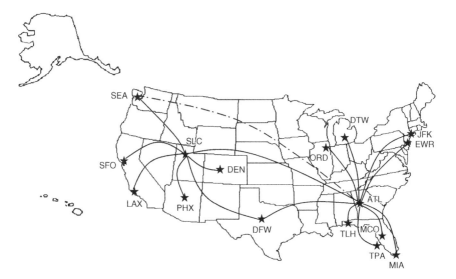

Figure 7.7 The proposed route ATL-SEA.

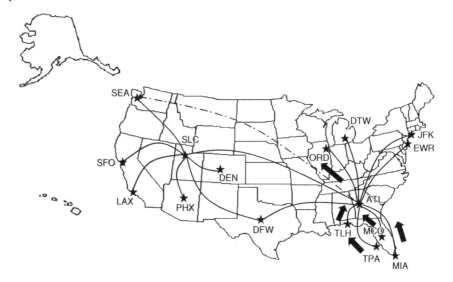

Figure 7.8 The possible feeding markets for the new route ATL-SEA.

in the local market: ATL-SEA. In addition, since ATL represents a hub for the airline, the proposed ATL-SEA flight is expected to serve connecting traffic. The route ATL-SEA could serve demand from MIA, MCO, TPA, and TLH to SEA through connecting at ATL. There is also a chance that some demand connects from other destinations (e.g. JFK, EWR, and DFW). However, this demand is expected to be minimum as the circuity and nondirectedness of the route make it unfavorable, compared with other directed itineraries offered by other airlines. Figure 7.8 shows all possible feeding markets for the new route, which include the local market ATL and the connecting markets MIA, MCO, TPA, and TLH.

7.2.3 Identifying the Size of the Demand Feeding Markets

Obviously, as the size (number of customers) of feeding markets increases, the proposed route could attract more passengers. For example, as the number of passengers traveling between ATL and SEA increases, there is a higher chance that a portion of this demand selects the proposed new route between ATL and SEA. Similarly, as the size of the feeding market MIA-SEA increases, there is a chance that some travelers between MIA and SEA use the proposed new route MIA-ATL-SEA, and hence connect on the new ATL-SEA flight.

The objective here is to obtain a good estimate of the size of each feeding market. The number of passengers traveling between any city-pair can be estimated using different techniques including, for example, time series analysis and causal modeling. Time series analysis, which is more common for this application, is a technique used to predict the future value of a variable using available historical data. In other

Table 7.5 The estimated markets sizes for the different demand feeding markets.

ATL-SEA	MIA-SEA	MCO-SEA	TPA-SEA	TLH-SEA
390	250	170	220	100

words, historical values of the market size variables are used to predict its value in the future. Causal modeling is an estimation approach that assumes the existence of a relationship that explains the value of an independent variable in terms of a set of independent variables. This relationship is derived using historical data of the dependent and independent variables and is used to predict the future value of the dependent variable using the predicted values of the independent variables. For example, assume the market size (demand) is a function of average population, ticket price, average income, etc. Given this function and the future value of these independent variables, the future demand can be estimated. For the purpose of continuing the discussion on this example, Table 7.5 gives the demand (market size) for the different feeding markets for the five feeding markets. For the purpose of this discussion, these values are assumed. However, in real-world applications, they can be accurately estimated using any of the two prediction techniques discussed above.

7.2.4 Analyzing Competition

It is unreasonable to assume that the entire demand from the feeding markets will be selecting itineraries that include the new flight ATL-SEA. Typically, in a competitive environment similar to that of the airline industry, travelers from each feeding market select from several travel options (itineraries). For example, let us assume that there are three other competing airlines (Airline 2, Airline 3, and Airline 4). Airline 2 has a hub at ORD as shown in Figure 7.9. Airline 3 operates using DFW as its main hub with a set of routes shown in Figure 7.10. Finally, Airline 4 has a hub at SEA with its routes shown in Figure 7.11. As such, customers traveling from any of the feeding markets could be selecting routes (itineraries) provided by any of these four airlines. For example, for customers traveling between MIA and SEA, several travel itineraries are available including:

MIA-ATL-SEA by Airline 1
MIA-ATL-SLC-SEA by Airline 1
MIA-ATL-DWF-SLC-SEA by Airline 1
MIA-ORD-SEA by Airline 2
MIA-DFW-SEA by Airline 3
MIA-SEA by Airline 4

Table 7.6 gives all itineraries available for the customers of each demand feeding market.

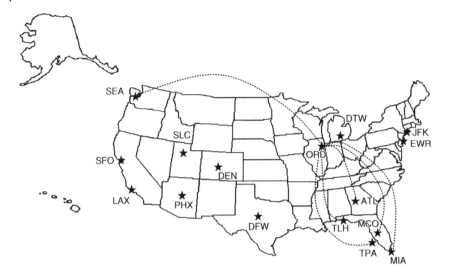

Figure 7.9 Routes of Airline 2.

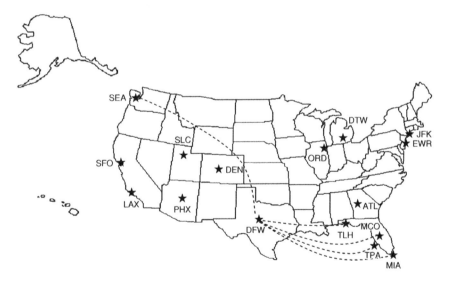

Figure 7.10 Routes of Airline 3.

7.2.5 Estimating Market Share

In this step, the objective is to estimate the percentage of demand selecting an itinerary that includes the proposed ATL-SEA flight. For example, considering the 250 passengers traveling from MIA-SEA, the percentage of customers

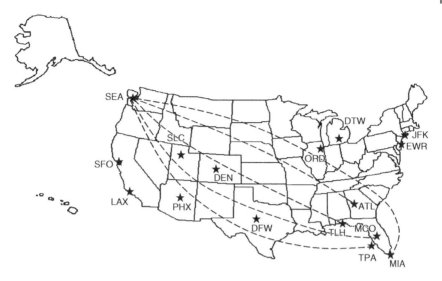

Figure 7.11 Routes of Airline 4.

that might select itinerary MIA-ATL-SEA by Airline 1 needs to be estimated. This percentage is sometimes referred to as the probability that an average individual traveling from MIA to SEA will select this itinerary. Several questions need to be considered here. First, what factors affect itinerary choice? Second, how this choice is affected by these factors? Third, can we estimate the probability of an itinerary choice? Fourth, how can it be estimated? Fifth, how accurate is the estimation? Finally, is this accuracy good enough for this preliminary feasibility study?

Several factors might affect the choice decision of an itinerary among a set of itineraries offered by all competing airlines. These factors are related to the traveler characteristics, the characteristics of the available itineraries, and the type of travel. Examples of the factors that are related to the traveler characteristics include income, age, gender, etc. For instance, high-income travelers are more likely to select most convenient itinerary regardless of price. In addition, females and aged individuals might be willing to avoid itineraries that have late-night arrival times at the destinations. Factors that are related to the characteristics of the itinerary include ticket price, departure and arrival times, number of connections, connection duration, airline reputation, etc. For instance, an itinerary that is less expensive, has convenient arrival and departure times, has no or short connection duration, and is operated by reputable airline is more likely to be selected by travelers. Finally, the type of travel is related to whether the travel is business or leisure trip, domestic or international, and short haul or long haul.

Table 7.6 Itineraries available for the customers of each demand feeding market.

City-pair	Available itineraries
ATL-SEA	ATL-SEA by Airline 1
	ATL-SLC-SEA by Airline 1
	ATL-DWF-SLC-SEA by Airline 1
	ATL-ORD-SEA by Airline 2
	ATL-DFW-SEA by Airline 3
	ATL-SEA by Airline 4
MIA-SEA	MIA-ATL-SEA by Airline 1
	MIA-ATL-SLC-SEA by Airline 1
	MIA-ATL-DWF-SLC-SEA by Airline 1
	MIA-ORD-SEA by Airline 2
	MIA-DFW-SEA by Airline 3
	MIA-SEA by Airline 4
TPA-SEA	TPA-ATL-SEA by Airline 1
	TPA-ATL-SLC-SEA by Airline 1
	TPA-ATL-DWF-SLC-SEA by Airline 1
	TPA-ORD-SEA by Airline 2
	TPA-DFW-SEA by Airline 3
	TPA-SEA by Airline 4
MCO-SEA	MCO-ATL-SEA by Airline 1
	MCO-ATL-SLC-SEA by Airline 1
	MCO-ATL-DWF-SLC-SEA by Airline 1
	MCO-ORD-SEA by Airline 2
	MCO-DFW-SEA by Airline 3
	MCO-SEA by Airline 4
TLH-SEA	TLH-ATL-SEA by Airline 1
	TLH-ATL-SLC-SEA by Airline 1
	TLH-ATL-DWF-SLC-SEA by Airline 1
	TLH-ORD-SEA by Airline 2
	TLH-DFW-SEA by Airline 3
	TLH-SEA by Airline 4

Historical data can be used to estimate the percentage of travelers who selected a given itinerary among several other itineraries in the choice set of each demand feeding market. For example, for the feeding market MIA-SEA, historical data could provide the percentage of travelers who selected each of the available itineraries. The problem of using these historical-based percentages is that the proposed new itinerary (MIA-ATL-SEA by Airline 1) was not historically included in the travelers' choice set. Thus, another technique is needed to estimate this percentage or market share, regardless of whether or not an itinerary is old or new. An analytical model can be estimated to provide the market share of an itinerary as a function of the variables listed above (i.e. traveler characteristics, itinerary characteristics, and trip type). One common form for this mathematical model is as follows (Coldren et al. 2003):

$$p_i^j = f\left(x, y_j, z\right)$$

where

p_i^j: Represents the probability that an itinerary i is selected by traveler j
x: Set of variables that describe the characteristics of the itineraries in the choice set of traveler j
y_j: Set of variables related to the characteristics of traveler j
z: Set of variables related to the trip purpose

The process of developing an itinerary choice model for market share forecasting is presented in detail in Chapter 8. Here, for the sake of this example, we assume that such an itinerary choice model is already developed. Figure 7.12 gives the probabilities that an average traveler from MIA to SEA selects each of the available itineraries. As shown in the figure, the model estimates that 80% of the travelers in the MIA-SEA market select the non-stop itinerary MIA-SEA by Airline 4. Another 8% select the connecting itinerary MIA-ATL-SEA by

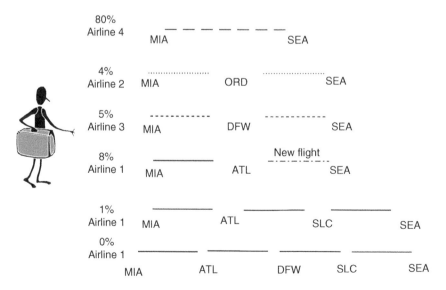

Figure 7.12 Assumed market share of each itinerary in the MIA-SEA city-pair.

Airline 1, and 4% select itinerary MIA-ORD-SEA by Airline 2. Itineraries MIA-DFW-SEA by Airline 3 and MIA-ATL-SLC-SEA by Airline 1 are selected by 5 and 1% of the travelers, respectively. Finally, the model estimates that no travelers select itinerary MIA-ATL-DWF-SLC-SEA by Airline 1. Again, the sum of these shares should add to 100%. According to these percentages (probabilities), out of the 250 travelers in the MIA-SEA market, 20 travelers are estimated to select the new itinerary MIA-ATL-SEA by Airline 1 (250 × 8%).

7.2.6 Estimating Total Flight Demand (Unconstrained Demand)

As explained above, the number of passengers from each feeding market to the new ATL-SEA flight can be estimated by multiplying the total demand of the feeding market by the estimated probability that an itinerary including this flight is selected. Table 7.7 gives all competing itineraries in each of the demand feeding markets. It also gives the estimated choice probability for

Table 7.7 Estimating the demand for each itinerary that includes the proposed flight ATL-SEA and total demand for the flight ATL-SEA.

City-pair	Demand	Competing itinerary	Itinerary market share (%)	Demand for the itinerary that includes the flight ATL-SEA
ATL-SEA	390	ATL-SEA by Airline 1	40	156
		ATL-SLC-SEA by Airline 1	9	
		ATL-DWF-SLC-SEA by Airline 1	0	
		ATL-ORD-SEA by Airline 2	11	
		ATL-DFW-SEA by Airline 3	10	
		ATL-SEA by Airline 4	40	
MIA-SEA	250	MIA-ATL-SEA by Airline 1	8	20
		MIA-ATL-SLC-SEA by Airline 1	1	
		MIA-ATL-DWF-SLC-SEA by Airline 1	0	
		MIA-ORD-SEA by Airline 2	4	
		MIA-DFW-SEA by Airline 3	5	
		MIA-SEA by Airline 4	80	

Table 7.7 (Continued)

City-pair	Demand	Competing itinerary	Itinerary market share (%)	Demand for the itinerary that includes the flight ATL-SEA
TPA-SEA	170	TPA-ATL-SEA by Airline 1	7	12
		TPA-ATL-SLC-SEA by Airline 1	1	
		TPA-ATL-DWF-SLC-SEA by Airline 1	0	
		TPA-ORD-SEA by Airline 2	4	
		TPA-DFW-SEA by Airline 3	6	
		TPA-SEA by Airline 4	82	
MCO-SEA	220	MCO-ATL-SEA by Airline 1	5	11
		MCO-ATL-SLC-SEA by Airline 1	1	
		MCO-ATL-DWF-SLC-SEA by Airline 1	0	
		MCO-ORD-SEA by Airline 2	8	
		MCO-DFW-SEA by Airline 3	5	
		MCO-SEA by Airline 4	81	
TLH-SEA	100	TLH-ATL-SEA by Airline 1	9	9
		TLH-ATL-SLC-SEA by Airline 1	1	
		TLH-ATL-DWF-SLC-SEA by Airline 1	0	
		TLH-ORD-SEA by Airline 2	5	
		TLH-DFW-SEA by Airline 3	5	
		TLH-SEA by Airline 4	80	
Total estimated number of passengers for the ATL-SEA flight			208	

each itinerary. According to these calculations, this new flight is expected to attract about 208 passengers. These 208 passengers represent the unconstrained demand for the ATL-SEA flight, as no information is provided yet on the capacity (number of seats) of the aircraft used to operate this flight.

In other words, the passenger demand of this flight is 208 passengers, assuming the capacity of the aircraft assigned to this flight is larger than this demand. However, if the capacity of the aircraft assigned to this flight is less than the demand, then the actual passenger demand of this flight will be constrained by this limited capacity. For example, if the aircraft assigned to this flight has only 170 seats, then the (constrained) passenger demand for this flight will be only 170 passengers. Thus, the estimated demand (208 passengers) is referred to as the unconstrained flight demand, which is the demand for the ATL-SEA flight assuming no capacity limitation (i.e. large aircraft). The actual demand is the least of the estimated demand and the capacity of the assigned aircraft. Considering the amount of estimated demand, the airline may be able to decide on the feasibility of having the new ATL-SEA flight. Of course, there are several other important factors that need to be carefully investigated before deciding on the feasibility of the new flight. These factors include operation cost, fares and revenue, potential growth, and competition resulting from introducing new itineraries. Once all these factors are examined, a final decision on initiating the new route can be finalized.

8

Market Share Models

The examples presented in the previous chapter emphasized the need for analytical tools (models) to estimate the market share of itineraries in the different city-pairs. When a city-pair is served by multiple itineraries, these models estimate the probability that each of these itineraries will be selected by the average traveler. These models are also known as the itinerary choice models, which replicate how customers select their itineraries. They are also referred to as quality service index or quality share index (QSI) models, which determine the relationship between the quality of the itinerary and its market share (Coldren et al. 2003; Wei and Hansen 2005).

In this chapter, the state-of-the-art models adopted by major airlines to estimate the itinerary market shares are presented. The chapter explains the basic model structure, main variables included, and models' application to estimate market shares. Before proceeding, we provide a brief review of the concept of model development (modeling) for unfamiliar readers.

8.1   What Is a Model?

A model is an implement that replicates or represents a system or a phenomenon in reality. A model is developed as it is usually easier and less expensive to experiment on a model than on the actual system. The model could be physical or analytical (mathematical). The analytical models could also be supported by graphics and computer animation. A physical model is usually in the form of a simplified assembly that replicates a large-scale complex system. For example, 3D architecture models that architects use to demonstrate their designs of buildings are examples of physical models. An analytical model is usually in the form of a mathematical equation or a collection of equations that replicate a certain phenomenon. Example of an analytical model is the mathematical expression that gives the relationship between the price of a house and the

Airline Network Planning and Scheduling, First Edition. Ahmed Abdelghany and Khaled Abdelghany.
© 2019 John Wiley & Sons, Inc. Published 2019 by John Wiley & Sons, Inc.

features of the house (area, number of rooms, number of bathrooms, swimming pool, etc.). Complex analytical models are sometimes supported by computer animations, which assist users to visualize and interpret results (Maruyama 1997; Fox 2015). For example, a simulation model that represents the movement of vehicles in a street intersection with traffic lights is an example of analytical models with animation. The model can show how vehicles approach the intersection in the different lanes and the relation with the traffic signal design (i.e. how much green light is given to each approach). The model can be used to estimate several measures of performance at the intersection including vehicular stopping time and queue lengths.

A model is usually developed to answer questions about the system that are typically difficult or expensive to answer by using the system itself. For example, an architect might design a 3D model for a building to demonstrate her/his design before construction. This 3D model can assist the building stakeholders to visualize the building and suggest design changes. Having the 3D model saves time and money that could be incurred, if these changes were to be made on the real building after its construction. Similarly, a traffic engineer can use a traffic simulation model to study the impact of altering the traffic signal timing scheme on the intersection performance, before deploying this new scheme in the intersection. The model can assist the traffic engineer to understand how the intersection will perform under different traffic signal timing schemes and whether any of them provides improvement over the current situation. The use of the model avoids any deterioration in the intersection performance in case the model indicates that a proposed signal timing scheme does not provide any improvement.

8.2 Model and Historical Data

Models differ in their complexity. A complex real-world system with many interconnected components usually requires a sophisticated model to be able to replicate this system (Osborne 2008; Hardy and Bryman 2004). The analyst may select to develop a simple model. However, this simplification could reduce the model's ability to replicate all aspects of the system under study. Developing a good model for a complex real-world system requires significant collaboration among individuals who have knowledge about the system (domain experts) and analysts who can translate this knowledge into an analytical model. Developing a model without having proper knowledge of the system usually results in an inoperable model that does not represent reality. One form of this knowledge of the system is historical data. Historical data provides a history of the behavior of the system under different conditions. A rich historical data records the system behavior under a wide variety of conditions.

8.3   Model Development Example

Consider a real-estate agent who is frequently asked by customers to estimate the price for their houses before listing them for sale. Due to the large inventory of houses, when determining the price of any given house, the real-estate agent has to search for all similar listings and recent sales. This process is usually time consuming. Since the real-estate agent is asked to perform this task frequently, she/he decided to develop a mathematical model that can estimate the price of the house as a function of the house's main features. The real-estate agent is aware that main features that affect a house price include number of bedrooms, number of bathrooms, number of car garages, whether or not the house has a swimming pool, and whether or not the house is located in a good school district. Of course, several other features that might affect the house pricing could be included in the model. However, the agent might have decided to ignore them due to their limited effect on the price, and/or there is no historical data that correlates the value of these features to house prices. For simplicity, the agent assumes that the features mentioned above have a linear effect on the house price. In other words, a change in the value of any of these features results in a corresponding linear change in the house price. Therefore, the relationship between the house price and the house features can be represented using a multiple linear regression (MLR) model that can be represented mathematically as follows:

$$Y = a_0 + a_1 X_1 + a_2 X_2 + a_3 X_3 + a_4 X_4 + a_5 X_5$$

where

Y: House price
X_1: Number of bedrooms
X_2: Number of bathrooms
X_3: Number of car garages
X_4: A binary variable that is equal to 1 if the house has a swimming pool and zero otherwise
X_5: A binary variable that is equal to 1, if the house is located in a good school district and zero otherwise
$a_0, a_1, a_2, a_3, a_4,$ and a_5: Parameters to be estimated

The variables $X_1, X_2, X_3, X_4,$ and X_5 are referred to as the independent variables, while the variable Y is known as the dependent variable as its value depends on the independent variables.

To develop (or estimate) a model, a reliable set of historical data is needed (Montgomery et al. 2012). This data provides previous records that relates house prices to the relevant house features mentioned above. Figure 8.1 shows a sample of this data. The figure is a snapshot from a spreadsheet that gives records of the prices for recently sold houses (column F) and their corresponding features (columns A–E). For example, the highlighted record shows the records for a house that was recently sold for \$296,727. This house has 3 bedrooms, 2.5 bathrooms, 2 car garages, and a swimming pool and is not located in a good school district. As mentioned above, the house price represents the dependent variable, and the other variables represent the independent variables. There should be enough data records to develop a model. Using this data, a regression analysis tool can be used to develop the model. Developing the model corresponds to estimating the values of the different parameters in the equation given above (i.e. a_0, a_1, a_2, a_3, a_4, and a_5). As an example, Microsoft Excel 2013 is used to develop an MLR model. The application provides the capability to perform regression analysis using the data analysis tool under the data tab, as illustrated in Figure 8.2.

	A	B	C	D	E	F	G
1	Number of bedrooms	Number of bathrooms	Number of car garage	Swimming pool	School district	Price	
3	3	2.5	2	1	0	296,727	
5	3	2	2	0	1	259,774	
6	3	2	2	0	0	238,388	
7	4	3	3	1	1	370,865	
8	5	3	3	1	1	467,729	
9	3	2.5	2	1	1	343,203	
10	4	3	3	1	0	384,078	
11	3	2.5	2	1	0	306,797	
12	4	2.5	2	0	1	315,724	
13	3	2	2	0	1	236,885	
14	3	2	2	0	0	223,585	
15	4	3	3	1	1	386,917	
16	5	3	3	1	1	431,401	
17	3	2.5	2	1	1	310,268	
18	4	3	3	1	0	369,160	
19	3	2.5	2	1	0	283,528	
20	4	2.5	2	0	1	330,684	
21	3	2	2	0	1	253,984	
22	3	2	2	0	0	234,363	
23	4	3	3	1	1	401,712	
24	5	3	3	1	1	447,906	
25	3	2.5	2	1	1	338,113	
26	4	3	3	1	0	355,263	
27	3	2.5	2	1	0	284,516	
28	4	2.5	2	0	1	321,197	
29	3	2	2	0	1	258,018	
30	3	2	2	0	0	223,035	

Figure 8.1 Example of historical data for the real-estate model.

Figure 8.2 Data analysis tool in Excel 2013.

The following steps are used to develop an MLR model using Microsoft Excel 2013.

Step 1: Activate the Data Analysis Box and select Regression as shown in Figure 8.3.
Step 2: Select the column that has the house price (column F) as the input for the Y Range, and select the other variables (columns A–E) as input for the X Range. If the labels in the first row are selected, check the corresponding box of Labels, as shown in Figure 8.4. In the output options of the window, select the output to be generated in a new worksheet.

Figure 8.5 gives the output of the MLR model as generated by Excel. The highlighted section gives the values of the different parameters of the model. In addition, it provides different statistics that describe the overall model quality. However, reviewing these statistics and explaining their meaning are beyond the scope of this chapter. Interested readers should refer to a fundamental statistics textbook.

Given the values of the estimated parameters, the model can be written as follows:

$$Y = 26{,}586 + 38{,}323\,X_1 + 31{,}882\,X_2 + 25{,}543\,X_3 + 55{,}087\,X_4 + 30{,}766\,X_5$$

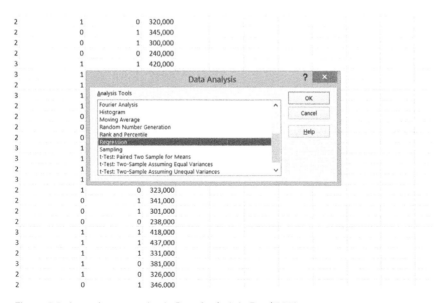

Figure 8.3 Accessing regression in Data Analysis in Excel 2013.

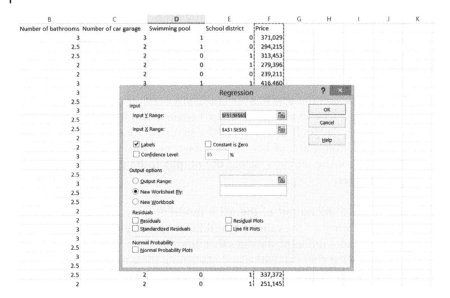

Figure 8.4 Model input.

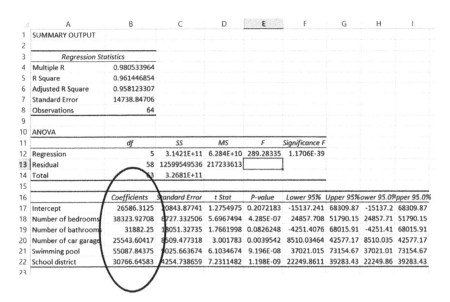

Figure 8.5 Model output.

The model indicates that the house price increases as the number of bedrooms increases. This trend is concluded because the parameter a_1 associated with the number of bedrooms variable has a positive sign. In addition, the value of the parameter a_1 indicates that the price of the house increases on average by \$38,323 for each additional bedroom. Similarly, the values of a_2 and a_3 indicate that the price of the house increases by \$31,882 and \$25,543 for each additional bathroom and car garage space, respectively. Furthermore, a swimming pool adds \$55,087 to the house value. Finally, having the house in a good school district adds \$30,766 to the house value. On average, the value of the empty lot is \$26,586, which corresponds to the value of a_0.

The model can now be used to predict the value of any house available for sale. For example, consider a house with 4 bedrooms, 2.5 bathrooms, and 3 car garages. The house has a swimming pool and exists in a good school district. The price of the house can be calculated by plugging in these numbers into the model as follows:

$$
\begin{aligned}
Y &= 26{,}586 + 38{,}323\left(4\right) + 31{,}882\left(2.5\right) + 25{,}543\left(3\right) + 55{,}087\left(1\right) + 30{,}766\left(1\right) \\
&= \$422{,}072
\end{aligned}
$$

8.4 Categorical Dependent Variable

The multiple regression analysis presented above assumes that the dependent variable has to be a continuous variable (similar to the price of the house). However, in some cases, the dependent variable representing a certain system/phenomenon is in the form of a categorical variable (Agresti 2003; Long and Freese 2006). Categorical variables take on values that are names or labels. The type of a car to be purchased by a household (e.g. minivan, sedan, truck, or sport utility vehicle) or airline to fly with (e.g. American Airlines, United Airlines, Delta Air Lines, or Southwest Airlines) are examples of categorical dependent variables. If the dependent variable is categorical, logistic regression model is recommended instead of MLR. Logistic regression measures the relationship between the categorical dependent variable and independent variables by estimating the probability that each category would be selected. When the dependent variable takes only two values (e.g. pass/fail, yes/no, win/lose, etc.), the binary (binomial) logistic regression is used. Cases with more than two categories (i.e. multiple choices) are referred to as multinomial logistic regression, and when the multiple categories are ordered (e.g. dependent variables are responses to questions in the form of low, medium, and high), it is referred to as ordinal logistic regression.

For example, a car dealer might be interested to know which type of vehicle to be purchased by a household as a function of the characteristics of the household and the characteristics of the available vehicle types. In this case, possible buyers can select from minivan, sedan, truck, and sport utility vehicle. The characteristics of the household might include income, size, number of children, commute distance, etc. The characteristics of the vehicle might include price, number of seats, horsepower, etc. In this case, the dependent variable is categorical, which is the type of car to be purchased. Thus, logistic regression can be used to estimate the probability to purchase each vehicle type. Similarly, an airline analyst might be interested to determine which airline itinerary is to be selected by a traveler. Logistic regression can be used to estimate the probability that each itinerary is selected as a function of a group of independent variables that represent the measurable attributes of the available itineraries and the attributes of the traveler.

8.5   Introduction to Discrete Choice Models

In many cases, there is a need to model a categorical dependent variable that is in the form of a choice among multiple options. For example, as mentioned above, an analyst might be interested to model which travel itinerary to be selected by a traveler among a set of itineraries offered by several competing airlines. In this case, the dependent variable is not a continuous variable. It represents one of the alternatives from the choice set (Ben-Akiva and Lerman 1985; Train 2009). Figure 8.6 depicts a typical choice process, where individual selects among N available alternatives constituting this individual's choice set.

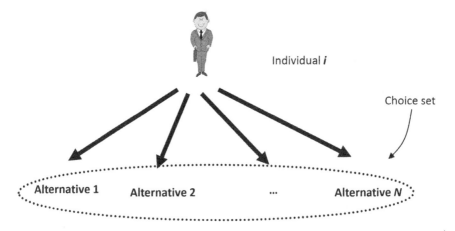

Figure 8.6 Choice set of an individual.

When rational individuals are in the situation of making a choice from multiple alternatives, they typically follow the following process:

- Identifying all available alternatives (i.e. the choice set).
- Determining relevant attributes that affect the attractiveness of the alternatives.
- Scoring these alternatives based on the values of their attributes.
- Ranking the alternatives based on their score.
- Making a choice, where the alternative with the highest score is most likely selected.

For example, for a rational individual selecting a travel itinerary, the following process is assumed to be followed:

- Identify all available itineraries from her/his origin to her/his destination on the day of travel offered by all airlines.
- Determine the relevant attributes of the itineraries that affect their attractiveness including price, number of connections (if any), departure time, arrival time, connection duration, circuity, etc.
- Score the available itineraries based on the values of these defined attributes.
- Rank the itineraries based on their score.
- Make a choice, where the itinerary with highest score is most likely selected.

Discrete choice models are used to model categorical dependent variables that represent a choice among several alternatives. The multinomial logit model (MLM) is a typical example of discrete choice models used to estimate the probability that an average individual selects each of the alternatives in her/his choice set. As explained hereafter, this probability is estimated as a function of a linear combination of several observed attributes (independent variables or explanatory variables) and problem-specific parameters. The values of these parameters are estimated using appropriate historical data that represents similar historical choice scenarios.

The multinomial logit relies on an important assumption known as independence of irrelevant alternatives (IIA). This assumption, which is a core hypothesis in rational choice theory, states that the probability of preferring one alternative to another does not depend on the presence or absence of other "irrelevant" alternatives. For example, the relative probabilities of choosing an itinerary scheduled by a low-cost airline or a legacy airline for travel do not change in case driving-your-own-car alternative is added as an additional alternative. In this case, driving your own car is considered as "irrelevant" alternative, and accordingly, its presence or absence does not change the probabilities of selecting the itineraries provided by the low-cost airline and the legacy airline. The IIA assumption is not always desirable, as it is frequently violated in many choice-making situations. An example of violating the IIA assumption is when the choice set of itineraries includes additional itinerary by the low-cost airline and/or the legacy airline. Suppose the choice odds ratio between the two itinerary alternatives offered by the low-cost airline and the legacy airline is 1.0–1.0. In case another itinerary scheduled by the legacy airline is added to the set of available itineraries, a traveler might be indifferent between the two itineraries offered by the legacy airline. In this case, he/she may exhibit odds ratio of 1.0, 0.5, and 0.5 among the itineraries offered by the low-cost airline, the first itinerary offered by the legacy airline, and the second itinerary offered by the legacy airline, respectively. Thus, when a second itinerary is added, the ratio between the itinerary of the low-cost airline and any of the two itineraries offered by the legacy airline is changed to 1.0–0.5. This change in the choice odds is contributed to the fact that the two itineraries offered by the legacy airline can be viewed as substitute of each other. In case the IIA is violated, especially in analysis that aims to predict how choices would change with the addition or removal of alternatives from the choice set, other models such as the nested logit can be used.

To explain how the MLM is developed and used, consider the example of an airline analyst, who is interested to evaluate competition among several itineraries (represented by the set *J*) in a given city-pair. The objective is to estimate the probability p_{ij} that an itinerary $j \in J$ is selected by an individual i. Next, a linear predictor function is considered to estimate the score of each alternative. This function is referred to as the utility function, which measures the attractiveness of each alternative. The utility function is a weighted sum of the explanatory variables. This function can be written as follows:

$$U_{i,j} = \beta_{0,j} + B_{1,j} x_{1,i} + B_{2,j} x_{2,i} + B_{3,j} x_{3,i} + B_{4,j} x_{4,i} + B_{5,j} x_{5,i} + \ldots$$

where

$U_{i,j}$: The utility (score) associated with individual i is choosing alternative j
X_i: Vector of explanatory variables describing observations associated with individual i
β_j: Vector of regression coefficients corresponding to outcome j

A set of data points that represent previous choice scenarios among the available alternative are required. The data set is used to estimate the different coefficients of the utility function. Specifically, assume a data set, of N observations, is collected. Each data point consists of the corresponding values of a set of M explanatory variables (independent variables or predictor variables) denoted as $x_{1,i}, x_{2,i}, x_{3,i}, x_{4,i}, \ldots, x_{m,i}$ and an associated categorical outcome (dependent variable) j_i, where $j_i \in J$ represents the choice made by individual i. In other words, each data point represents a choice scenario that is made by individual i and the values of explanatory variables that lead to this choice (outcome). The goal of the multinomial logistic regression is to construct a model that explains the relationship between the explanatory variables (independent variables) and the outcome (dependent variable). In other words, the model explains the relative effect of different explanatory variables on the choice. Developing such model, the outcome of a new choice scenario can be predicted as a function of the values of the explanatory variables.

The data points are fed into specialized statistical packages to estimate the values of the coefficients of the utility function. An estimation method known as the maximum likelihood estimation (MLE) is typically used for this purpose. MLE corresponds to many well-known estimation methods in statistics such as the ordinary least square method that is used for MLR. The description of the MLE is beyond the scope of this presentation and can be found elsewhere (Ben-Akiva and Lerman 1985).

The MLM is used to estimate the probability p_{ij} that an alternative $j \in J$ is selected by an individual i as shown below. The equation indicates that the probability of selecting an alternative $j \in J$ by an individual i is proportional to the attractiveness (utility or score) of this alternative:

$$p_{ij} = \frac{e^{U_{i,j}}}{\sum_{j \in J} e^{U_{i,j}}}$$

where e is the exponential constant; $e = 2.712$.

8.6 Itinerary Choice Models

In the 1990s, discrete choice models gained acceptance in modeling the choice behavior in many applications including economics, politics, marketing, and transportation. Researchers and practitioners in the airline industry had found discrete choice models as a suitable technique to model and understand how travelers select among their travel itineraries. The goal is to understand the main measurable variables that affect the itinerary choice and the relative importance of these variables and to investigate whether these discrete choice models can predict the market share of competing itineraries. Among the first effort to develop itinerary choice models is the work of Coldren et al. (2003), who developed itinerary choice models for a major US airline to represent competition among airlines in the US domestic market. The research team worked closely with this airline to design the model structure and select main variables to be included in the model in order to explain the itinerary choice behavior.

Variables that affect itinerary choice could be grouped under three main categories as follows: (i) itinerary characteristics, (ii) traveler characteristics, and (iii) trip characteristics. The itinerary-related characteristics include the number of connections, connection duration, travel distance (or circuity), departure and arrival times, price, aircraft type, and airline brand (i.e. low cost or legacy). It is generally expected that travelers prefer nonstop itineraries or itineraries that have less number of connections and with adequate connection duration. In addition, for connecting itineraries, travelers prefer itineraries with minimum distance circuity. Travelers also prefer convenient departure/arrival times, avoiding very early-morning or late-night itineraries. They also prefer to pay less for a good service on comfortable aircraft. Several traveler characteristics might impact itinerary choice including, for example, income, age, and gender. It is expected that high-income travelers might be less sensitive to itinerary price and accordingly select most convenient itineraries that satisfy their travel plan. Specific age and gender groups might have preferences to specific itineraries with convenient departure and arrival times, connections, or special airline service. Finally, trip characteristics include trip type (business vs. leisure), domestic or international, and short haul or long haul. It is common that travelers' itinerary choice behavior varies based on whether they travel for leisure or business trip. For example, leisure travelers are generally more price sensitive than business travelers. Travelers on long-haul and international trips might perceive the quality of available itineraries differently than that when they travel for short-haul and domestic trips.

Coldren et al. (2003) argue that there is no single model that can represent the travelers' choice behavior in all city-pairs. The main issue is that different city-pairs have different time zones. A combination of departure–arrival times for an itinerary might be convenient for travelers in a particular time zone but not the other. For example, the travel from a destination in the east coast of the United States to another destination in the west coast (and vice versa) is about five hours, and there are three hours of time-zone difference between the two destinations. A flight departure at 6:00 p.m. (east coast local time) from the east coast will arrive at about 8:00 p.m. (west coast local time), which is a convenient departure–arrival time combination. On the other hand, the travel from the west coast to the east coast will have different consideration. For example, a flight departure from the west coast at the same departure time that is 6:00 p.m. (West Coast local time) will arrive at 2:00 a.m. (East Coast local time), which is not a convenient departure–arrival time combination. These two examples show that an itinerary with a departure time at 6:00 p.m. is acceptable for an east-to-west trip, while the same departure time is undesirable for a west-to-east trip.

Thus, an approach is adopted in which several itinerary choice models were developed for the domestic US market considering the different time-zone combinations. Following this approach, the domestic US market is divided following the four standard time zones covering most parts of the United States, which are Eastern Standard Time (E), Central Standard Time (C), Mountain Standard Time (M), and Western Standard Time (W). Depending on the location of the origin and the destination of a city-pair, a distinct itinerary choice model is developed. Thus, given all origin–destination combinations, sixteen different models were developed, as given in Table 8.1. For example, if the

Table 8.1 The different time zones for the US-based domestic itinerary choice models.

	Eastern (E)	Central (C)	Mountain (M)	Western (W)
Eastern (E)	E-E	E-C	E-M	E-W
Central (C)	C-E	C-C	C-M	C-W
Mountain (M)	M-E	M-C	M-M	M-W
Western (W)	W-E	W-C	W-M	W-W

origin of the city-pair is in the Central time zone (e.g. Chicago or Dallas) and the destination is in the Western time zone (e.g. San Francisco or Seattle), a model that corresponds to the Central–Western (C–W) combination is used. In addition, several other models were considered to model travel from the mainland to the states of Alaska and Hawaii, which are out of these four considered time zones.

Several variables have been selected to be included in a linear-shaped utility (score) function. These variables can be classified into five categories including level of service, connection quality, carrier attributes, aircraft type, and departure time. These variables were mainly selected based on their measurability and availability. As mentioned above, historical data is needed to estimate the parameters of the different variables included in the utility function. This historical data is usually available from historical booking data. Historical booking data includes most itinerary characteristics. However, this data typically lacks detailed traveler characteristics (e.g. income) and trip characteristics (e.g. trip purpose, leisure or business). In addition, when the model is developed, the variables included in the utility function should be measurable for all itineraries, such that they can be plugged into the itinerary choice model to estimate and predict the market share of the itineraries. Accordingly, most of the variables included by Coldren et al. (2003) in the itinerary choice models are the itinerary characteristics variables. The list of the variables and their definitions are given in Table 8.2.

The level-of-service variable is related to the number of connections of the itinerary. Four different types of itineraries were considered in the domestic US market. These types include nonstop, single-connect, and double-connect, in addition to a direct itinerary, which is a single-connect itinerary that both flights in the itinerary are assigned to the same aircraft. Thus, direct itineraries have the advantage over the single-connect itinerary in that connecting passengers on direct itinerary do not have to connect between two different aircraft at two different gates. The level-of-service variable is considered as a dummy variable and is considered in relation to the best level of service in the city-pair. The best level of service in the city-pair corresponds to the level of service of the best itinerary serving the city-pair. For example, a city-pair such

Table 8.2 Definitions of explanatory variables included in the itinerary choice models.

Variable	Description
Level of service	Dummy variable representing the level of service of the itinerary (nonstop, direct, single connect, or double connect) with respect to the best level of service available in the city-pair
Second-best connection	For connection itineraries sharing a common leg, a dummy variable indicating that the itinerary is not the best connection (with respect to ground time) for the given incoming or outgoing leg at a transfer city
Second-best connection time difference	If the second-best connection indicator equals one, this variable measures the ground time difference between the itinerary and the best connection itinerary
Best connection time difference	Elapsed time difference between an itinerary involving a stop or connection and the fastest itinerary involving a stop or connection for each city-pair independent of transfer cities
Distance ratio	Itinerary distance divided by the shortest itinerary distance for the city-pair multiplied by 100
Point-of-sale-weighted city presence	Carrier origin and destination presence (determined by percentage of departures) weighted by industry city-pair point-of-sale percentages divided by 100 to give units of percent from 0 to 100
Fare ratio	Carrier average fare divided by the industry average fare for the city-pair multiplied by 100
Carrier	Dummy variable representing carriers having more than 0.5% of itineraries in the entity. All other carriers are combined together in a single category
Code share	Dummy variable indicating whether any leg of the itinerary was booked as a code share
Regional jet	Dummy variable indicating whether the smallest aircraft on any part of the itinerary is a regional jet
Propeller aircraft	Dummy variable indicating whether the smallest aircraft on any part of the itinerary is a propeller aircraft
Mainline jet seats	If an itinerary involves neither a regional jet nor a propeller aircraft leg, this variable measures the number of seats on the smallest aircraft for the itinerary
Regional jet seats	If an itinerary includes a regional jet leg (but no propeller aircraft leg), this variable measures the number of seats on the smallest regional jet aircraft for the itinerary
Propeller aircraft seats	If an itinerary includes a propeller aircraft leg, this variable measures the number of seats on the smallest propeller aircraft for the itinerary
Time of day	Dummy variable for each hour of the day (based on the local departure time of the first leg of the itinerary)

as Chicago O'Hare (ORD) to Dallas–Fort Worth (DFW) is considered a nonstop city-pair because there is at least one airline that provides nonstop service in this city-pair. A city-pair such as ORD to Daytona Beach (DAB) is considered as a single-connect city-pair because this is the best itinerary service provided in this city-pair.

The second-best connection and the second-best connection time difference are two related variables that are used only for connecting itineraries. The first variable is a dummy variable that indicates whether the flights of the connecting itinerary are the best connection at the transfer airport with respect to the ground time. This dummy variable is equal to 1, if the flights of the connecting itinerary are not the best connection, and zero otherwise. In case the dummy variable is equal to 1 for any itinerary, the second variable, the best connection time difference, measures the time difference between the connection duration of this itinerary and the connection duration of the best connecting itinerary. The best connection time difference variable measures the elapsed time difference between an itinerary involving a stop or connection and the fastest itinerary involving a stop or connection for each city-pair, regardless of transfer cities. Distance ratio measures the itinerary distance with respect to the shortest itinerary distance for the city-pair.

The point-of-sale-weighted city presence reflects the fact that itineraries offered by airlines that have more presence in the origin and/or the destination of the trip are more likely to be selected by travelers. For example, travelers leaving DFW are more likely to select itineraries offered by American Airlines, since this airline has more city presence in the airport. This variable measures the airline's city presence at the origin and destination of the itinerary. The fare ratio measures the airline's average fare with respect to the industry's average fare for the city-pair. It should be noted that this fare ratio variable is not measured at the itinerary level. In other words, it does not represent the ticket price of each itinerary. It rather measures, in general, whether the airline is a low-fare airline that offers inexpensive itineraries in the city-pair or not. The carrier variable is a dummy variable to indicate whether travelers have any airline preference. Travelers might have a particular preference to one airline over the other airlines. Thus, if identical itineraries are offered by the different airlines, the itinerary offered by the preferred airline is more likely to be selected. The code-share variable is a dummy variable that indicates whether the itinerary is a code-share itinerary, where the marketing airline is different from the operating airline for one or more of the flights of the itinerary.

Regional jet is a dummy variable indicating whether the smallest aircraft used on any flight of the itinerary is a regional jet. Propeller aircraft is a dummy variable indicating whether the smallest aircraft used on any flight of the itinerary is a propeller aircraft. The variable mainline jet seats measures the number of seats on the smallest aircraft for the itinerary if the itinerary involves neither a regional jet nor a propeller aircraft leg. The variable regional jet seats measures the number of seats on the smallest regional jet aircraft for the itinerary, if an itinerary includes a regional jet leg (but no propeller aircraft leg). The variable propeller aircraft seats measures the number of seats on the smallest propeller aircraft for the itinerary, if an itinerary includes a propeller aircraft leg. Time of day is a dummy variable for each hour of the day, based on the local departure time of the first leg of the itinerary (Coldren et al. 2003).

Historical airline booking data was used to estimate the values of the different parameters of the utility function for the different considered time zones. Table 8.3 gives the values of these parameters for five models including E-E, E-C, E-M, E-W, and W-E (Coldren et al. 2003). The other models were not published by Coldren et al. (2003). Here, the parameters of the carrier constant were eliminated because of proprietary reasons.

Table 8.3 The values of the parameters for the itinerary choice models.

Explanatory variables	E-E	E-C	E-M	E-W	W-E
Level of service					
Nonstop itinerary in nonstop market	0	0	0	0	0
Direct itinerary in nonstop market	−1.6819	−1.5253	−1.7168	−1.6615	−1.5349
Single-connect itinerary in nonstop market	−3.1213	−2.876	−2.7087	−2.9729	−2.8858
Double-connect itinerary in nonstop market	−7.7557	−7.297	−7.5112	−7.213	−6.5395
Direct itinerary in direct market	0	0	0	0	0
Single-connect itinerary in direct market	−0.7901	−0.7768	−1.0989	−1.0615	−0.9847
Double-connect itinerary in direct market	−4.9442	−4.5565	−4.3308	−4.4475	−4.5869
Single-connect itinerary in single-connect market	0	0	0	0	0
Double-connect itinerary in single-connect market	−3.0953	−3.1205	−2.5431	−2.8209	−2.7105
Connection quality					
Second-best connection	−0.556	−0.5058	−0.5322	−0.7269	−0.4377
Second-best connection time difference	−0.0157	−0.019	−0.0178	−0.0162	−0.018
Best connection time difference	−0.0108	−0.0094	−0.0093	−0.0104	−0.0088
Distance ratio	−0.0125	−0.0116	−0.021	−0.0173	−0.0207

Table 8.3 (Continued)

Explanatory variables	E-E	E-C	E-M	E-W	W-E
Carrier attributes					
Point-of-sale-weighted city presence	0.0024	0.01	0.0078	0.0071	0.0077
Fare ratio	−0.0018	−0.0038	−0.0045	−0.0035	−0.004
Carrier constants (proprietary)	xxxxxx	xxxxxx	xxxxxx	xxxxxx	xxxxxx
Code share	−1.5911	−2.1255	−1.922	−2.1658	−2.2082
Aircraft size and type					
Mainline	0	0	0	0	0
Propeller	−1.242	−1.0599	−0.8081	−0.9988	−0.9269
Regional jet	−0.7046	−0.8317	−0.8844	−0.7079	−0.411
Propeller seats	0.0246	0.0146	0.0157	0.0184	0.0212
Regional jet seats	0.0117	0.0101	0.0144	0.008	0.0052
Mainline seats	0.0041	0.0037	0.0047	0.0032	0.0036
Time of day					
Midnight to 5 a.m.	−1.1653	−1.1577	−1.2341	−1.2037	−0.7228
5–6 a.m.	−0.4653	−0.4577	−0.5341	−0.5037	0.1482
6–7 a.m.	0	0	0	0	0
7–8 a.m.	0.2865	0.2693	0.2179	0.2297	0.057
8–9 a.m.	0.2836	0.413	0.3983	0.3168	0.2354
9–10 a.m.	0.2046	0.3204	0.4156	0.3413	0.0311
10–11 a.m.	0.1219	0.2571	0.4539	0.3425	0.0592
11–12 noon	0.1022	0.2539	0.3842	0.267	0.0196
12–1 p.m.	0.1452	0.3137	0.4252	0.2977	0.057
1–2 p.m.	0.1732	0.3676	0.2298	0.2594	0.0219
2–3 p.m.	0.2622	0.4742	0.1964	0.2462	−0.0988
3–4 p.m.	0.3438	0.5254	0.2136	0.2557	0.1175
4–5 p.m.	0.423	0.5853	0.2476	0.2097	0.0873
5–6 p.m.	0.4667	0.5959	0.2419	0.2324	0.1779
6–7 p.m.	0.4054	0.5516	0.2491	0.1889	−0.2437
7–8 p.m.	0.2047	0.4627	−0.0099	0.0472	0.2914
8–9 p.m.	−0.0362	0.0866	−0.0984	−0.2243	0.1417
9–10 p.m.	−0.3565	−0.2917	−0.6878	−0.3924	0.0541
10 p.m. to midnight	−0.6468	−0.7454	−0.7637	−0.317	−0.0573

As shown in the table, for the five models, the signs and values of the parameters of the different variables included in the utility function are logical. For example, for the E-E model, the parameters of the level-of-service variables are negative, and the values of these parameters decrease as the number of connections increases. For example, for the nonstop city-pairs (markets), the nonstop itinerary is considered the reference, and the value of its corresponding parameter is equal to 0. The values of the parameters that correspond to the single-connection itinerary, the direct itinerary, and the double connection itinerary are −1.68, −3.12, and −7.75, respectively.

When the itinerary is not the best connection (i.e. the second-best connection), the itinerary score is reduced by 0.556 (the parameter value of the second-best connection dummy variable is −0.556). For every minute difference between the second-best connection itinerary and the best connection itinerary, the itinerary score is reduced by 0.0157. In addition, for every minute of elapsed time difference between an itinerary involving a connection and the fastest itinerary involving a stop or connection for each city-pair independent of transfer cities, the itinerary score is reduced by 0.0108. Furthermore, the itinerary score decreases as its distance increases. The score of the itinerary decreases by 0.0125 for each unit increase in distance ratio.

As the city presence for an airline in the city-pair increases, the attractiveness of its itineraries increases. Expensive airlines in the city-pair are less likely to be selected, as the score of its itinerary decreases by 0.0018 for every unit increase in the fare ratio. Airlines do not prefer code-share itineraries. The parameter of the code-share dummy variable is −1.59. With code-share itineraries, travelers usually get confused about which airline to actually check in with. Passengers prefer itineraries operated by mainline aircraft compared with regional jets and propeller. In addition, for the same aircraft type, passengers prefer larger aircraft that has more seats compared with the smaller ones. As a side note, this result might be controversial for aircraft operated by ultralow-cost airlines that tend to increase the number of seat rows on the aircraft to maximize their available capacity on the expenses of spacing for legroom between seat rows. These aircraft might be less preferred by customers.

Finally, the departure time parameters associated with the departure time dummy variables indicate the relative preference of the different departure times. For example, by assuming that 6–7 a.m. is the reference, the best-preferred departure interval in the E-E city-pairs is 5–6 p.m., which corresponds to the highest parameter value of 0.4667. This departure time is usually preferred to business travelers after finishing their workday.

8.7 Applying Itinerary Choice Models: An Example

This section explains how the itinerary choice model can be used to estimate the unconstrained market share of a given itinerary (i.e. the probability that it will be selected, given that this itinerary and its competing itineraries have no limitation on capacity). For this purpose, consider the example given in Table 8.4. It gives the characteristics of four competing itineraries (itinerary 1, itinerary 2, itinerary 3, and itinerary 4) in a nonstop city-pair (market) in the E-E region. For example, itinerary 3 is a single-connection itinerary that has a distance ratio of 120%, which means that it is 20% longer than the least-distance itinerary in this city-pair. The airline that operates itinerary 3 has 25% city presence, and it offers itineraries that are 10% cheaper than the market average (the fare ratio is 90%). At least one of the flights of this itinerary is operated by a regional jet that has 120 seats. Finally, the departure time of this itinerary is within the 7–8 a.m. interval.

The E-E model is used to calculate the systematic utility of each itinerary ($v1$, $v2$, $v3$, and $v4$), as shown in Table 8.4. For each itinerary, the value of the variable is multiplied by the corresponding parameter estimated for the E-E model, and all terms are added up to estimate the total utility (score) of each of the four itineraries. Accordingly, itinerary 1, itinerary 2, itinerary 3, and itinerary 4 have a utility of –0.6708, –0.3960, –3.7374, and –4.1717, respectively.

The utility values are plugged into the probability equation to estimate the probability of selecting each itinerary by an average rational traveler. The probability of selecting the four competing itineraries is given below. One should notice that the sum of these four probabilities should equal to 100%:

$$p_{\text{itin }1} = \frac{e^{-0.6708}}{e^{-0.6708} + e^{-0.396} + e^{-3.7374} + e^{-4.1717}} = 41.79\%$$

$$p_{\text{itin }2} = \frac{e^{-0.396}}{e^{-0.6708} + e^{-0.396} + e^{-3.7374} + e^{-4.1717}} = 55.0\%$$

$$p_{\text{itin }3} = \frac{e^{-3.7374}}{e^{-0.6708} + e^{-0.396} + e^{-3.7374} + e^{-4.1717}} = 1.95\%$$

$$p_{\text{itin }4} = \frac{e^{-4.1717}}{e^{-0.6708} + e^{-0.396} + e^{-3.7374} + e^{-4.1717}} = 1.26\%$$

Assume one of the competing itineraries is sold out. This itinerary has to be removed for the choice set of the next customer. For example, if itinerary 1 is sold out, the choice set of the next customers includes only itinerary 2, itinerary 3,

Table 8.4 Example applying itinerary choice models for four itineraries.

Explanatory variables	E-E	Itin 1	Itin 2	Itin 3	Itin 4	v1	v2	v3	v4
Level of service									
Nonstop itinerary in nonstop market	0	1	1			0	0	0	0
Direct itinerary in nonstop market	−1.6819					0	0	0	0
Single-connect itinerary in nonstop market	−3.1213			1	1	0	0	−3.1213	−3.1213
Double-connect itinerary in nonstop market	−7.7557					0	0	0	0
Direct itinerary in direct market	0					0	0	0	0
Single-connect itinerary in direct market	−0.7901					0	0	0	0
Double-connect itinerary in direct market	−4.9442					0	0	0	0
Single-connect itinerary in single-connect market	0					0	0	0	0
Double-connect itinerary in single-connect market	−3.0953					0	0	0	0
Connection quality									
Second-best connection	−0.556					0	0	0	0
Second-best connection time difference	−0.0157					0	0	0	0
Best connection time difference	−0.0108					0	0	0	0
Distance ratio	−0.0125	100	100	120	110	−1.25	−1.25	−1.5	−1.375
Carrier attributes									
Point-of-sale-weighted city presence	0.0024	25	25	25	25	0.06	0.06	0.06	0.06
Fare ratio	−0.0018	110	105	90	88	−0.198	−0.189	−0.162	−0.1584

	Coefficient									
Carrier constants (proprietary)	xxxxxx									
Code share	-1.5911			1		0	0	0	0	0
Aircraft size and type										
Mainline	0	1				0	0	0	0	0
Propeller	-1.242					0	0	0	0	0
Regional jet	-0.7046		1	1		0	-0.7046	-0.7046	-0.7046	0
Propeller seats	0.0246				120	0	0	0	0	0
Regional jet seats	0.0117		120	120		0	1.404	1.404	1.404	0
Mainline seats	0.0041	150				0.615	0	0	0	0
Time of day										
Midnight to 5 a.m.	-1.1653					0	0	0	0	0
5–6 a.m.	-0.4653					0	0	0	0	0
6–7 a.m.	0					0	0	0	0	0
7–8 a.m.	0.2865		1			0	0	0.2865	0	0
8–9 a.m.	0.2836			1		0	0.2836	0	0	0
9–10 a.m.	0.2046					0	0	0	0	0
10–11 a.m.	0.1219					0	0	0	0	0
11–12 noon	0.1022				1	0.1022	0	0	0	0
12–1 p.m.	0.1452					0	0	0	0	0
1–2 p.m.	0.1732					0	0	0	0	0
2–3 p.m.	0.2622					0	0	0	0	0
3–4 p.m.	0.3438					0	0	0	0	0

(Continued)

Table 8.4 (Continued)

Explanatory variables	E-E	Itin 1	Itin 2	Itin 3	Itin 4	v1	v2	v3	v4
4–5 p.m.	0.423				1	0	0	0	0.423
5–6 p.m.	0.4667					0	0	0	0
6–7 p.m.	0.4054					0	0	0	0
7–8 p.m.	0.2047					0	0	0	0
8–9 p.m.	−0.0362					0	0	0	0
9–10 p.m.	−0.3565					0	0	0	0
10 to midnight	−0.6468					0	0	0	0
V						−0.6708	−0.396	−3.7374	−4.1717
Exp(V)						0.5113	0.6730	0.0238	0.0154
Probability						41.79%	55.00%	1.95%	1.26%

Table 8.5 Example applying itinerary choice models for three itineraries.

Explanatory variables	E-E	Itin 2	Itin 3	Itin 4	v2	v3	v4
Level of service							
Nonstop itinerary in nonstop market	0	1			0	0	0
Direct itinerary in nonstop market	−1.6819				0	0	0
Single-connect itinerary in nonstop market	−3.1213		1	1	0	−3.1213	−3.1213
Double-connect itinerary in nonstop market	−7.7557				0	0	0
Direct itinerary in direct market	0				0	0	0
Single-connect itinerary in direct market	−0.7901				0	0	0
Double-connect itinerary in direct market	−4.9442				0	0	0
Single-connect itinerary in single-connect market	0				0	0	0
Double-connect itinerary in single-connect market	−3.0953				0	0	0
Connection quality							
Second-best connection	−0.556				0	0	0
Second-best connection time difference	−0.0157				0	0	0
Best connection time difference	−0.0108				0	0	0
Distance ratio	−0.0125	100	120	110	−1.25	−1.5	−1.375
Carrier attributes							
Point-of-sale-weighted city presence	0.0024	25	25	25	0.06	0.06	0.06
Fare ratio	−0.0018	105	90	88	−0.189	−0.162	−0.1584
Carrier constants (proprietary)	xxxxxx				0	0	0
Code share	−1.5911				0	0	0

(Continued)

Table 8.5 (Continued)

Explanatory variables	E-E	Itin 2	Itin 3	Itin 4	v2	v3	v4
Aircraft size and type							
Mainline	0			1	0	0	0
Propeller	−1.242				0	0	0
Regional jet	−0.7046	1	1		−0.7046	−0.7046	0
Propeller seats	0.0246				0	0	0
Regional jet seats	0.0117	120	120		1.404	1.404	0
Mainline seats	0.0041				0	0	0
Time of day							
Midnight to 5 a.m.	−1.1653				0	0	0
5–6 a.m.	−0.4653				0	0	0
6–7 a.m.	0				0	0	0
7–8 a.m.	0.2865		1		0	0.2865	0
8–9 a.m.	0.2836	1			0.2836	0	0
9–10 a.m.	0.2046				0	0	0
10–11 a.m.	0.1219				0	0	0
11–12 noon	0.1022				0	0	0
12–1 p.m.	0.1452				0	0	0
1–2 p.m.	0.1732				0	0	0
2–3 p.m.	0.2622				0	0	0
3–4 p.m.	0.3438				0	0	0

4–5 p.m.	0.423		0	0.423
5–6 p.m.	0.4667		0	0
6–7 p.m.	0.4054	1	0	0
7–8 p.m.	0.2047		0	0
8–9 p.m.	−0.0362		0	0
9–10 p.m.	−0.3565		0	0
10 to midnight	−0.6468		0	0
V	−0.396		−3.7374	−4.1717
Exp(V)	0.67300		0.02384	0.01542
Probability	94.49%		3.34%	2.17%

and itinerary 4. Table 8.5 gives the utility of each of these three itineraries. The probabilities of these three itineraries are calculated as shown below. As one anticipates, the probability of selecting each of the remaining three itineraries increases with the removal of itinerary 1 from the choice set:

$$P_{\text{itin } 2} = \frac{e^{-0.396}}{e^{-0.396} + e^{-3.7374} + e^{-4.1717}} = 94.49\%$$

$$P_{\text{itin } 3} = \frac{e^{-3.7374}}{e^{-0.396} + e^{-3.7374} + e^{-4.1717}} = 3.34\%$$

$$P_{\text{itin } 4} = \frac{e^{-4.1717}}{e^{-0.396} + e^{-3.7374} + e^{-4.1717}} = 2.17\%$$

9

Profitability Forecasting Models

9.1 Introduction

The previous chapter presents an example of itinerary choice models that estimate the probability of selecting a given itinerary among a set of competing itineraries in a given city-pair. This probability corresponds to what is known as the unconstrained market share of the itinerary. The unconstrained market share of a given itinerary represents the percentage of travelers that select the itinerary assuming unlimited seat capacity. In other words, it represents the passenger market share of the itinerary when the seat capacity of its flight(s) can accommodate all passengers who select the itinerary. However, when the seat capacity of the itinerary is less than its demand, some of this demand is spilled to other itineraries serving this city-pair. In this case, the actual market share of this itinerary is determined by its seat capacity. In addition, the actual market share of each itinerary in the city-pair would depend partially on the number of passengers spilled from other itineraries and recaptured on this itinerary.

To estimate the actual market share of the itineraries, an analytical tool is needed to capture the interaction among the different itineraries at the network level. This tool replicates how demand is spilled and recaptured among the different itineraries of the different competing airlines and their corresponding flights (Abdelghany and Abdelghany 2012). It also tracks the number of passengers and computes the corresponding passenger load factor on each flight. If fares are known, the total revenue of the schedule can be estimated. In addition, if the cost of operating the schedule is known, the revenue and the cost are used to estimate the schedule profitability, and thus the model is referred to as the profitability forecasting model (PFM).

PFM is used primarily to evaluate a schedule of an airline considering the competition with other airlines. It evaluates how any proposed schedule for an airline would sustain competition with other airlines. The model can also

Airline Network Planning and Scheduling, First Edition. Ahmed Abdelghany and Khaled Abdelghany.
© 2019 John Wiley & Sons, Inc. Published 2019 by John Wiley & Sons, Inc.

answer many what-if scenarios that reflect changes in the schedules of the target and the competing airlines or the ticket prices in the different markets.

Another way to look at the PFM is that it replicates the travelers' ticket booking behavior in the different city-pairs. Along the booking horizon of the schedule, travelers in each city-pair identify available itineraries and select the itinerary that maximizes their utility. Once an itinerary is selected by a traveler, a seat is reserved, and seat availability on the flight(s) of the itinerary is updated. The process continues until the entire travel demand is assigned. During the process, when a flight is marked as full (i.e. sold out), this flight and its corresponding itineraries are removed from the choice set of subsequent travelers.

This chapter presents a description of the PFM used by major airlines to evaluate schedule competitiveness and profitability. Assume the availability of information on the travels demand in different city-pair and the flight schedules of all competing airlines. Also, assume that the behavioral models that describe how travelers select among itineraries are known. The objective of PFMs is to estimate the market share for each airline at the network, market, and flight levels. Figure 9.1 illustrates the overall framework of PFM presented in this chapter. The model is composed of the input module, itinerary builder, itinerary valuation (scoring) and choice models, and the demand assignment module. The description of each of these modules is given hereafter.

9.2 Model Input

As mentioned above, PFMs are used to evaluate the schedule of an airline considering the competition with other airlines. Thus, the input of the PFMs consists of the total demand in the different city-pairs, the proposed flight schedule of the target airline, and the flight schedules of all competing airlines. The schedule of an airline (timetable) is typically presented in the form of a list of flights, where each flight is defined by its flight number, origin, destination, scheduled departure time, scheduled arrival time, and seat capacity. Figure 9.2 shows an example of a flight schedule data set in a tabular format, where each row represents a flight record. Each record includes information on the marketing airline, operating airline, origin, destination, flight number, departure time, arrival time, flight distance, code-share indicator, and aircraft type. Each flight record also includes seven binary indicators to designate days of the week in which this flight is scheduled to operate.

The target airline has full information on the flight schedule that needs to be evaluated. On the contrary, the flight schedules of the competing airlines are not necessarily known, as competing airlines may decide to change their schedule for any future period. Airlines usually publish most of their flight schedules 9–12 months in advance so that passengers can plan their trips and book their tickets. However, this published schedule is confirmed only for the

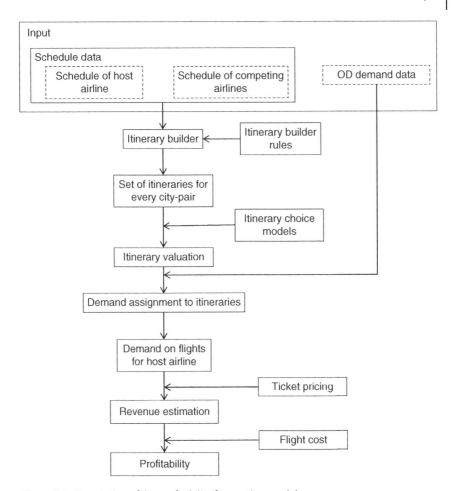

Figure 9.1 Description of the profitability forecasting model.

future 3–6 months. The flight schedules for the future months beyond this horizon are periodically updated over time. As such, if the host airline is interested to evaluate its schedule for a future month that is beyond the 6-month period, some assumptions need to be made regarding the schedule of the competitors. In this case, the host airline might decide to depend on historical schedule data. For instance, if the target airline is studying the competition for a schedule for the month of January of next year, it may use the competitors' schedules of the month of January of the current year to substitute for any missing information in the future schedule. In addition, the target airline might also depend on any published news on the competitors' plans to add or remove routes in the different city-pairs.

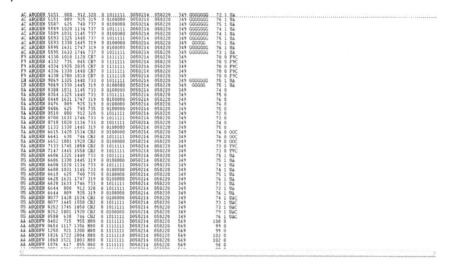

Figure 9.2 Example of schedule data in text format.

In the US domestic market, the number of passengers traveling in every city-pair (origin–destination pair or OD pair) is recorded (Devriendt et al. 2009). This data is typically collected by the Department of Transportation and is referred to as the Airline Origin and Destination Survey (DB1B). In this survey, a 10% sample of airline tickets from reporting airlines is collected by the Office of Airline Information of the Bureau of Transportation Statistics. This database is used to determine passenger demands in the different city-pairs, air traffic patterns, and the airline market shares. DB1B includes three main tables, which are DB1B Ticket, DB1B Market, and DB1B Coupon. DB1B Ticket contains summary of each domestic itinerary on the Origin and Destination Survey including the reporting airline, itinerary fare, number of passengers, originating airport, round-trip indicator, and miles flown. DB1B Market contains information on directional market characteristics of each domestic itinerary of the Origin and Destination Survey such as the reporting airline, origin and destination airport, prorated market fare, number of market coupons, market miles flown, and airline change indicators. DB1B Coupon provides coupon-specific information for each domestic itinerary of the Origin and Destination Survey such as the operating airline, origin and destination airports, number of passengers, fare class, coupon type, trip break indicator, and trip distance.

To obtain the number of passengers in the different city-pairs, for any future scenarios, the DB1B Market data is usually extrapolated for any future period taking into consideration demand seasonality, growth trend, and any known special event that might affect travel demand in a particular city-pair.

9.3 Itinerary Builder Module

As mentioned above, the flight schedule data is given in the form of a list of flights for each airline. In many cases, travelers select connecting itineraries for their travel, especially between city-pairs that do not have nonstop flight service. Accordingly, the list of possible nonstop and connecting itineraries in the different city-pairs needs to be determined. For this purpose, an itinerary builder module is included in the model. The purpose of this module is to generate all feasible itineraries using the flights of each airline and its partner airlines (e.g. alliance and code-share agreements) (Seredyński et al. 2014). This itinerary builder is based on several rules to generate feasible itineraries. These rules are designed to make sure that realistic itineraries are generated by the model. These rules are:

- The origin of the first flight should be the origin of the itinerary.
- The destination of the last flight should be the destination of the itinerary.
- A threshold on maximum number of connections (flights) should be considered.
- The origin of a flight leg should be the destination of its preceding flight leg.
- The destination of a flight leg should be the origin of its next flight leg.
- The departure time of a flight leg in the itinerary should be greater than the arrival time of its preceding flight leg.
- The minimum connection time for passengers should be satisfied.
- A maximum connection time should be considered.
- Itineraries can be constructed between partner airlines and those that implement code-share agreements.
- The ratio between the itinerary distance and the shortest itinerary in the city pair should not exceed a pre-defined threshold to avoid itineraries with circuity.

Once the itineraries for every city-pair are defined, the available choice set of each traveler is determined. Then, the itinerary choice models are used to score these itineraries and determine the probability of selecting each itinerary in the choice set.

9.4 How the Model Works?

PFM replicates how travelers book their seats. To elaborate, consider a daily schedule of an airline to be evaluated against its competitors. The following steps explain how the model works (Abdelghany and Abdelghany 2008, 2012).

- Step 1: Demand generation – This step replicates demand generation. First, one city-pair is selected for demand generation. Demand is generated from the different city-pairs in a random order in proportion to the number of passengers (market size) in the city-pair. For example, if the demand of the city-pair A–B is twice the demand of the city-pair C–D, it is twice more likely that the next generated traveler belongs to the A–B city-pair. In other words, city-pairs with large demands are more likely to be selected for demand generation than those with smaller demands.

- Step 2: Itinerary generation – The itinerary builder is activated to generate all feasible itineraries for the selected city-pair.
- Step 3: Itinerary choice – For this city-pair, the corresponding itinerary choice model is used to estimate the probability of selecting each itinerary in the choice set. Given the value of these probabilities, one itinerary is selected randomly from the choice set in proportion to the probability associated with each itinerary.
- Step 4: Flight capacity update – Once this itinerary is selected, seat availability on all flights of this itinerary is updated by eliminating one seat on each flight of the itinerary.
- Step 5: Flight capacity check – If a flight becomes full, this flight is removed from the schedule and from all itineraries for subsequent travelers.
- Step 6: Demand assignment check – If the entire demand in all city-pairs is generated and assigned to itineraries, stop and generate measure of performance.

9.5 Load Factor, Market Share, and Market Concentration

PFM calculates several measures of performance including load factor for each flight, market share, and metrics related to service concentration. The load factor represents the percentage of revenue-generating seats of each flight. The market share is calculated for every city-pair. It gives the percentage of passengers selecting the itineraries of the host airline as a percentage of the entire demand in this city-pair. The market share can be aggregated for a given hub, regional area, or for the whole schedule. It can also be aggregated for a given time period (e.g. morning, afternoon, etc.). The market share is also referred to as the quality service index or QSI (or quantity share index) because it measures how the quality of the itinerary (schedule) attracts customers and contributes to the airline's market share.

The market concentration metrics measure the level of competition between airlines in a given city-pair. The Herfindahl–Hirschman Index (HHI) is one of the common measures that is used for this purpose (Stavins 2001). HHI is calculated by summing the squares of the market share (or QSI) of each airline competing in the city-pair as follows:

$$HHI = \sum_{n=1}^{N} QSI_n^2$$

where

n: Index representing an airline competing in the city-pair
N: Number of airlines competing in the city-pair
HHI: Measure of concentration in a given city-pair (or market) represented by HHI
QSI_n: Market share of airline n providing service in the city-pair

For example, consider competition among three airlines (Airline I, Airline II, and Airline III) in two different city-pairs (A–B and C–D) with different market shares as shown in Table 9.1. In the city-pair (A–B), the three airlines are competing with comparable market shares of 35, 32, and 33%. In the city-pair (C–D), Airline II has a dominant market share of 90%, compared with Airlines I and III with limited market share of 4 and 6%, respectively.

The HHI for the two city-pairs can be calculated as follows:

$$HHI_{A-B} = 0.35^2 + 0.32^2 + 0.33^2 = 0.3338$$

$$HHI_{C-D} = 0.04^2 + 0.90^2 + 0.06^2 = 0.8152$$

Low values of the HHI indicate that there is strong competition in the city-pair, as in the case of the city-pair A–B, where airlines serving in this city-pair have comparable market shares. On the contrary, a high HHI value indicates that the market share is concentrated, and probably this city-pair is dominated by one airline. Theoretically, the minimum value of the HHI is 0, where large number (infinity) of competitors serve the market and each one has small market share (close to 0). The maximum value of the HHI is 1.0, representing a pure monopoly scenario, in which there is only one airline serving the city with 100% market share.

Table 9.1 Example for calculating the HHI.

	Market share	
	City-pair A–B (%)	City-pair C–D (%)
Airline I	35	4
Airline II	32	90
Airline III	33	6

In case an airline decided to expand its network, an important question for this airline to answer is which city-pairs to consider. Should the airline consider adding service in market with low or high HHI? The answer to this question is not straightforward. One could think that city-pairs with low HHI are characterized by strong competition among several airlines. This strong competition could lead to what is known as a fare war among airlines serving in this city-pair resulting in low yield. Thus, an airline might not be able to provide profitable service in the city-pair unless it has a competitive cost structure (i.e. low cost per available seat mile (CASM)). On the other hand, city-pairs with higher HHI values are usually characterized by the existence of a dominant airline. This dominant airline most likely controls capacity and pricing in the city-pair. This dominant airline might apply tactical pricing and capacity adjustments to make it difficult for other airlines to enter the market and compete.

As explained earlier, airlines are competing to attract passengers and maximize their market share. However, this is not always the right strategy. Increasing passengers and load factor does not necessarily guarantee profitability, especially if the yield in the city-pair is low. Historically, airlines compete for passengers based on frequency of service, price, and quality of products offered.

Airlines had always believed that market share is proportional to capacity and frequency provided in the city-pair. Adding flight frequency gives more itinerary options to travelers. These itineraries are eventually selected by travelers, which increases the market share of the airline. In addition, if an airline decided to increase service frequency, it represents a barrier for other airlines to enter the market. However, it should be clear that capacity is inversely proportional to ticket prices. As suggested by most airline revenue management models, when capacity increases, prices charged to customers decrease. Hence, the average yield in the market decreases, and the service in this city-pair might not be profitable. However, when an airline dominates the market with the adequate capacity, it usually increases the ticket prices to take advantage from its dominance in the city-pair.

In the domestic US market, Southwest Airlines is an example of airlines that competes based on frequency. Southwest Airlines serves less number of destinations compared with other major US airlines including Delta Air Lines, American Airlines, and United Airlines. However, in any city-pair that Southwest Airlines serves, it tries to dominate by providing high frequency. This dominance represents entry barrier in the city-pair for other airlines. When Southwest Airlines dominates the service in the city-pair, it may adjust ticket price to profit from its dominance.

Airlines adjust their fares to be competitive to attract customers and increase market share. The risk here is that the average yield in the city-pair decreases and the route might not be profitable. Airlines that compete based on low ticket price should have low CASM to remain profitable. Typically, low-cost airline (and ultralow-cost airlines) such as Frontier Airlines and Spirit Airlines are good examples of airlines in the domestic US market that depend on low fares to attract customers. These airlines typically have low CASM and can provide sustainable service with low fares. The low fares also represent entry barrier for airlines that plan to compete in city-pairs served by the low-cost airlines. In many cases, when airlines adopt a low-cost business model, they usually sacrifice the quality of the product provided. Most major airlines still maintain high product quality to attract high revenue customers (typically business travelers). The product quality includes all additional services and amenities provided with the flight. It includes ticket restrictions, airport lounges, inflight service, seat legroom, baggage fees, priority boarding, inflight internet connection, flight entertainment, meals and drinks, benefits from mileage plan accounts, etc. Most major airlines invest in product quality especially in city-pairs that have significant business travel.

10

Partnership Agreements

10.1 Introduction

Several options are available to airlines to increase demand and expand its network coverage domestically and internationally through partnership agreements. For example, an airline can extend its network by establishing operational partnerships with regional airlines that can serve small markets to feed the mainline airline. An airline also could engage in a code-share agreement (code-sharing) with another airline. Code-sharing is a capacity purchase agreement in which an airline can sell tickets on some pre-identified flights of its code-share partner airline. Airlines can also participate in an alliance that includes several airlines. An airline alliance is an arrangement between two or more airlines to cooperate on different aspects including code-share agreements. An alliance typically works as a global airline to facilitate inter-airline connections worldwide. Another important aspect of how airlines can boost their sales and market penetration is through expanding ticket distribution channels. Airline managers should understand how the different itineraries designed by their network planning teams are competing in the different ticket distribution channels. They should examine the trade-off between using distribution channels that are characterized with large market, expensive fees, and high competition among subscribed airlines (e.g. Expedia and Orbitz) versus distribution channels with small market, less expensive, and low/no airlines competition (e.g. the airline website). Finally, airlines adopt loyalty programs to increase the number of returning customers. A loyalty program is a rewards program offered by a company to frequent flying customers. A loyalty program may give a customer advanced access to new products, special sales coupons, or free merchandise. Airlines can also increase its demand by having agreements with large corporations. In these agreements, an airline offers fare discount in return of having the employees of the corporation use the airline for their business travel. In the following subsection, we discuss the practices in which airlines enter into partnership agreements to expand its services and sales.

Airline Network Planning and Scheduling, First Edition. Ahmed Abdelghany and Khaled Abdelghany.
© 2019 John Wiley & Sons, Inc. Published 2019 by John Wiley & Sons, Inc.

10.2   Regional Airlines

Small communities do not present sufficient demand to attract mainline airlines to provide regular service to these communities. Instead, these communities are served by airlines operating small-size aircraft. These small-size aircraft typically have limited range, and hence they can only provide service regionally, connecting small communities to close major hub(s). Thus, considering the nature of their regional services, airlines using these small aircraft are referred to as regional airlines. Although different classifications are used to categorize airlines based on their revenues or aircraft sizes, we use the term regional airlines or regional carriers to represent airlines that carry traffic from small communities to a close major hub(s) using small- or mid-size aircraft. Following the Airline Deregulation Act of 1978, the US federal government has continued to support the regional airline sector to ensure air travel access to small and isolated rural communities (Forbes and Lederman 2007; Levine 2011; Fageda and Flores-Fillol 2012).

As such, mainline airlines enter partnership agreements with regional airlines to secure demand from these small communities to their major hub(s) or focus cities (Truitt and Haynes 1994; Graham 1997; French 1998). According to these partnership agreements, regional airlines operate as an extension of the mainline airline. They serve the demand of the small communities through providing service to the mainline airline's hub(s). Figure 10.1 gives an example of a hypothetical network of a mainline airline that extends its network with flights operated by a regional airline. In this figure, the routes of the mainline airline are shown by solid lines, and the routes of the regional airline are shown by dashed lines.

Three main approaches are reported to establish partnership between a regional airline and a mainline airline. First, the regional airline operates as an independent airline under its own brand, mostly providing service to connect small and isolated communities to a major hub. This form of partnership, which is the least common, is often in the form of a code-share agreement (i.e. see next section for more details on code-share agreements) or a capacity purchase arrangement with the regional airline. The regional airline performs all planning and operation decisions (flight scheduling, pricing, and operation irregularity management). Second, the regional airline is affiliated with the mainline airline. In this case, the regional airline contracts with a mainline airline to operate under its brand name or pseudo brand name (e.g. United Airlines and United Express). The aircraft of the regional airline are painted with the brand and logo of the mainline airline. The regional airline, in this case, delivers passengers to the mainline airline's hub(s) from/to surrounding small communities. Regional airlines might also serve on the mainline airline's short-haul routes during low demand seasons that do not warrant the use of large aircraft. In these two approaches, the mainline airline can perform

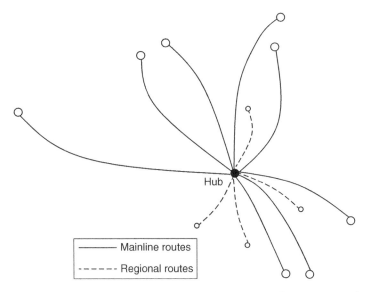

Figure 10.1 Example of a network of a mainline airline with agreement with a regional airline.

partnership with multiple regional airlines as needed. Large regional airlines can establish partnership with more than one mainline airline, if they have adequate resources (i.e. aircraft and crew). Third, the regional airline operates as a subsidiary of a parent company that holds both the mainline and the regional airlines. The holding company controls investment decisions of the regional airline such as service expansion, aircraft purchase, etc. An example of this approach is the regional airline Envoy Air, which is fully owned by American Airlines Group. In the latter two methods, the mainline airline has more control on the planning and operations of flights operated by the regional airline. The regional airline operates under the mainline airline's two-letter flight designator code. The mainline airline is in charge of determining the city-pairs to be flown, flight frequency and schedule, time banking, pricing strategies, and marketing and sales. The mainline airline is also in charge of operation-related decisions such as flight delays and cancellations.

10.3 Code-share Agreements

Code-sharing (or code-share agreement) is a partnership agreement between two airlines: the operating airline and the marketing airline. The operating airline allows the marketing airline to promote and sell seats on some of its flights. The code-share agreement brings benefit to both the marketing and the

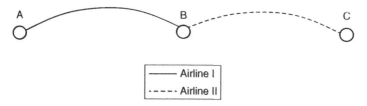

Figure 10.2 Example of two routes operated by two airlines participating in a code-share agreement.

operating airlines of the agreement. It allows the marketing airline to increase network coverage by gaining access to the routes (flights) of the code-share partner. In addition, code share provides relatively seamless travel to the marketing airline's customers, when their travel requires connecting between the flights of different airlines. The operating airline uses the agreement to increase its flights' load factor and provides an additional stream of revenue (Park and Zhang 2000; Brueckner 2001; Ito and Lee 2007; Abdelghany et al. 2009; Morandi et al. 2015; Zou and Chen 2017).

The code-share agreement is typically a two-way agreement, in which the two participating airlines play the role of the marketing airline with respect to flights operated by its partner. For example, Figure 10.2 shows two routes operated by two airlines participating in a code-share agreement. The first airline (Airline I) operates flights in city-pair A-B, while the second airline (Airline II) operates flights in city-pair B-C. When these two airlines engage in a code-share agreement, Airline I can market the flights in city-pair B-C, which is operated by Airline II. Meanwhile, Airline II can market the flights in city-pair A-B, which are operated by Airline I. Both airlines can serve the A-C market using a code-share connecting itinerary. It is possible that the marketing airline participates simultaneously in multiple code-share agreements with different operating airlines to extend its network in the different geographical areas. Similarly, the operating airline may participate simultaneously in multiple code-share agreements with different marketing airlines. Thus, several marketing airlines promote the flights of the operating airline simultaneously.

Three common types of code-share agreements are adopted in the airline industry: the hard block, the soft block, and the free sale agreements. In the hard block agreement, the marketing airline purchases a predetermined number of seats at a predetermined price from the operating airline. Then, the marketing airline is free to market and sell these seats at whatever price it chooses. In the soft block agreement, the size of the block is not determined in advance and is expected to vary based on the operating airline's anticipated demand level. Finally, in the free sale (also known as free flow) agreement, the operating airline gives the marketing airline access to its seat inventory, which the marketing airline can promote under its own code. In some cases, the operating airline uses a seat inventory control mechanism to block certain booking classes from being accessed by the marketing airline.

Code-share agreements require schedule planning and capacity allocation coordination between the marketing and the operating airlines. In the example above, when Airline I and Airline II participate in a code-share agreement, in order to develop attractive itineraries to serve the demand of city-pair A-C, the schedule of the flights in city-pair B-C should be aligned with the schedule of the flights in the city-pair A-B. In addition, for any connecting itinerary in the city-pair A-C, the seat capacity of the flights in city-pairs A-B and B-C should be adequate to carry local traffic as well as the connecting code-share traffic. Similarly, to serve the demand of city-pair C-A, the schedule of the flights in city-pair B-A should be aligned with the schedule of flights in the city-pair C-B. Also, for any connecting itinerary in the city-pair C-A, the seat capacity of flights in city-pairs C-B and B-A should be adequate to carry the local and connecting code-share demand.

Regardless of the type of the code-share agreement, each participating airline must answer several questions as follows: Which airlines would be most beneficial to enter into a code-share agreement with? Which flights would be accessible to each code-share partner? How many seats could be sold by the different partners on each flight and at what price? How many seats should be requested on the flights of the partners and the price of these seats?

In early implementations of code-share agreements, it was believed that extending the code-share agreement in terms of number of flights increases the revenue of the operating airline. Most code-share agreements are implemented based on the assumption that the agreement will generate additional demand that fills empty seats. However, in some cases, the additional (discounted) code-share demand would displace non-code-share passengers on the operating airline's flights. Accordingly, it is crucial that each airline evaluates the trade-off between the incremental revenue from the code-share agreement and the revenue loss due to displacing non-code-share passengers. This problem becomes more convoluted when the airline participates in multiple code-share agreements with several marketing airlines. In this case, the operating airline must find the optimal seat allocation among the different partners to minimize the chances of displacing non-code-share passengers. Most airlines adopt ad hoc quantitative rules to evaluate the profitability of their code-share agreements. In most cases, the terms of the agreement are determined after cycles of negotiation between the operating airline and the marketing airline, where each airline seeks to maximize its own expected revenues. One main drawback of the current code-sharing practice is that most airlines overlook the possible revenue loss from displacing non-code-share passengers. In addition, the different code-share agreements are evaluated individually, ignoring the trade-off among the different proposed agreements and its overall impact on profitability (Abdelghany et al. 2009).

10.4 Airline Alliances

Alliance is a general form of collaboration between two or more airlines (Fan et al. 2001; Gudmundsson and Rhoades 2001; Kleymann and Seristö 2001; Tiernan et al. 2008; Wang 2014; Lordan et al. 2015). Code-sharing is one example of how participating airlines can collaborate in an alliance. However, alliance could incorporate other forms of collaboration including ground operations, sharing gates, maintenance, etc. Airline alliances have become powerful tools for expanding networks, enhancing revenue, and reducing costs. Three major worldwide alliances currently exist, which represent about 60% of the world air traffic. These alliances are Star Alliance, SkyTeam, and Oneworld. These three alliances may be seen as three global airlines that are competing in the different city-pairs at the global level. Throughout every region of the world, airlines have embraced global alliances as a mean to access new markets, legitimize their brand in an increasingly global market, and provide incremental growth and profitability through the cooperative mechanisms within the alliance. Many airlines have also entered joint ventures to operate more closely on applicable routes. Airlines benefit from the alliance by extending their network and strengthening city presence by adding flight frequency with partners, often realized through code-share agreements. Airlines also realize possible cost reduction by sharing sales offices, maintenance facilities, operational facilities, catering facilities, computer systems, and operational staff (e.g. ground handling personnel at check-in and boarding desks). However, it should be noted that the ability of an airline to join an alliance may be restricted by its ability to meet the service standards of the alliance. It could also be restricted by laws and regulations of the airlines' host country or subject the authorities' approval.

10.5 Distribution Channels and Point of Sale

Selling a product requires marketing it to potential customers. Airlines depend on a wide range of ticket distribution channels to market their services to travelers. Ticket distribution channels are the media through which airlines communicate travel service information, including available itineraries and prices, to their prospective customers (Shaw 2016). These distribution channels include traditional travel agents, online travel agents, airline's own website, and airline's phone reservation systems. These channels differ in terms of volume of customers, level of competition with other airlines in the same channel, and distribution and commission cost charged by the channel's administrator. Airlines seek to find the optimal distribution strategy to maximize their sales and market share while at the same time minimizing their ticket marketing and distribution cost. Airlines usually subscribe to one or more global ticket distribution systems (GDS) (e.g. Sabre, Travelport, and Amadeus) to distribute their tickets. These GDS access real-time seat availability and ticket price information from the airlines' internal computer reservation systems. This information is then made available to travel agents and wholesalers who can effectively reach more customer groups on behalf of the airlines. In return, GDS charge airlines a pre-specified fee for each confirmed booking (Alamdari and Mason 2006). Airlines might also pay commissions to travel agents to promote sales of their tickets in highly competitive markets. It has been estimated that the US airlines paid over $7 billion in 2012 to distribute their tickets to consumers. This increasing distribution cost represents additional detrimental cost to the airlines, which strive to minimize their operation cost in a competitive and revenue-deteriorating environment.

Advances in computing and communication technologies along with the widespread of the Internet usage over the past two decades have created many business opportunities in the ticket distribution industry. Several Internet-based ticket distribution channels have emerged that differ in their underlying technologies and capabilities. These channels include the airline websites (e.g. www.airlineName.com) and Orbitz that is a travel technology company developed by a consortium of major US airlines. Both channels depend on technologies that bypass global distribution systems to directly use the airlines' internal reservation systems. New channels also include many Internet-based travel agencies (e.g. Expedia, Priceline, and Travelocity – a subsidiary of one of the global distribution systems that is sold to another online travel agent). These Internet-based travel agencies work as an interface for global distribution systems over the World Wide Web; however, they cost the airlines less than the cost charged by traditional travel agencies.

Distribution channels can be classified into two main categories. The first represents the airline's representatives, which promote ticket sales for this airline only. The second represents travel agents that simultaneously promote ticket sales for more than one airline. The two channels provide several ways of communication with customers including the Internet, call-in telephone lines, and walk-in sales offices. These distribution channels differ in their cost, market penetration, type of customers, easiness to use, and the amount of information presented to customers.

Nevertheless, the emergence of several new ticket distribution channels has raised an important question to airlines on how to determine the optimal distribution portfolio of their tickets among available channels. Given the different characteristics of these channels in terms of market volume, charged booking fees, and level of competition with other airlines in the channel, each airline needs to determine which channels to use and how many tickets should be allocated to each channel. Clearly, the goal is to maximize sold tickets and minimize associated distribution cost. Airline managers should understand how the different itineraries designed by their network planning teams are competing in the different distribution channels. They should examine the trade-off between using two common types of ticket distribution channels by airlines: (i) distribution channels with high market power and high competition among subscribed airlines (e.g. Expedia and Orbitz) versus (ii) distribution channels with low market power and low airlines competition (e.g. airline. com such as aa.com and continental.com) (Abdelghany and Abdelghany 2007a). On the other hand, ticket distribution businesses are required to study operation characteristics that maximize their revenues. These include, for example, market shares, subscribed airlines, information display, service fees, etc. Finally, regulators and policy makers need to evaluate different regulations/deregulations strategies that might be imposed on the ticket distribution industry to guarantee fair market competition. In other words, they should incorporate the characteristics of the ticket distribution channels while capturing airlines competition.

10.6 Loyalty Programs

Loyalty programs are generally defined as programs to retain the current customers of the company. A loyalty program is a reward program offered to frequent-flyer customers. In this program, customers receive rewards in the form of advanced access to new products, service upgrades, special sales coupons, or free merchandise. Customer retention is an important strategy to maintain a strong market share. Airlines recognize the fact that it does no good to the business spending significant resources attracting new customers only to lose them later to other competitors (Mimouni-Chaabane and Volle 2010; Dolnicar et al. 2011).

Most airlines adopt loyalty programs for customer retentions. Customers are asked to sign in a loyalty program in which their status in the program grows, as they make more travel with the airline. Customers accumulate rewards in their loyalty accounts, which are proportional to the miles they traveled or the monetary values they spend on the airline's products. Customers can redeem these rewards as free flight, flight upgrade to premium class, priority boarding on flights, free baggage, use of lounges at airports, etc. Loyalty programs are effective in retaining customers in case all competing airlines provide comparable service with the same value.

10.7 Corporate Travel

Airlines rely on a large number of sales personnel to promote their flights and itineraries in the different markets. The objective is to market flights of the airline and increase its market share. Among the groups that are targeted by sales personnel are corporate travelers. Each airline seeks to establish travel agreements with corporations with excessive business travel in city-pairs served by the airline. The objective of these agreements is to attract business travelers to the airline and lower the chance that these travelers select to fly on competing airlines. In these agreements, airlines offer fare discount to corporations in return of securing a certain market share of the corporation's travel (Abdelghany and Abdelghany 2007b).

Section 3

11

Basic Fleet Assignment Model (FAM)

11.1   Introduction

Most major airlines operate more than one aircraft (fleet) type. The fleet heterogeneity gives airlines the flexibility to serve markets (city-pairs) with different characteristics related to trip distance, demand, and operations cost and revenue. For example, airlines can serve long-haul markets using long-range aircraft and serve short-haul markets with short-range aircraft. They can also serve high-demand markets using large aircraft and serve low-demand markets using small aircraft. It also allows airlines to adopt different strategies to strengthen its competition in the different markets. For example, an airline could decide to serve a city-pair through scheduling a few flights using a large aircraft or scheduling frequent flights using a small aircraft. Nonetheless, fleet heterogeneity is expected to increase the operations cost of the airline as the cost elements associated with maintenance, crew training, and flight recovery during irregular operations are expected to increase (Gu et al. 1994; Subramanian et al. 1994; Hane et al. 1995; Rushmeier and Kontogiorgis 1997; Rexing et al. 2000; Sherali et al. 2006).

When airlines operate more than one aircraft fleet type, they need to decide on the most suitable fleet type for each flight. This problem is referred to as the fleet assignment problem and is applicable only to airlines that have more than one fleet type (or subfleet, derivatives of the same fleet). Figure 11.1 gives an illustration of the fleet assignment problem for a hypothetical airline that operates three fleet types. As shown in the figure, the flights of this airline are assigned to three fleet types.

Many factors define the solution of the fleet assignment problem. These factors are related to the characteristics of the fleet and the flights constituting the schedule. The factors related to fleet include number of aircraft in every fleet, aircraft seat capacity, cargo capacity, aircraft range, length of landing runway, landing fees, ability to fly over water, level of noise, fuel burning rate, maintenance

Airline Network Planning and Scheduling, First Edition. Ahmed Abdelghany and Khaled Abdelghany.
© 2019 John Wiley & Sons, Inc. Published 2019 by John Wiley & Sons, Inc.

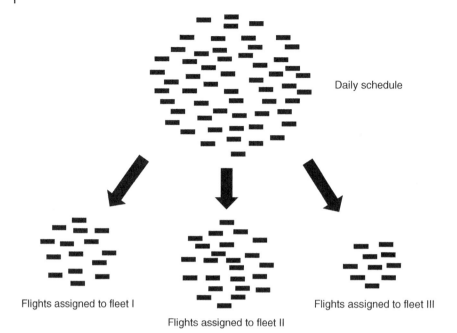

Figure 11.1 Fleet assignment considering three fleet types.

requirements, crew cost, and other operating cost. Factors related to the flights include passenger and cargo demand, flight length, revenue and yield, departure/arrival airport (e.g. runway and infrastructure suitability for the aircraft at the origin and destination airports), curfews at origin and destination, and noise restrictions. Overall, airlines determine a fleet assignment that maximizes its revenue from transporting passengers and cargo and minimizes its operations cost.

In practice, two major approaches are used for the fleet assignment problem, namely, the warm start approach and the cold start approach. The warm start approach assumes that a solution to the fleet assignment problem of a previous similar schedule exists, and hence only minor adjustments to this solution are required. These adjustments respond to changes in fleet, demand, or competition. In this approach, the fleeting solution of many flights remains unchanged. Airlines adopt the warm start approach to avoid significant fleet assignment changes as slight modifications can be performed over time. On the other hand, the cold start approach assumes no previous fleeting solution. Thus, it can produce a fleeting solution that is significantly different from those developed for any previous schedule. Airlines adopt the cold start approach, when there is a significant change in the schedule, demand, resources, and/or competition. They also use such approach to benchmark their fleet assignment solutions in order to ensure that the year-to-year adjustments maintain a near-optimal solution.

The fleet assignment problem is an important component of the schedule planning process. It is typically solved after determining a preliminary flight schedule. Figure 11.2 shows the structure of the schedule planning process adopted by most major airlines. As shown in the figure, the process follows an iterative sequential approach. The figure also shows that the fleet assignment represents the core of the schedule planning process, where it interacts significantly with other components including flight scheduling, revenue management, and aircraft routing. The interrelations among the fleet assignment and other planning decisions are discussed in more detail in Chapter 3. In this chapter, the basic fleet assignment model is introduced. Chapters 12 and 13 provide more detailed examples on the basic fleet assignment problem and its applications.

The basic fleet assignment problem is a simplified version of the problem solved by the airlines. However, it provides a good foundation to understand the real-world problem. The basic fleet assignment problem adopts three major assumptions that differentiate it from the real-world one. These assumptions are (i) the flight schedule is predetermined and fixed, (ii) the demand at the flight level is known and fixed, and (iii) the schedule repeats itself on a daily basis. The first two assumptions simplify the problem. As discussed in Chapter 3, the flight schedule and fleet assignment decisions are interdependent. For example, an airline might decide to serve a city-pair using a large aircraft and schedule small number of flights (i.e. small frequency) or serve the same city-pair considering frequent flights using a small aircraft. In addition,

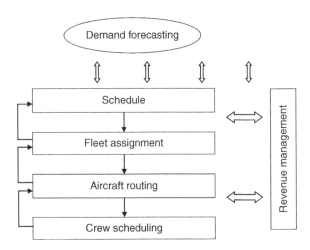

Figure 11.2 The airline schedule planning process.

the arrival time of a flight (which is essential information in the flight schedule) depends on the speed of the aircraft assigned to the flight. Furthermore, the minimum ground time for an aircraft at any airport depends on the aircraft type. Typically, large aircraft requires long service time to prepare the aircraft for its next flight. The ground time of the aircraft affects the arrival and departure times of the aircraft's inbound and outbound flights, respectively.

In addition, as discussed in Chapter 3, the assumption that the flight demand is known and fixed at the flight level is also not realistic. For example, for a city-pair that is served by more than one flight, the passenger demand of any flight depends on the fleet type selected for all other flights serving this city-pair. The third assumption that the schedule repeats itself daily is also not realistic because most airlines have different schedules that vary by day of week. Airlines adjust the scheduled flights and departure times to match the daily demand variations.

11.2 Graphical Representation: Time-staggered Diagram

In this section, the graphical representation of the fleet assignment problem is presented. The fleet assignment solution is presented using the time-staggered (time–space) diagram. In this diagram, the different stations are shown as vertical lines representing the timeline at each station. The top of the vertical line represents the start of the day. Each flight is represented by an arrow connecting two vertical lines that represent the flight's origin and destination stations, respectively. As mentioned earlier, the departure time of the flight is typically known. However, the arrival time of the flight depends on the speed of its assigned aircraft. Figure 11.3 illustrates a graphical representation of a flight connecting stations A and B. The flight could be assigned to one of three fleet types. The departure time of the flight is given, while its arrival time at station B is determined based on the fleet type.

Figures 11.4 and 11.5 extend the figure above to represent the ground arc and the overnight arc. The ground arc represents the time to get the aircraft ready for its next flight. The overnight represents an aircraft staying overnight at a station. In Figure 11.4, flights F1 and F2 are connected by a ground arc. In Figure 11.5, flights F3 and F4 are connected by an overnight arc, where the aircraft that operates flight F4 arrives by the end of the day at station B, and it is scheduled in the morning to operate flight F3.

Similar to the flight arc, the ground arc and the overnight arc depend on the fleet assignment of their corresponding inbound flights. Figure 11.6 shows an example of the time-staggered diagram for four hypothetical flights, namely, F5, F6, F7, and F8. The diagram shows the schedule at seven stations named

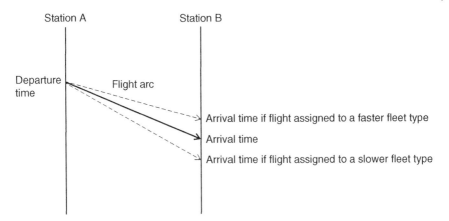

Figure 11.3 A graphical representation of a flight connecting stations A and B.

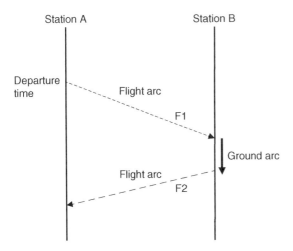

Figure 11.4 Example of a ground arc.

A through G. In this example, each of the four flights could be assigned to one of two fleet types. The fleet types are indicated by the solid (fleet type I) and the dashed (fleet type II) lines, respectively. As shown in the figure, the lengths of the flight arcs, the ground arcs, and the overnight arcs depend on the fleet type assigned to the flights. For example, when flight F6 is assigned to the fleet type II, the ground arc at station G between flights F6 and F7 is shorter than that when flight F6 is assigned to fleet type I. Figure 11.7 shows two time-staggered diagrams that represent the assignment of flights F5, F6, F7, and F8 to the two fleet types I and II.

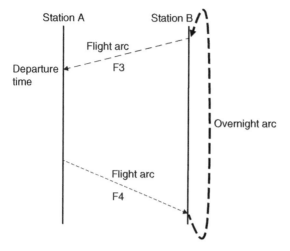

Figure 11.5 Example of an overnight arc.

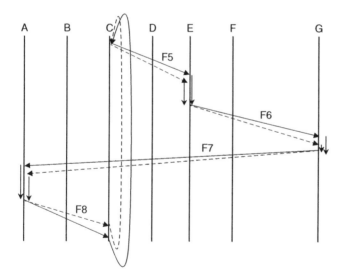

Figure 11.6 A time-staggered diagram for four flights and two fleet types.

In the fleet assignment solution, each flight is assigned to only one fleet. For instance, in the example above, flights F5 and F6 could be assigned to fleet type I, while flights F7 and F8 could be assigned to fleet type II. This fleet assignment solution can be represented using the time-staggered diagrams as shown in Figure 11.8.

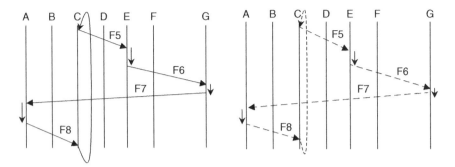

Figure 11.7 Two time-staggered diagrams for two different fleet types.

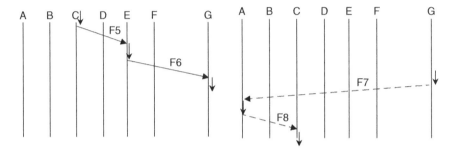

Figure 11.8 An example of fleet assignment solution with four flights and two fleet types.

The time-staggered diagrams can also be used to identify number of aircraft needed to cover all flights. This is done by drawing a horizontal line at any time instance and counting the number of flight arcs, ground arcs, and overnight arcs that intersect with this horizontal line. For example, Figure 11.9 shows the time-staggered diagram for a hypothetical airline. A horizontal line can be drawn to count the number of aircraft required to operate these flights. This line can be drawn late in the night, when there are minimum operations and all aircraft are on the ground for overnight turns. As shown in the figure, a dashed horizontal line is drawn toward the end of the day, which indicates that at least four aircraft are needed for the solution. These aircraft include two aircraft at station A, one aircraft at station C, and one aircraft at station F.

11.3 Problem Input

The basic fleet assignment problem entails five major input elements, including (i) the daily flight schedule, (ii) the expected passenger and cargo demand for each flight, (iii) ticket pricing at the flight level, (iv) fleet information including cost of flying, and (v) operation constraints.

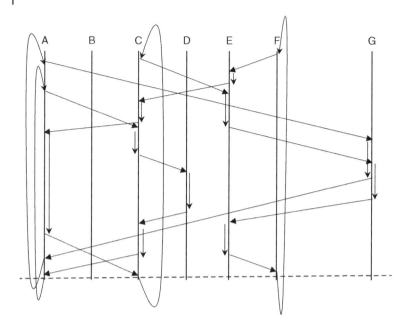

Figure 11.9 Determining number of aircraft needed for the fleet assignment solution.

The daily flight schedule represents the list of flights that are proposed in the schedule and need to be assigned to the different fleet types. Each flight has to be defined by a flight number, origin, destination, and scheduled departure time. As mentioned above, the scheduled arrival time cannot be determined without knowing the speed of the aircraft assigned to the flight, which is not known yet.

The flight demand is the number of passengers and cargo that need to be transported by the flight. As mentioned above, the demand at the flight level cannot be determined with accuracy. The passenger demand of any flight depends on the fleet type assigned to other flights. Thus, it is not realistic to assume that demand is accurately known for a given flight. In addition, for airlines that adopt a hub-and-spoke network with considerable connecting traffic, the demand is typically predicted at the origin–destination (OD) level. Thus, a mechanism is needed to convert the connecting OD demand into its corresponding flight-based demand. Finally, to properly estimate the demand, airlines should have adequate information on the schedule and tactics of its competitors. Any unexpected change in schedule or tactics by competitors can significantly affect the expected demand.

Information on the ticket prices is required to determine the expected revenue for each flight. In its simplest format, the average expected fare could be used for all filled seats. The flight revenue can be calculated by multiplying the expected demand of the flight by the average fare. The revenue from cargo can be calculated in a similar way. However, as more detailed pricing information becomes available indicating that several fares are charged, the demand can be clustered based on the charged fares into booking classes. The revenue for each booking class is calculated by multiplying the number of passengers in the booking class by its fare. The total revenue is calculated by adding the revenues from all booking classes of the flight. It should be noted that pricing data is usually available at the OD level and not at the flight level. Thus, a mechanism is needed to prorate the OD fares to estimate the fares at the flight level.

The solution of the fleet assignment problem requires detailed information on the fleet, which includes the number of aircraft available by each fleet or subfleet. Data obtained for each fleet type includes the number of seats per cabin class (i.e. first, business, and coach), operations cost (fuel, crew, landing fees, etc.), cargo capacity, aircraft range, speed, location of the maintenance facility for each fleet type, noise level and curfews, and required airport infrastructure (i.e. runway, gates, etc.).

Finally, any operation constraints related to fleet assignments should be considered. One major constraint is the through-flights constraint. Through flights is a connection (typically at the hub) between an inbound flight and an outbound flight, where both flights are to be operated by the same aircraft. Through flights are included in the schedule to facilitate connectivity for passengers using both flights. When there is considerable demand that is connecting between two major cities through the hub, the airline provides a through flight. If two connecting flights are to be operated as a through flight, the fleet assignment solution has to ensure that both flights are assigned to the same fleet type such that both flights can be assigned later to one aircraft. Figure 11.10 shows

Figure 11.10 Example of through flights.

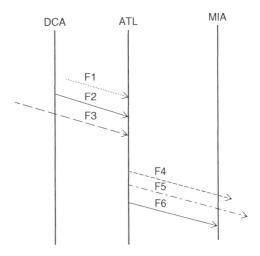

an example of a partial schedule serving three stations DCA, ATL, and MIA. As there is a considerable connecting traffic between DCA and MIA, the airline facilitates the connection at the hub (ATL) by scheduling flight F2 (DCA-ATL) and flight F6 (ATL-MIA) to operate using the same aircraft. Thus, the fleet assignment solution has to ensure that flights F2 and F6 are assigned to the same fleet type. The fleet assignment problem has additional constraints for maintenance and crew that will be ignored at this stage.

11.4 Problem Definition and Formulation

For larger airlines, the basic fleet assignment problem is formulated as an optimization problem. The objective of this optimization problem is to determine the optimal fleet assignment solution that maximizes the airline's net revenue. This objective is achieved by optimally matching the demand (number of passenger and cargo) and the supply (fleet seat and cargo capacity) while maximizing the profitability of the schedule. To formulate the fleet assignment problem as an optimization problem, its decision variables, objective function, and constraints need to be defined.

We consider a hypothetical airline that operates a set of flights F serving a set of stations S, where each flight $f \in F$ is defined by its flight number, origin, destination, and scheduled departure time. This airline operates several fleet types defined by the set E (e.g. Boeing 737-300, Airbus A320, etc.). Each fleet $e \in E$ is defined by number of aircraft available, cabins configuration, number of seats per cabin class (i.e. first, business, and coach), operations cost (fuel, crew, landing fees, etc.), cargo capacity, aircraft range, speed, location(s) of maintenance facility for each fleet type, noise level and curfews, and required airport infrastructure (i.e. runway, gates, etc.).

The decision variables of the problem are represented mathematically by a set of binary variables X, where each element $x_{fe} \in X$ is equal to 1 if flight $f \in F$ is assigned to equipment $e \in E$, and zero otherwise. The number of elements in the set X is equal to the multiplication of the number of flights $|F|$ in the schedule by the number of fleet types $|E|$. The solution of the fleet assignment problem is to determine the value of x_{fe} $\forall f \in F$ and $\forall e \in E$.

If flight $f \in F$ is assigned to fleet type $e \in E$, the airline gains revenue and incurs cost. The values of the revenue and cost are specific to the flight-fleet combination. In other words, if the same flight is assigned to a different fleet type, different revenue and cost values should be considered. For instance, if a flight with large demand is assigned to a small aircraft, the revenue is expected to be less than the revenue that is expected if the same flight is assigned to a large aircraft that can capture more demand. The revenue depends on the passenger and cargo demand of flight $f \in F$, the ticket fare(s), the cargo tariff(s), and the seat/cargo capacity of fleet $e \in E$. In addition, the cost of operating a flight depends on the type of the aircraft (i.e. fleet type). This cost is a function of aircraft fuel consumption rates, crew salary rates, maintenance cost, landing fees at the destination airport of the flight, and other operations cost elements.

We use r_{fe} to denote the unconstrained revenue associated with assigning flight $f \in F$ to any fleet type $e \in E$. The unconstrained revenue is a constant and does not depend on the aircraft type. It is defined as the maximum attainable revenue for a flight regardless of assigned capacity to this flight. The unconstrained revenue is calculated as a function of the unconstrained demand of the flight and its average fare. The unconstrained demand is the maximum demand that the airline is able to capture. The term unconstrained is used to denote that the amount of demand is determined without taking into consideration any flight capacity restrictions.

We also use c_{fe} to denote the cost, associated with assigning flight $f \in F$ to fleet type $e \in E$. This cost is a function of the assigned fleet type. It consists of two main components: (i) the flight operating cost and (ii) the spill cost. The flight operating cost is related to fuel cost, crew cost, landing fees, etc. The spill cost of a flight is the revenue lost in case the assigned aircraft (fleet type) cannot accommodate all passengers (i.e. the unconstrained demand). The difference between the unconstrained revenue r_{fe} and the cost c_{fe} is the flight net revenue (or net income or profit) and is denoted by p_{fe}, which can be computed mathematically as follows:

$$p_{fe} = r_{fe} - c_{fe}$$

For example, assume a flight serving a city-pair with a total demand of 300 passengers and an average fare value of \$200. Thus, the total unconstrained revenue of this flight is \$60,000. Assume also that a certain fleet type with a seat capacity of 180 seats is assigned to this flight. Thus, 120 passengers cannot be served by this flight. Accordingly, the spill cost is \$24,000. Assume that the operating cost of this flight using this fleet type is \$20,000. Then, the total cost is \$44,000 (\$20,000 plus \$24,000), and the net revenue is \$16,000 (\$60,000 minus \$44,000).

The value of the flight profit p_{fe} associated with assigning flight $f \in F$ to fleet type $e \in E$ is incurred by the airline only if the flight is actually assigned to the fleet, which is represented by setting the decision variable x_{fe} equal to 1. Thus, the actual incurred profit associated with assigning flight $f \in F$ to fleet type $e \in E$ is the multiplication of p_{fe} by x_{fe} (i.e. $p_{fe} \cdot x_{fe}$). If the flight $f \in F$ is not assigned to the fleet $e \in E$, the value of the decision variable x_{fe} is equal to 0, and hence the incurred profit is also equal to 0.

Accordingly, the total profit for the airline can be calculated by summing the profits resulting from all fleet assignment possibilities for all flights that can be written mathematically as follows:

$$P = \sum_{f \in F} \sum_{e \in E} p_{fe} \cdot x_{fe}$$

or

$$P = \sum_{f \in F} \sum_{e \in E} \left(r_{fe} - c_{fe} \right) . x_{fe}$$

In solving the fleet assignment problem, the objective is to maximize the profit P. Since the unconstrained revenue r_{fe} is constant (i.e. does not change by altering the fleet assignment to the flights), the unconstrained revenue term r_{fe} can be removed from the objective function as follows:

$$\sum_{f \in F} \sum_{e \in E} \left(-c_{fe} \right) . x_{fe}$$

If the negative sign is removed from the term above, we obtain the total assignment cost of the schedule. The total cost of the schedule is calculated by summing the costs resulting from all fleet assignment decisions for all flights. Removing the negative sign and converting the objective function into minimization instead of maximization, the objective is to minimize the total cost:

$$C = \sum_{f \in F} \sum_{e \in E} c_{fe} . x_{fe}$$

11.5 The Constraints of the Basic Fleet Assignment Problem

The basic feet assignment problem includes four main sets of constraints known as the coverage constraints, the resources constraints, the through flights constraints, and the balance constraints. These four sets of constraints are explained below.

11.5.1 The Coverage Constraints

The coverage constraints are pertinent to making sure that each flight in the schedule is assigned to one and only one fleet type. If a flight is not assigned to any fleet, the flight is marked as uncovered and cannot be included in the schedule (or otherwise the problem is infeasible to solve). In addition, a flight cannot be assigned to more than one fleet type. The coverage constraints can be written mathematically as follows:

$$\sum_{e \in E} x_{fe} = 1, \forall f \in F$$

The constraint indicates that the sum of the assignment decision variables for a flight over all the fleet types must be equal to 1. For example, if there are four fleet feasible types for flight $f \in F$, the coverage constraint is written as follows:

$$x_{f1} + x_{f2} + x_{f3} + x_{f4} = 1$$

This constraint forces only one of the decision variables x_{f1}, x_{f2}, x_{f3}, and x_{f4} to be equal to 1 and forces the remaining variables to be equal to zeros. Thus, the flight $f \in F$ is assigned to the fleet type that its corresponding decision variable is equal to 1. For instance, if $x_{f3} = 1$, this means that the flight f is assigned to fleet type III, and x_{f1}, x_{f2}, and x_{f4} must be equal to 0.

11.5.2 Resources Constraints

The resources constraints ensure that there is a sufficient number of aircraft from each fleet type that can satisfy the fleet assignment to the flights. For the fleet assignment solution, at any point in time throughout the day, the number of aircraft in the air plus the number of aircraft on the ground at all stations should not exceed the number of aircraft belonging to that fleet. As explained above, the time-staggered diagram can be used to count the number of aircraft needed for the fleet assignment solution. The mathematical constraints can be written as follows:

$$\sum_{s \in S} y_{se} \leq N_e, \forall e \in E$$

where y_{se} represents the maximum number of arcs (flight and ground) assigned to fleet type $e \in E$ at station $s \in S$ and N_e represents the number of available aircraft of type $e \in E$.

11.5.3 The Through-flights Constraints

As mentioned above, an airline might decide to connect an inbound flight at a given station with an outbound flight at the same station using the same aircraft. In other words, one aircraft is used for both flights. In this case, the fleet assignment solution must ensure that both flights are assigned to the same fleet type. Simply, the fleet assignment of the inbound flight has to be the same as the fleet assignment of the outbound flight. For each through-flights $f - f'$, the fleet assignment decision of the inbound flight f and fleet assignment decision of the outbound flight f' have to be equal. This constraint can be written mathematically as follows:

$$x_{fe} = x_{f'e} \quad \forall e \in E$$

11.5.4 The Balance Constraints

The balance constraints are essential to correctly define the structure of the fleet assignment problem. The balance constraints entail that, at every station $s \in S$, the number of inbound flights that are assigned to a given fleet type is equal to the number of outbound flights that are assigned to that fleet type. For example, Figure 11.11 shows an example of two inbound flights and two outbound flights at a station. In case this airline operates two fleet types such as Airbus A320 and Airbus A319, there are several balance scenarios at this station:

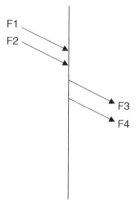

Figure 11.11 Example of balancing fleet types at a station.

Balance 1: All the four flights are assigned to Airbus A320.

Balance 2: All the four flights are assigned to Airbus A319.

Balance 3: Flights F1 and F3 are assigned to Airbus A320, while flights F2 and F4 are assigned to Airbus A319.

Balance 4: Flights F1 and F4 are assigned to Airbus A320, while flights F2 and F3 are assigned to Airbus A319.

Balance 5: Flights F1 and F3 are assigned to Airbus A319, while flights F2 and F4 are assigned to Airbus A320.

Balance 6: Flights F1 and F4 are assigned to Airbus A319, while flights F2 and F3 are assigned to Airbus A320.

Flights that are inbound to a given station are outbound flights from other station(s). Thus, the balance constraints have to be satisfied simultaneously for all stations. In addition, ensuring the balance of the resources is more binding than the demand–supply matching. For example, it could be the case that the two inbound flights F1 and F2 are more suitable to be operated by Airbus A320 and the two outbound flights F3 and F4 are more suitable to be operated by Airbus A319. However, this assignment plan cannot be adopted because it violates the balance constraints. The balance constraints can be written mathematically as follows:

$$\sum_{f \in IN_s} x_{fe} = \sum_{f \in OUT_s} x_{fe}, \forall e \in E, \forall s \in S$$

where IN_s and OUT_s represent the set of inbound and outbound flights for the airline at station $s \in S$.

12

A Walk-through Example of the Basic Fleet Assignment Model

12.1 Problem Definition

In this chapter, an example is presented to further explain the formulation of the basic fleet assignment model, which is presented in Chapter 11. For this purpose, we assume a small hypothetical airline. This airline has a hub at Chicago O'Hare Airport (ORD) and serves four airports including Boston Logan International Airport (BOS), John F. Kennedy International Airport (JFK), Los Angeles International Airport (LAX), and Seattle–Tacoma International Airport (SEA). Each of these four cities is served by two round trips to and from the hub. Thus, the schedule includes 16 flights that are either originating from or ending at the hub. Assume that the airline has two fleet types, namely, $e1$ and $e2$. The number of aircraft in each fleet type is assumed to be known. The scheduled departure time for these 16 flights is also given. The arrival time of each flight depends on whether the flight is assigned to fleet $e1$ or $e2$. Figure 12.1 shows the network structure of the hypothetical airline.

Figures 12.2 and 12.3 show two time-staggered diagrams that represent the schedule of the 16 flights when assigned to fleets $e1$ and $e2$, respectively. Each flight is represented by an arrow that connects between two stations. The flight number is given next to each arrow. The departure time of each flight is represented by the intersection of the arrow's tail with the timeline at its origin station. The arrival time of each flight is represented by the intersection of the arrow's head with the timeline at its destination station. The main difference between the two schedules is as follows. When flight 6 is assigned to fleet $e1$, it arrives at SEA before the departure of flight 7 (Figure 12.2). However, when flight 6 is assigned to fleet $e2$, it arrives at SEA after the departure of flight 7 (Figure 12.2).

The assignment cost c_{fe} (the flight operating cost plus the spill cost) associated with assigning flight f to fleet type e is assumed to be known. Table 12.1 gives the notations used to represent the assignment cost for the 16 flights and the two fleet types. For example, the term $c_{12,2}$ represents the cost associated with assigning flight 12 to fleet $e2$.

Airline Network Planning and Scheduling, First Edition. Ahmed Abdelghany and Khaled Abdelghany.

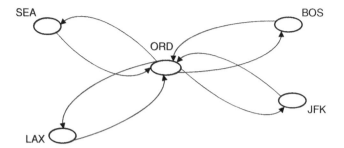

Figure 12.1 The network structure of a hypothetical airline with a hub at ORD.

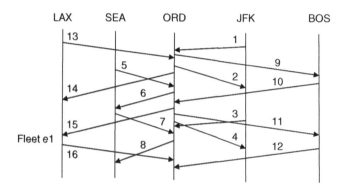

Figure 12.2 Time-staggered diagram for fleet type $e1$.

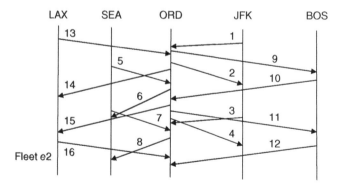

Figure 12.3 Time-staggered diagram for fleet type $e2$.

The problem includes 32 assignment decision variables, where each of the 16 flights could be assigned to one of the two fleet types. Table 12.2 gives the notations used to represent these decision variables. For example, $x_{15,1}$ is equal to 1 if flight 15 is assigned to fleet $e1$, and zero otherwise.

Table 12.1 Notations for the assignment cost.

Flight f	Fleet type $e1$	Fleet type $e2$
1	$c_{1,1}$	$c_{1,2}$
2	$c_{2,1}$	$c_{2,2}$
3	$c_{3,1}$	$c_{3,2}$
4	$c_{4,1}$	$c_{4,2}$
5	$c_{5,1}$	$c_{5,2}$
6	$c_{6,1}$	$c_{6,2}$
7	$c_{7,1}$	$c_{7,2}$
8	$c_{8,1}$	$c_{8,2}$
9	$c_{9,1}$	$c_{9,2}$
10	$c_{10,1}$	$c_{10,2}$
11	$c_{11,1}$	$c_{11,2}$
12	$c_{12,1}$	$c_{12,2}$
13	$c_{13,1}$	$c_{13,2}$
14	$c_{14,1}$	$c_{14,2}$
15	$c_{15,1}$	$c_{15,2}$
16	$c_{16,1}$	$c_{16,2}$

Table 12.2 The symbolic representation of the decision variables of the fleet assignment.

Flight f	Fleet type $e1$	Fleet type $e2$
1	$x_{1,1}$	$x_{1,2}$
2	$x_{2,1}$	$x_{2,2}$
3	$x_{3,1}$	$x_{3,2}$
4	$x_{4,1}$	$x_{4,2}$
5	$x_{5,1}$	$x_{5,2}$
6	$x_{6,1}$	$x_{6,2}$
7	$x_{7,1}$	$x_{7,2}$
8	$x_{8,1}$	$x_{8,2}$
9	$x_{9,1}$	$x_{9,2}$
10	$x_{10,1}$	$x_{10,2}$
11	$x_{11,1}$	$x_{11,2}$
12	$x_{12,1}$	$x_{12,2}$
13	$x_{13,1}$	$x_{13,2}$
14	$x_{14,1}$	$x_{14,2}$
15	$x_{15,1}$	$x_{15,2}$
16	$x_{16,1}$	$x_{16,2}$

12.2 The Objective Function

The objective function of the problem is to minimize the total assignment cost C. The total cost can be written as follows:

$$
\begin{aligned}
C = {} & c_{1,1} \cdot x_{1,1} + c_{1,2} \cdot x_{1,2} + c_{2,1} \cdot x_{2,1} + c_{2,2} \cdot x_{2,2} + c_{3,1} \cdot x_{3,1} + c_{3,2} \cdot x_{3,2} + c_{4,1} \cdot x_{4,1} \\
& + c_{4,2} \cdot x_{4,2} + c_{5,1} \cdot x_{5,1} + c_{5,2} \cdot x_{5,2} + c_{6,1} \cdot x_{6,1} + c_{6,2} \cdot x_{6,2} + c_{7,1} \cdot x_{7,1} \\
& + c_{7,2} \cdot x_{7,2} + c_{8,1} \cdot x_{8,1} + c_{8,2} \cdot x_{8,2} + c_{9,1} \cdot x_{9,1} + c_{9,2} \cdot x_{9,2} + c_{10,1} \cdot x_{10,1} \\
& + c_{10,2} \cdot x_{10,2} + c_{11,1} \cdot x_{11,1} + c_{11,2} \cdot x_{11,2} + c_{12,1} \cdot x_{12,1} + c_{12,2} \cdot x_{12,2} \\
& + c_{13,1} \cdot x_{13,1} + c_{13,2} \cdot x_{13,2} + c_{14,1} \cdot x_{14,1} + c_{14,2} \cdot x_{14,2} + c_{15,1} \cdot x_{15,1} \\
& + c_{15,2} \cdot x_{15,2} + c_{16,1} \cdot x_{16,1} + c_{16,2} \cdot x_{16,2}
\end{aligned}
$$

12.3 The Constraints

As mentioned above, the basic fleet assignment problem includes four main sets of constraints, which are the coverage constraints, the resources availability constraints, the through flights constraints, and the balance constraints. The coverage constraints entail that each flight is to be assigned to one fleet type only. The problem has sixteen coverage constraints, one for each flight. These coverage constraints are written as follows:

$$x_{1,1} + x_{1,2} = 1$$

$$x_{2,1} + x_{2,2} = 1$$

$$x_{3,1} + x_{3,2} = 1$$

$$x_{4,1} + x_{4,2} = 1$$

$$x_{5,1} + x_{5,2} = 1$$

$$x_{6,1} + x_{6,2} = 1$$

$$x_{7,1} + x_{7,2} = 1$$

$$x_{8,1} + x_{8,2} = 1$$

$$x_{9,1} + x_{9,2} = 1$$

$$x_{10,1} + x_{10,2} = 1$$

$$x_{11,1} + x_{11,2} = 1$$

$$x_{12,1} + x_{12,2} = 1$$

$$x_{13,1} + x_{13,2} = 1$$

$$x_{14,1} + x_{14,2} = 1$$

$$x_{15,1} + x_{15,2} = 1$$

$$x_{16,1} + x_{16,2} = 1$$

The resources availability constraints ensure that the number of used aircraft in the fleet assignment solution does not exceed the available fleet resources. As mentioned in Chapter 11, the resources constraints are applied by counting the number of aircraft that are used at any time epoch. It is more practical to count the aircraft later in the night when most of the aircraft are parking at the different stations. For this purpose, we define the decision variable $RON_{s\,e}$ to represent the number of aircraft of type e that remain overnight at station s (RON is an abbreviation for remaining overnight). For example, the variable $RON_{LAX\,e1}$ represents the number of aircraft of type $e1$ that remain overnight at LAX.

For instance, in the example given above, assume that there are only three aircraft of type $e1$ and four aircraft of type $e2$. The aircraft of type $e1$ that are used simultaneously in the fleet assignment solution can be counted by counting the number of aircraft of type $e1$ that remain overnight at the different stations. Figure 12.4 shows the number of aircraft of type $e1$ that remain overnight at each station. Thus, the resources constraint for fleet type $e1$ is written as follows:

$$RON_{LAX\,e1} + RON_{SEA\,e1} + RON_{ORD\,e1} + RON_{JFK\,e1} + RON_{BOS\,e1} \leq 3$$

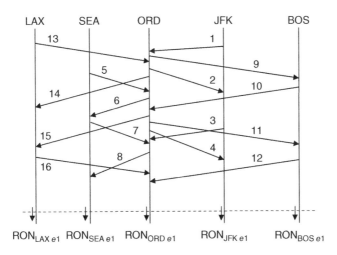

Figure 12.4 Counting aircraft remaining overnight at the different stations for fleet type $e1$.

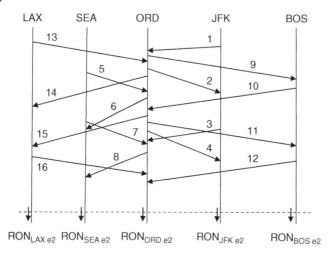

Figure 12.5 Counting aircraft remaining overnight at the different stations for fleet type *e2*.

Similarly, Figure 12.5 shows the number of aircraft of type *e2* that remain overnight at each station. Thus, the resources constraint for fleet type *e2* is written as follows:

$$\text{RON}_{\text{LAX}\,e2} + \text{RON}_{\text{SEA}\,e2} + \text{RON}_{\text{ORD}\,e2} + \text{RON}_{\text{JFK}\,e2} + \text{RON}_{\text{BOS}\,e2} \leq 4$$

The through-flight constraints ensure that the inbound and outbound flights of each through flight are assigned the same fleet type. Thus, they can be assigned to the same aircraft. For example, if there is significant traffic between BOS and LAX, the airline might decide to develop a smooth connection for passengers at ORD by arriving on flight 10 and connecting to flight 15 without changing the aircraft. This requires flights 10 and 15 to be assigned to the same fleet type. The through-flight constraints are written as follows:

$$x_{10,1} = x_{15,1}$$

or

$$x_{10,2} = x_{15,2}$$

This constraint implies that if flight 10 is assigned to fleet type *e1*, flight 15 should also be assigned to fleet type *e1*. Otherwise, if flight 10 is assigned to fleet type *e2*, flight 15 should also be assigned to fleet type *e2*.

The balance constraints guarantee that at each station, the number of inbound flights of a certain fleet should be equal to the number of outbound flights assigned to the same fleet. Accordingly, the balance constraints are considered for each fleet type and at each station. In this problem, there are five stations and two fleet types, resulting in ten balance constraints.

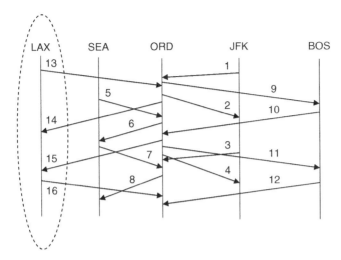

Figure 12.6 The inbound and outbound flights at LAX operated by fleet $e1$.

For example, considering fleet type $e1$ at LAX, the number of inbound flights assigned to fleet $e1$ must be equal to the number of outbound flights assigned to the same fleet. As shown in Figure 12.6, at LAX, flights 14 and 15 are inbound flights, while flights 13 and 16 are outbound flights. Thus, the balance constraint at LAX for fleet type $e1$ can be written as follows:

$$x_{14,1} + x_{15,1} = x_{13,1} + x_{16,1}$$

If the left-hand side of this equation, for example, is equal to 0, it means that none of flights 14 and 15 are assigned to fleet $e1$. Hence, the balance constraint ensures that the right-hand side of the equation is 0, indicating that none of flights 13 or 16 are assigned to fleet $e1$. Similarly, if the left-hand side of this equation, for example, is equal to 2, it means that both flights 14 and 15 are assigned to fleet $e1$. Thus, the balance constraint ensures that the right-hand side of the equation is equal to 2, forcing flights 13 and 16 to be assigned to fleet $e1$. Finally, if the left-hand side of this equation is equal to 1, it means that either flight 14 or flight 15 is assigned to fleet $e1$. The balance constraint ensures that the right-hand side of the equation is equal to 1, indicating that either flight 13 or flight 16 is assigned to fleet $e1$.

A similar balance constraint is written for fleet $e2$ at LAX as follows:

$$x_{14,2} + x_{15,2} = x_{13,2} + x_{16,2}$$

Figure 12.7 illustrates the inbound and outbound flights operated by fleet $e1$ at SEA. Considering the two inbound flights 6 and 8 and the two outbound flights 5 and 7, the balance constraint at SEA for fleet $e1$ is written as follows:

$$x_{6,1} + x_{8,1} = x_{5,1} + x_{7,1}$$

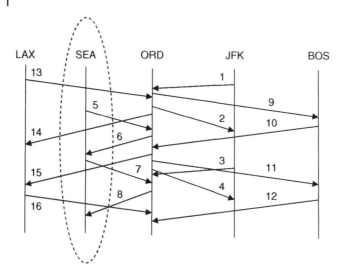

Figure 12.7 The inbound and outbound flights at SEA operated by fleet type e1.

Similarly, the balance constraint at SEA for fleet $e2$ is written as follows:

$$x_{6,2} + x_{8,2} = x_{5,2} + x_{7,2}$$

Similarly, the balance constraints at ORD for fleets $e1$ and $e2$ can be written as follows:

$$x_{1,1} + x_{10,1} + x_{3,1} + x_{12,1} + x_{13,1} + x_{5,1} + x_{7,1} + x_{16,1}$$
$$= x_{9,1} + x_{2,1} + x_{11,1} + x_{4,1} + x_{14,1} + x_{6,1} + x_{15,1} + x_{8,1}$$

$$x_{1,2} + x_{10,2} + x_{3,2} + x_{12,2} + x_{13,2} + x_{5,2} + x_{7,2} + x_{16,2}$$
$$= x_{9,2} + x_{2,2} + x_{11,2} + x_{4,2} + x_{14,2} + x_{6,2} + x_{15,2} + x_{8,2}$$

The balance constraints at JFK can be written as follows:

$$x_{2,1} + x_{4,1} = x_{1,1} + x_{3,1}$$

$$x_{2,2} + x_{4,2} = x_{1,2} + x_{3,2}$$

Finally, the balance constraints at BOS can be written as follows:

$$x_{9,1} + x_{11,1} = x_{10,1} + x_{12,1}$$

$$x_{9,2} + x_{11,2} = x_{10,2} + x_{12,2}$$

12.4   Interconnection Nodes

The balance constraints described above might result in inefficient solutions characterized by long fleet ground times. To explain this inefficiency, consider the flight operations at BOS with two inbound flights (9 and 11) and two outbound flights (10 and 12) (Figure 12.8).

The balance constraints described above could result in six different fleet assignment solutions at BOS as follows:

1) Flights 9, 10, 11, and 12 are assigned to $e1$.
2) Flights 9, 10, 11, and 12 are assigned to $e2$.
3) Flights 9 and 10 are assigned to $e1$ and flights 11 and 12 are assigned to $e2$.
4) Flights 9 and 10 are assigned to $e2$ and flights 11 and 12 are assigned to $e1$.
5) Flights 9 and 12 are assigned to $e1$ and flights 10 and 11 are assigned to $e2$.
6) Flights 9 and 12 are assigned to $e2$ and flights 10 and 11 are assigned to $e1$.

Solutions one through four entail that flights 9 and 10 are operated by the same fleet type and flights 11 and 12 are also operated by the same fleet type. These fleet assignments result in rotations (flight 9 to flight 10 and flight 11 to flight 12) with short aircraft ground times. Thus, when aircraft rotations are determined to take advantage of the short-time rotations, one aircraft is assigned to operate flights 9 and 10, and another aircraft is assigned to operate flights 11 and 12, as shown in Figure 12.9. On the contrary, solutions five and six entail that flights 9 and 10 are operated using two different fleet types. Also, flights 11 and 12 are operated by different fleet types. However, flights 9 and 12 are operated by the

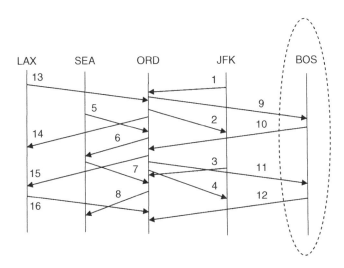

Figure 12.8 The inbound and outbound flights at BOS.

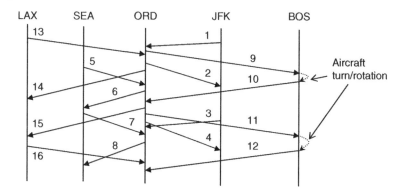

Figure 12.9 Efficient fleet balance at BOS with short ground time for the aircraft.

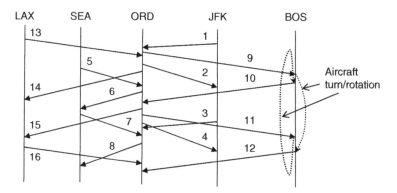

Figure 12.10 Fleet balance at BOS with undesirable long ground time for the aircraft.

same fleet type. Also, flights 10 and 11 are operated by the same fleet type. Accordingly, when one aircraft is selected to operate flights 9 and 12, this aircraft will have relatively undesirable long ground time at BOS. Similarly, when one aircraft is assigned to operate flights 11 and 10, this aircraft has to stay overnight at BOS. These two assignments are shown in Figure 12.10.

To avoid such inefficiencies, the balance constraints defined above are modified. The modification entails balancing flights assigned to a certain fleet type over shorter time periods (intervals) instead of balancing the flights over the entire day. Thus, the timeline of each station is divided into what is known as interconnection nodes. An interconnection node is defined by a set of flight arrivals followed by a set of flight departures. The number of interconnection nodes at each station depends on the number of scheduled flights (arrivals and departures) at this station.

To further explain the concept of interconnection nodes, consider the schedule of flights operated by fleet $e1$, given in Figure 12.11. The interconnection

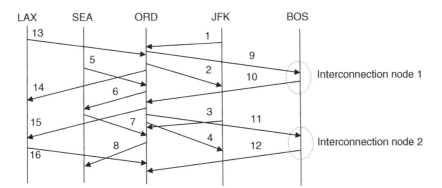

Figure 12.11 Interconnection nodes at BOS for schedule operated by fleet type *e*1.

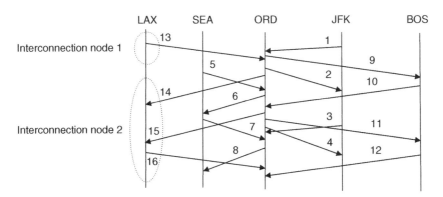

Figure 12.12 Interconnection nodes at LAX for schedule operated by fleet type *e*1.

nodes are defined at each station. For example, at BOS, the schedule starts by one arrival (flight 9), followed by one departure (flight 10), before the next arrival (flight 11). Thus, flights 9 and 10 are included in the first interconnection node at BOS. Similarly, flights 11 and 12 are included in the second interconnection node at BOS. Figure 12.11 shows the two interconnection nodes at BOS. Similarly, at LAX, the schedule starts by a departure (flight 13), before the next arrival (flight 14). Thus, the first node includes only one departure. The second node includes two arrivals (flights 14 and 15) and one departure (flight 16). The two interconnection nodes at LAX are shown in Figure 12.12. The interconnection nodes at the five different stations for the schedule if operated by fleet type *e*1 are given in Figure 12.13. Figure 12.14 gives the interconnection nodes for the five stations assuming fleet type *e*2 is used. One should notice the difference in the interconnection nodes at SEA when fleet type *e*2 is used instead of fleet type *e*1. The difference is due to the different arrival times of flight 6 in these two cases.

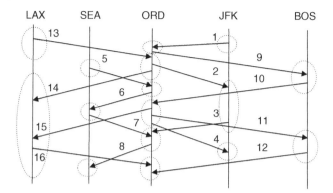

Figure 12.13 Interconnection nodes for the schedule operated by fleet type *e1*.

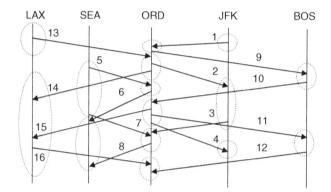

Figure 12.14 Interconnection nodes for the schedule operated by fleet type *e2*.

The balance constraints should be applied for each interconnection node and for each fleet type. Therefore, a new set of ground and overnight links are considered to maintain aircraft counts between successive interconnection nodes at the same station. For example, at LAX and for fleet type *e1*, there are two interconnection nodes. Thus, a ground link is used to connect these two nodes (node 1 and node 2). Also, a variable denoted as $y_{(1-2)\,LAX\,e1}$ is used to count the number of aircraft of type *e1* on the ground between nodes 1 and 2 at LAX. In addition, an overnight link is introduced, and a variable $RON_{LAX\,e1}$ is considered to count the number of aircraft of type *e1* that remain overnight at LAX. Figure 12.15 shows the two interconnection nodes at LAX for fleet type *e1* along with the ground and overnight links. It should be noted that the number of aircraft of a fleet type starting the day at a station is equivalent to the number of aircraft from the same fleet that remain overnight at the same station. This condition entails the assumption that the flight schedule is repeating itself every day.

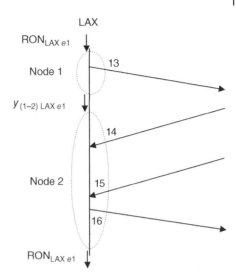

Figure 12.15 The interconnection nodes at LAX for fleet type $e1$ illustrating the ground and overnight links.

Then, two balance constraints for fleet type $e1$ are considered at LAX: one constraint for each interconnection nodes. The balance of the first node is as follows:

$$\mathrm{RON}_{\mathrm{LAX}\,e1} = x_{13,1} + y_{(1-2)\,\mathrm{LAX}\,e1}$$

This constraint indicates that the number of aircraft of type $e1$ that start the day at LAX is equal to the value of the decision variable $x_{13,1}$ (which could be one or zero) plus the number of aircraft of type $e1$ that remain on the ground and to be used later in the day.

The balance of the second node at LAX for fleet type $e1$ is as follows:

$$y_{(1-2)\mathrm{LAX}\,e1} + x_{14,1} + x_{15,1} = x_{16,1} + \mathrm{RON}_{\mathrm{LAX}\,e1}$$

This constraint ensures that the number of aircraft of type $e1$ that remain on the ground after the departure of flight 13 plus the values of the decision variables $x_{14,1}$ and $x_{15,1}$ (which could be one or zero) is equal to the value of the decision variables $x_{16,1}$ plus the number of aircraft of type $e1$ that remain overnight at LAX.

Another example of the balance constraints is shown in Figure 12.16 for SEA considering fleet type $e1$. As shown in this figure, there are three interconnection nodes at SEA. One ground link is used to connect nodes 1 and 2. The variable $y_{(1-2)\,\mathrm{SEA}\,e1}$ is used to represent the number of aircraft that remain on the ground for this ground link. Similarly, another ground link is used to connect nodes 2 and 3. The variable $y_{(2-3)\,\mathrm{SEA}\,e1}$ is used to represent the number of aircraft that remain on the ground for this link. In addition, the variable $\mathrm{RON}_{\mathrm{SEA}\,e1}$, associated with the overnight link, is used to count the number of aircraft of type $e1$ that remain overnight at SEA.

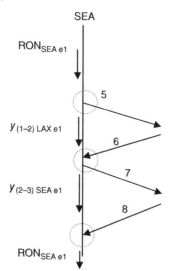

SEA

Figure 12.16 The interconnection nodes at SEA for fleet type e1 illustrating the ground and overnight links.

Three balance constraints are considered for fleet type $e1$ at SEA as follows:

$$\text{RON}_{\text{SEA } e1} = x_{5,1} + y_{(1-2)\text{SEA } e1}$$

$$y_{(1-2)\text{SEA } e1} + x_{6,1} = x_{7,1} + y_{(2-3)\text{SEA } e1}$$

$$y_{(2-3)\text{SEA } e1} + x_{8,1} = \text{RON}_{\text{SEA } e1}$$

Based on the discussion, to complete the problem formulation, the following decision variables are considered (Table 12.3).

The full list of the balance constraints for the two fleet types at the different nodes is as follows:

Fleet type $e1$ at LAX

$$\text{RON}_{\text{LAX } e1} = x_{13,1} + y_{(1-2)\text{LAX } e1}$$

$$y_{(1-2)\text{LAX } e1} + x_{14,1} + x_{15,1} = x_{16,1} + \text{RON}_{\text{LAX } e1}$$

Fleet type $e1$ at SEA

$$\text{RON}_{\text{SEA } e1} = x_{5,1} + y_{(1-2)\text{SEA } e1}$$

$$y_{(1-2)\text{SEA } e1} + x_{6,1} = x_{7,1} + y_{(2-3)\text{SEA } e1}$$

$$y_{(2-3)\text{SEA } e1} + x_{8,1} = \text{RON}_{\text{SEA } e1}$$

Table 12.3 The list of additional decision variables for the fleet assignment example.

$RON_{LAX\,e1}$	Number of aircraft of type $e1$ that remain overnight at LAX
$RON_{SEA\,e1}$	Number of aircraft of type $e1$ that remain overnight at SEA
$RON_{ORD\,e1}$	Number of aircraft of type $e1$ that remain overnight at ORD
$RON_{JFK\,e1}$	Number of aircraft of type $e1$ that remain overnight at JFK
$RON_{BOS\,e1}$	Number of aircraft of type $e1$ that remain overnight at BOS
$y_{(1-2)\,LAX\,e1}$	Number of aircraft of type $e1$ on the ground link that connects interconnection nodes 1 and 2 at LAX
$y_{(1-2)\,SEA\,e1}$	Number of aircraft of type $e1$ on the ground link that connects interconnection nodes 1 and 2 at SEA
$y_{(2-3)\,SEA\,e1}$	Number of aircraft of type $e1$ on the ground link that connects interconnection nodes 2 and 3 at SEA
$y_{(1-2)\,ORD\,e1}$	Number of aircraft of type $e1$ on the ground link that connects interconnection nodes 1 and 2 at ORD
$y_{(2-3)\,ORD\,e1}$	Number of aircraft of type $e1$ on the ground link that connects interconnection nodes 2 and 3 at ORD
$y_{(3-4)\,ORD\,e1}$	Number of aircraft of type $e1$ on the ground link that connects interconnection nodes 3 and 4 at ORD
$y_{(4-5)\,ORD\,e1}$	Number of aircraft of type $e1$ on the ground link that connects interconnection nodes 4 and 5 at ORD
$y_{(5-6)\,ORD\,e1}$	Number of aircraft of type $e1$ on the ground link that connects interconnection nodes 5 and 6 at ORD
$y_{(1-2)\,JFK\,e1}$	Number of aircraft of type $e1$ on the ground link that connects interconnection nodes 1 and 2 at JFK
$y_{(2-3)\,JFK\,e1}$	Number of aircraft of type $e1$ on the ground link that connects interconnection nodes 2 and 3 at JFK
$y_{(1-2)\,BOS\,e1}$	Number of aircraft of type $e1$ on the ground link that connects interconnection nodes 1 and 2 at BOS
$RON_{LAX\,e2}$	Number of aircraft of type $e2$ that remain overnight at LAX
$RON_{SEA\,e2}$	Number of aircraft of type $e2$ that remain overnight at SEA
$RON_{ORD\,e2}$	Number of aircraft of type $e2$ that remain overnight at ORD
$RON_{JFK\,e2}$	Number of aircraft of type $e2$ that remain overnight at JFK
$RON_{BOS\,e2}$	Number of aircraft of type $e2$ that remain overnight at BOS
$y_{(1-2)\,LAX\,e2}$	Number of aircraft of type $e2$ on the ground link that connects interconnection nodes 1 and 2 at LAX
$y_{(1-2)\,SEA\,e2}$	Number of aircraft of type $e2$ on the ground link that connects interconnection nodes 1 and 2 at SEA
$y_{(1-2)\,ORD\,e2}$	Number of aircraft of type $e2$ on the ground link that connects interconnection nodes 1 and 2 at ORD

(*Continued*)

Table 12.3 (Continued)

$y_{(2-3)\text{ ORD }e2}$	Number of aircraft of type $e2$ on the ground link that connects interconnection nodes 2 and 3 at ORD
$y_{(3-4)\text{ ORD }e2}$	Number of aircraft of type $e2$ on the ground link that connects interconnection nodes 3 and 4 at ORD
$y_{(4-5)\text{ ORD }e2}$	Number of aircraft of type $e2$ on the ground link that connects interconnection nodes 4 and 5 at ORD
$y_{(5-6)\text{ ORD }e2}$	Number of aircraft of type $e2$ on the ground link that connects interconnection nodes 5 and 6 at ORD
$y_{(1-2)\text{ JFK }e2}$	Number of aircraft of type $e2$ on the ground link that connects interconnection nodes 1 and 2 at JFK
$y_{(2-3)\text{ JFK }e2}$	Number of aircraft of type $e2$ on the ground link that connects interconnection nodes 2 and 3 at JFK
$y_{(1-2)\text{ BOS }e2}$	Number of aircraft of type $e2$ on the ground link that connects interconnection nodes 1 and 2 at BOS

Fleet type $e1$ at ORD

$$RON_{\text{ORD}\,e1} + x_{1,1} = x_{9,1} + y_{(1-2)\text{ORD}\,e1}$$

$$y_{(1-2)\text{ORD}\,e1} + x_{13,1} = x_{2,1} + x_{14,1} + y_{(2-3)\text{ORD}\,e1}$$

$$y_{(2-3)\text{ORD}\,e1} + x_{5,1} = x_{6,1} + y_{(3-4)\text{ORD}\,e1}$$

$$y_{(3-4)\text{ORD}\,e1} + x_{10,1} = x_{15,1} + x_{11,1} + x_{4,1} + y_{(4-5)\text{ORD}\,e1}$$

$$y_{(4-5)\text{ORD}\,e1} + x_{3,1} + x_{7,1} = x_{8,1} + y_{(5-6)\text{ORD}\,e1}$$

$$y_{(5-6)\text{ORD}\,e1} + x_{16,1} + x_{12,1} = RON_{\text{ORD}\,e1}$$

Fleet type $e1$ at JFK

$$RON_{\text{JFK}\,e1} = x_{1,1} + y_{(1-2)\text{JFK}\,e1}$$

$$y_{(1-2)\text{JFK}\,e1} + x_{2,1} = x_{3,1} + y_{(2-3)\text{JFK}\,e1}$$

$$y_{(2-3)\text{JFK}\,e1} + x_{4,1} = RON_{\text{JFK}\,e1}$$

Fleet type $e1$ at BOS

$$RON_{\text{BOS}\,e1} + x_{9,1} = x_{10,1} + y_{(1-2)\text{BOS}\,e1}$$

$$y_{(1-2)\text{BOS}\,e1} + x_{11,1} = x_{12,1} + RON_{\text{BOS}\,e1}$$

Fleet type *e2* at LAX

$$\text{RON}_{\text{LAX}\,e2} = x_{13,2} + y_{(1-2)\text{LAX}\,e2}$$

$$y_{(1-2)\text{LAX}\,e2} + x_{14,2} + x_{15,2} = x_{16,2} + \text{RON}_{\text{LAX}\,e2}$$

Fleet type *e2* at SEA

$$\text{RON}_{\text{SEA}\,e2} = x_{5,2} + x_{7,2} + y_{(1-2)\text{SEA}\,e2}$$

$$y_{(1-2)\text{SEA}\,e2} + x_{6,2} + x_{8,2} = \text{RON}_{\text{SEA}\,e2}$$

Fleet type *e2* at ORD

$$\text{RON}_{\text{ORD}\,e2} + x_{1,2} = x_{9,2} + y_{(1-2)\text{ORD}\,e2}$$

$$y_{(1-2)\text{ORD}\,e2} + x_{13,2} = x_{2,2} + x_{14,2} + y_{(2-3)\text{ORD}\,e2}$$

$$y_{(2-3)\text{ORD}\,e2} + x_{5,2} = x_{6,2} + y_{(3-4)\text{ORD}\,e2}$$

$$y_{(3-4)\text{ORD}\,e2} + x_{10,2} = x_{15,2} + x_{11,2} + x_{4,2} + y_{(4-5)\text{ORD}\,e2}$$

$$y_{(4-5)\text{ORD}\,e2} + x_{3,2} + x_{7,2} = x_{8,2} + y_{(5-6)\text{ORD}\,e2}$$

$$y_{(5-6)\text{ORD}\,e2} + x_{16,2} + x_{12,2} = \text{RON}_{\text{ORD}\,e2}$$

Fleet type *e2* at JFK

$$\text{RON}_{\text{JFK}\,e2} = x_{1,2} + y_{(1-2)\text{JFK}\,e2}$$

$$y_{(1-2)\text{JFK}\,e2} + x_{2,2} = x_{3,2} + y_{(2-3)\text{JFK}\,e2}$$

$$y_{(2-3)\text{JFK}\,e2} + x_{4,2} = \text{RON}_{\text{JFK}\,e2}$$

Fleet type *e2* at BOS

$$\text{RON}_{\text{BOS}\,e2} + x_{9,2} = x_{10,2} + y_{(1-2)\text{BOS}\,e2}$$

$$y_{(1-2)\text{BOS}\,e2} + x_{11,2} = x_{12,2} + \text{RON}_{\text{BOS}\,e2}$$

In the next chapter, we provide an example in which these set of constraints are used to solve a hypothetical fleet assignment problem.

13

Application of the Basic Fleet Assignment Model

13.1 Introduction

This chapter presents an application of the fleet assignment problem presented in the previous chapter. Given the values of the different parameters describing the problem, Microsoft Excel Solver is used to determine the optimal fleet assignment to the 16 flights considered in the schedule (Baker 2012). The network structure of the hypothetical airline is given in the previous chapter as shown in Figure 12.1. In addition, the flight schedules assuming the use of fleet types $e1$ and $e2$ are also given in Figures 12.2 and 12.3, respectively. All cost and revenue values given in this chapter are hypothetical values provided for illustration purposes only. Values used in real-world applications are carefully estimated using thorough data collection and accounting methodologies.

13.2 Problem Input

As shown in the previous chapter, the hypothetical airline is assumed to operate 16 flights using two different fleet types. Figure 13.1 shows a snapshot of an Excel spreadsheet that describes the problem input. In this sheet, rows represent the 16 flights and columns represents the two fleet types, $e1$ and $e2$. In this sheet, the fleet assignment cost c_{fe} is given, which is the cost associated with assigning flight f to fleet type e $\forall f \in F$ and $\forall e \in E$. As mentioned earlier, the assignment cost includes two main components: (i) the flight operating cost and (ii) the spill cost. The flight operating cost includes fuel cost, crew cost, landing fees, etc. The spill cost is the lost revenue in case the aircraft (fleet type) assigned to the flight cannot accommodate all passengers. The table gives the cost for the 16 flights considering the two fleet types. For example, as shown in row 13, assume a cost of $31,200 is recorded in case

Airline Network Planning and Scheduling, First Edition. Ahmed Abdelghany and Khaled Abdelghany.
© 2019 John Wiley & Sons, Inc. Published 2019 by John Wiley & Sons, Inc.

Flight Cost		
	Fleet	
Flight	e1	e2
1	$ 31,230	$ 35,420
2	$ 37,200	$ 34,100
3	$ 35,460	$ 37,200
4	$ 33,270	$ 39,500
5	$ 36,000	$ 32,450
6	$ 32,500	$ 33,780
7	$ 37,400	$ 39,500
8	$ 33,650	$ 33,200
9	$ 38,500	$ 35,420
10	$ 31,200	$ 38,500
11	$ 30,500	$ 39,800
12	$ 34,500	$ 34,520
13	$ 33,680	$ 36,580
14	$ 39,800	$ 37,500
15	$ 36,750	$ 36,850
16	$ 32,700	$ 32,560

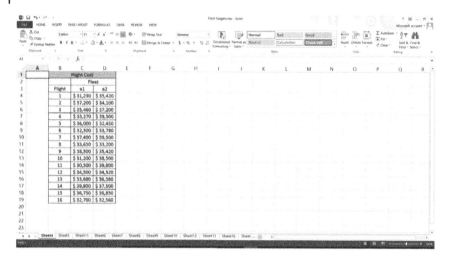

Figure 13.1 The fleet assignment cost for the 16 flights considering the two fleet types.

Flight Net Cost		
	Fleet	
Flight	e1	e2
1	$ 31,230	$ 35,420
2	$ 37,200	$ 34,100
3	$ 35,460	$ 37,200
4	$ 33,270	$ 39,500
5	$ 36,000	$ 32,450
6	$ 32,500	$ 33,780
7	$ 37,400	$ 39,500
8	$ 33,650	$ 33,200
9	$ 38,500	$ 35,420
10	$ 31,200	$ 38,500
11	$ 30,500	$ 39,800
12	$ 34,500	$ 34,520
13	$ 33,680	$ 36,580
14	$ 39,800	$ 37,500
15	$ 36,750	$ 36,850
16	$ 32,700	$ 32,560

Decision Variables		
	Fleet	
Flight	e1	e2
1	0	0
2	0	0
3	0	0
4	0	0
5	0	0
6	0	0
7	0	0
8	0	0
9	0	0
10	0	0
11	0	0
12	0	0
13	0	0
14	0	0
15	0	0
16	0	0

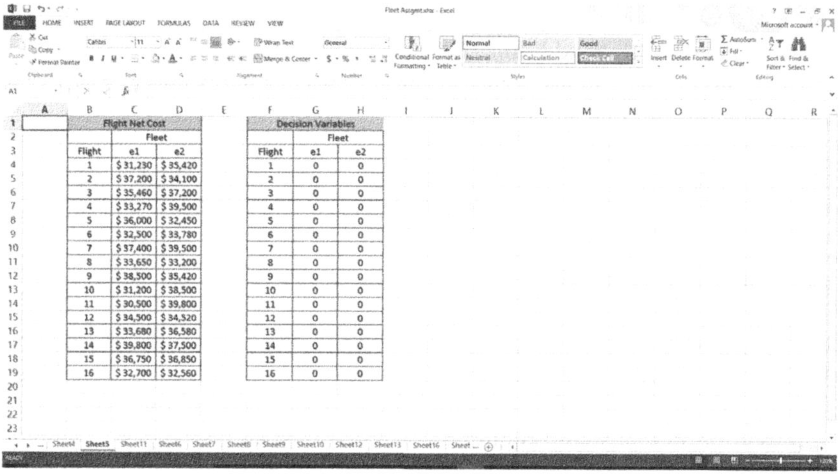

Figure 13.2 The fleet assignment decision variables for the 16 flights and the two fleet types.

flight 10 is operated by fleet type $e1$. If this flight is assigned to fleet $e2$, the cost increases to $38,500.

Figure 13.2 shows a snapshot of the Excel sheet of the problem after defining the fleet assignment decision variables. As explained in the previous chapter, this problem has 32 decision variables. All these decision variables are binary. Initially, these decision variables can be set to any values. After the problem is

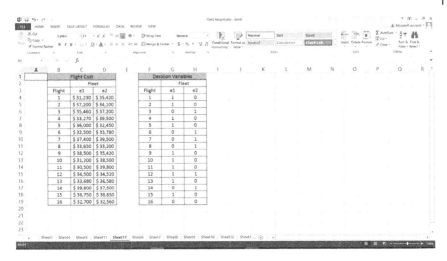

Figure 13.3 Initial values of the fleet assignment decision variables.

solved, the optimal values of these decision variables are determined. As shown in the figure, all decision variables are initially set to zero.

In Figure 13.3, initial values of the fleet assignment decision variables are not zero. For example, in row 6, flight 3 is initially assigned to fleet type $e2$. In row 7, flight 4 is initially assigned to fleet type $e1$. One can notice that in row 15, flight 12 is initially assigned to both fleets $e1$ and $e2$. This initial assignment is infeasible, as the flight has to be assigned to one fleet type only. However, adding the coverage constraints to the problem eliminates this infeasibility. As mentioned above, coverage constraints ensure that each flight is assigned to one fleet type only.

In Figure 13.4, the sum of the decision variables that correspond to each flight is calculated. Column "I" in the Excel sheet includes the sum of column "G" and column "H." Such calculation is required for the flight coverage constraints.

Figures 13.5 and 13.6 show another snapshot of the Excel sheet when the incurred cost is calculated, as given in columns K through M. Figure 13.5 shows the calculation of column L, while Figure 13.6 shows the calculation of column M. The cost of operating a flight using a certain fleet type is only incurred when the corresponding decision variable is equal to 1. Thus, the incurred cost is calculated by multiplying the operation cost of a flight by its corresponding decision variable. For example, in row 6, flight 3 is assigned to fleet type $e2$. Accordingly, the incurred cost of this flight is the cost of fleet type $e2$, which is $37,200.

The next step is to add the decision variables that correspond to the number of aircraft of each fleet type that remain overnight at the different stations and

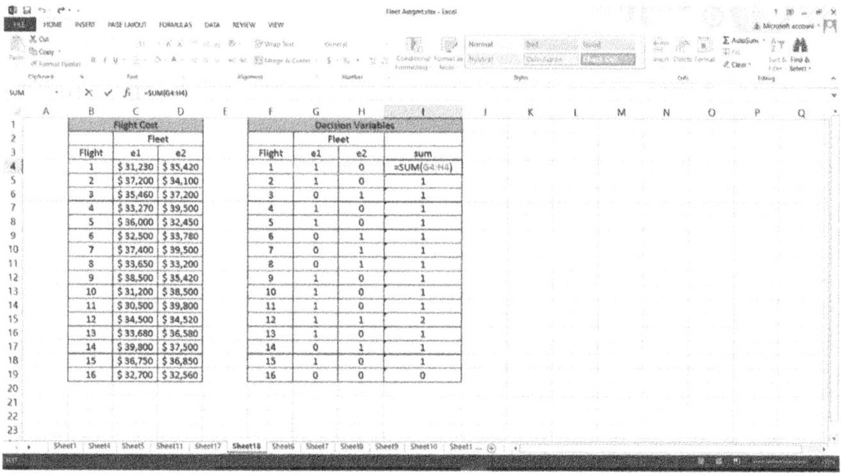

Flight	e1	e2	Flight	e1	e2	sum
1	$ 31,230	$ 35,420	1	1	0	=SUM(G4:H4)
2	$ 37,200	$ 34,100	2	1	0	1
3	$ 35,460	$ 37,200	3	0	1	1
4	$ 33,270	$ 39,500	4	1	0	1
5	$ 36,000	$ 32,450	5	1	0	1
6	$ 32,500	$ 33,780	6	0	1	1
7	$ 37,400	$ 39,500	7	0	1	1
8	$ 33,650	$ 33,200	8	0	1	1
9	$ 38,500	$ 35,420	9	1	0	1
10	$ 31,200	$ 38,500	10	1	0	1
11	$ 30,500	$ 39,800	11	1	0	1
12	$ 34,500	$ 34,520	12	1	1	2
13	$ 33,680	$ 36,580	13	1	0	1
14	$ 39,800	$ 37,500	14	0	1	1
15	$ 36,750	$ 36,850	15	1	0	1
16	$ 32,700	$ 32,560	16	0	0	0

Figure 13.4 The sum of the decision variables for each flight.

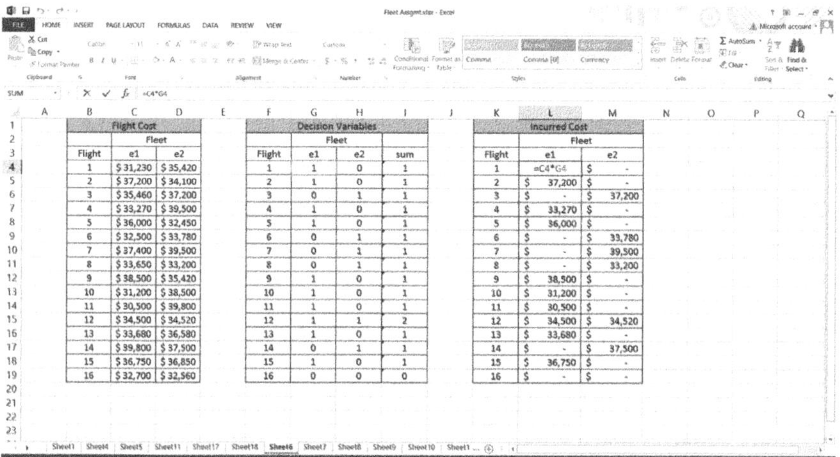

Flight	e1	e2	Flight	e1	e2	sum	Flight	e1	e2
1	$ 31,230	$ 35,420	1	1	0	1	1	=C4*G4	$ -
2	$ 37,200	$ 34,100	2	1	0	1	2	$ 37,200	$ -
3	$ 35,460	$ 37,200	3	0	1	1	3	$ -	$ 37,200
4	$ 33,270	$ 39,500	4	1	0	1	4	$ 33,270	$ -
5	$ 36,000	$ 32,450	5	1	0	1	5	$ 36,000	$ -
6	$ 32,500	$ 33,780	6	0	1	1	6	$ -	$ 33,780
7	$ 37,400	$ 39,500	7	0	1	1	7	$ -	$ 39,500
8	$ 33,650	$ 33,200	8	0	1	1	8	$ -	$ 33,200
9	$ 38,500	$ 35,420	9	1	0	1	9	$ 38,500	$ -
10	$ 31,200	$ 38,500	10	1	0	1	10	$ 31,200	$ -
11	$ 30,500	$ 39,800	11	1	0	1	11	$ 30,500	$ -
12	$ 34,500	$ 34,520	12	1	1	2	12	$ 34,500	$ 34,520
13	$ 33,680	$ 36,580	13	1	0	1	13	$ 33,680	$ -
14	$ 39,800	$ 37,500	14	0	1	1	14	$ -	$ 37,500
15	$ 36,750	$ 36,850	15	1	0	1	15	$ 36,750	$ -
16	$ 32,700	$ 32,560	16	0	0	0	16	$ -	$ -

Figure 13.5 The fleet assignment cost for flights assigned to fleet type *e*1.

the number of aircraft of each fleet type that remain on the ground between the different interconnection nodes. Figure 13.7 shows a snapshot of the Excel sheet after adding these decision variables. As shown in this figure, rows 20–24 define the decision variables for the number of aircraft that remain overnight at the five stations for the two fleet types. Similarly, rows 25–35 define the decision variables for the number of aircraft that remain on the ground between the different interconnection nodes, as defined in the previous chapter.

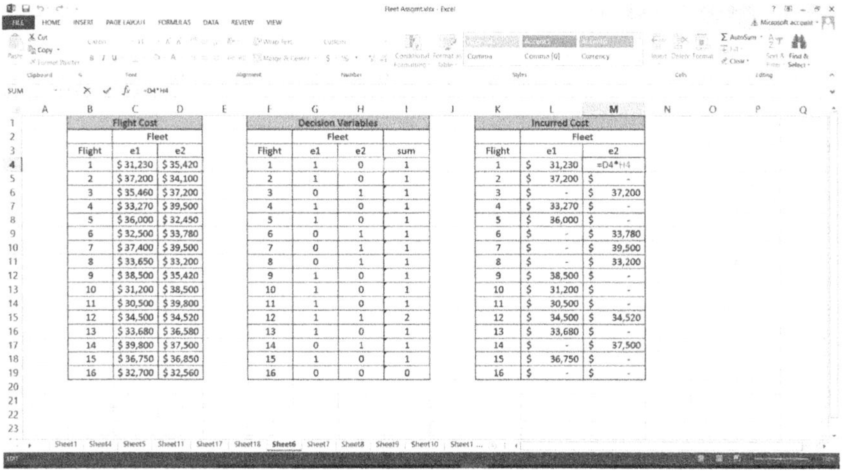

Flight Cost				Decision Variables				Incurred Cost		
	Fleet				Fleet				Fleet	
Flight	e1	e2		Flight	e1	e2	sum	Flight	e1	e2
1	$ 31,230	$ 35,420		1	1	0	1	1	$ 31,230	=D4*H4
2	$ 37,200	$ 34,100		2	1	0	1	2	$ 37,200	$ -
3	$ 35,460	$ 37,200		3	0	1	1	3	$ -	$ 37,200
4	$ 33,270	$ 39,500		4	1	0	1	4	$ 33,270	$ -
5	$ 36,000	$ 32,450		5	1	0	1	5	$ 36,000	$ -
6	$ 32,500	$ 33,780		6	0	1	1	6	$ -	$ 33,780
7	$ 37,400	$ 39,500		7	0	1	1	7	$ -	$ 39,500
8	$ 33,650	$ 33,200		8	0	1	1	8	$ -	$ 33,200
9	$ 38,500	$ 35,420		9	1	0	1	9	$ 38,500	$ -
10	$ 31,200	$ 38,500		10	1	0	1	10	$ 31,200	$ -
11	$ 30,500	$ 39,800		11	1	0	1	11	$ 30,500	$ -
12	$ 34,500	$ 34,520		12	1	1	2	12	$ 34,500	$ 34,520
13	$ 33,680	$ 36,580		13	1	0	1	13	$ 33,680	$ -
14	$ 39,800	$ 37,500		14	0	1	1	14	$ -	$ 37,500
15	$ 36,750	$ 36,850		15	1	0	1	15	$ 36,750	$ -
16	$ 32,700	$ 32,560		16	0	0	0	16	$ -	$ -

Figure 13.6 The fleet assignment cost for flights assigned to fleet type *e*2.

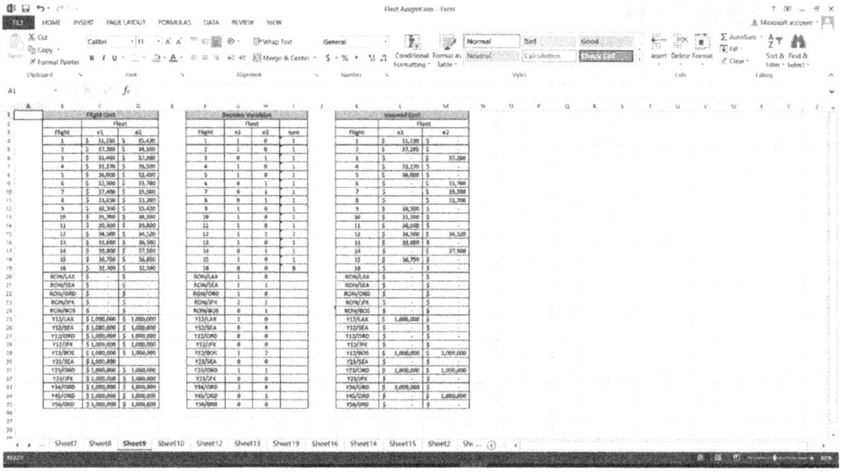

Figure 13.7 Adding the remaining overnight and ground decision variables.

Columns C and D give the corresponding cost. In this problem, it is assumed that there is no cost associated with an aircraft that remains overnight at any of the five stations (i.e. zero values are given in the table). On the other hand, a high cost value is considered (i.e. $1,000,000) for each aircraft to remain on the ground between two successive interconnection nodes. This high cost is used in the model to avoid solutions that do not utilize the aircraft efficiently during the day. Columns G and H give initial values for the decision variables. It should be noted that these decision variables are integer variables that are greater than or equal to 0. For example, in row 23, the decision variables for the number of aircraft remaining overnight at JFK for fleet type $e1$ and fleet type $e2$ are assumed equal to 2 and 1, respectively. Similarly, in rows 25–35, initial values are assumed for the number of aircraft of each fleet type to remain on the ground during the day, between the different interconnection nodes, at the different stations. For example, in row 31, the decision variables for the number of aircraft remaining overnight at ORD between interconnection nodes 2 and 3 for fleet type $e1$ and fleet type $e2$ are both assumed to be equal to 1. Finally, columns L and M give the incurred cost, which is calculated by multiplying the cost values in columns C and D by their corresponding decision variables.

Next, the total incurred cost of the problem is calculated by summing up the different cost elements. In Figure 13.8, the total cost of the fleet assignment problem is given in cell O3. This cost is calculated by summing up all cost elements in columns L and M. Figure 13.9 shows the numerical value of the incurred cost, which is $10,558,530. This cost represents the objective function of the mathematical program formulated for the problem. The optimal fleet assignment solution is obtained when this objective function is minimized.

Next, the balance constraints are defined in the Excel sheet. As explained in the previous chapter, these constraints are considered for all interconnection nodes at all stations for the two fleet types. For this purpose, two tables are introduced in the Excel sheet as shown the Figure 13.10. The first table (FLEET 1), which begins at cell B42, is used for fleet type $e1$. The second table (FLEET 2), which begins at cell B53, is used for fleet type $e2$. In these tables, for each

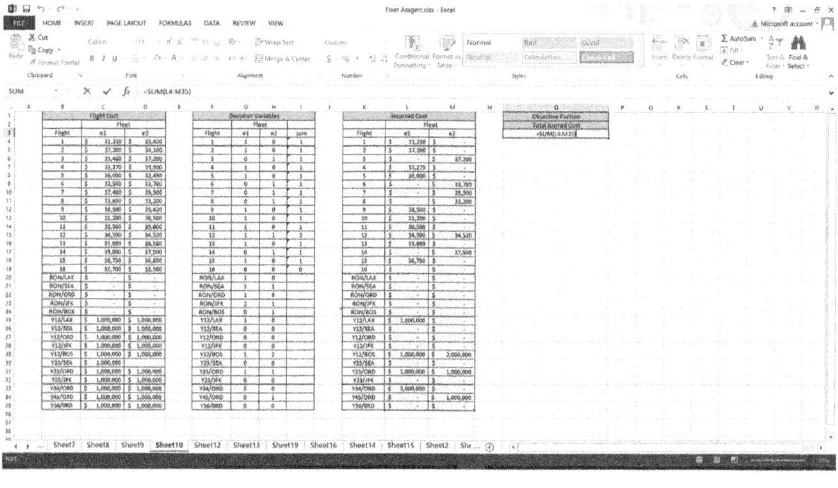

Figure 13.8 Calculating the total cost of the fleet assignment problem.

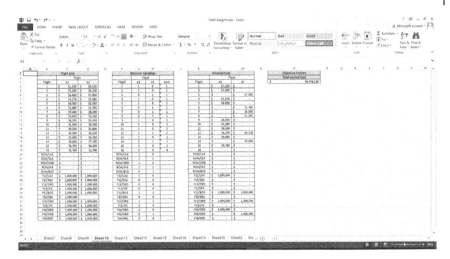

Figure 13.9 The value of the total incurred cost in cell O3.

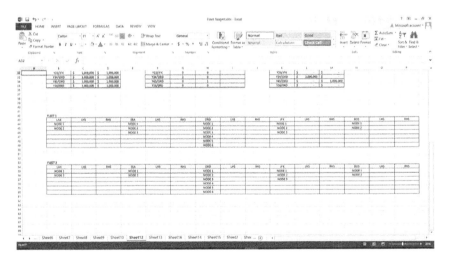

Figure 13.10 Tabular format for the balance constraints.

station, a row is given to represent the balance constraint of an interconnection node. For example, considering LAX and fleet type $e1$, two interconnection nodes are defined. Thus, two rows (44 and 45) are defined in the Excel sheet by NODE 1 and NODE 2. Similarly, for JFK and fleet type $e2$, three interconnection nodes are defined. Thus, three rows are defined (55, 56, and 57) in the sheet by NODE 1, NODE 2, and NODE3. For each interconnection node, two columns are defined as LHS and RHS, which represent the left-hand side and the right-hand side of the balance constraint of this interconnection node.

Figures 13.11 and 13.12, respectively, show how the LHS and the RHS cells are defined for the first interconnection node (i.e. NODE 1) at LAX for fleet type $e1$. The balance constraint for this node is given in the previous chapter as follows:

$$\text{RON}_{\text{LAX}\,e1} = x_{13,1} + y_{(1-2)\,\text{LAX}\,e1}$$

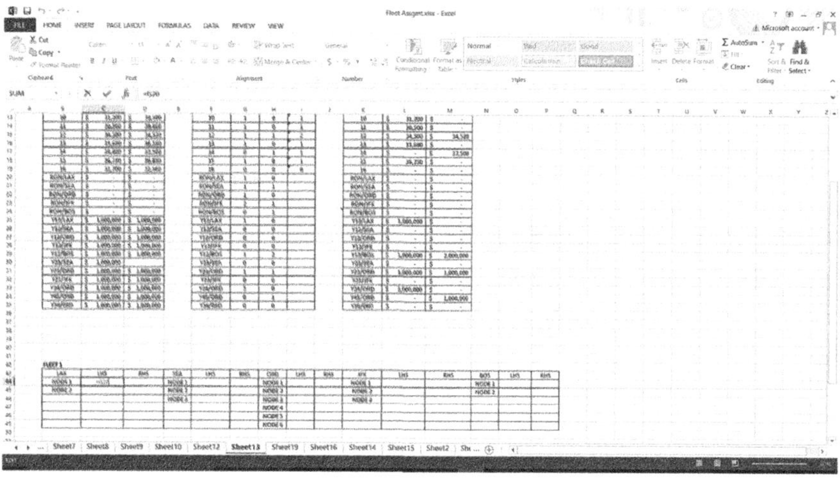

Figure 13.11 Defining the LHS of the balance constraint for the first interconnection node at LAX for fleet type $e1$.

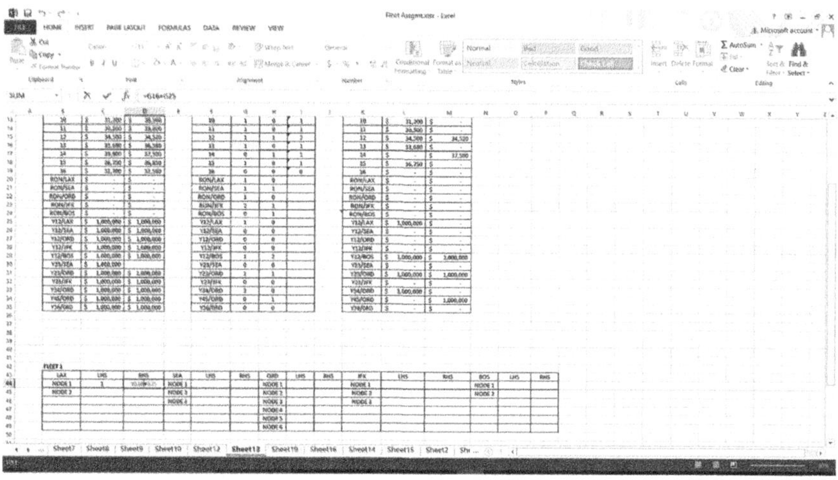

Figure 13.12 Defining the RHS of the balance constraint for the first interconnection node at LAX for fleet type $e1$.

LHS represents the number of aircraft remaining overnight at LAX for fleet type $e1$. This information is provided in cell G20. Thus, the value of cell C44 is set to be equal to the value of cell G20. Similarly, RHS represents the sum of the value of the decision variable that corresponds to whether or not flight 13 is assigned to fleet type $e1$ (cell G16) plus the number of aircraft of fleet type $e1$ that remain on the ground in the period between interconnection nodes 1 and 2 (cell G25). Thus, cell D44 defining the RHS of this constraint is set to be equal to the sum of cells G16 and G25.

Another example to set up the balance constraints is given for the second interconnection node (NODE 2) at LAX for fleet type $e1$. Figures 13.13 and 13.14, respectively, show how the LHS and the RHS cells are defined for this interconnection node. The balance constraint for this node was given in the previous chapter as follows:

$$y_{(1-2)\text{LAX }e1} + x_{14,1} + x_{15,1} = x_{16,1} + \text{RON}_{\text{LAX }e1}$$

LHS represents the sum of the number of aircraft of fleet type $e1$ that remain on the ground in the period between interconnection nodes 1 and 2 (cell G25), the value of the decision variable that corresponds to whether or not flight 14 is assigned to fleet type $e1$ (Cell G17), and the value of the decision variable that corresponds to whether or not flight 15 is assigned to fleet type $e1$ (cell G18). Thus, cell C45 is set to be equal to the sum of cells G25, G17, and G18. Similarly, RHS represents the sum of the value of the decision variable that corresponds to whether or not flight 16 is assigned to fleet type $e1$ (cell G19) plus the

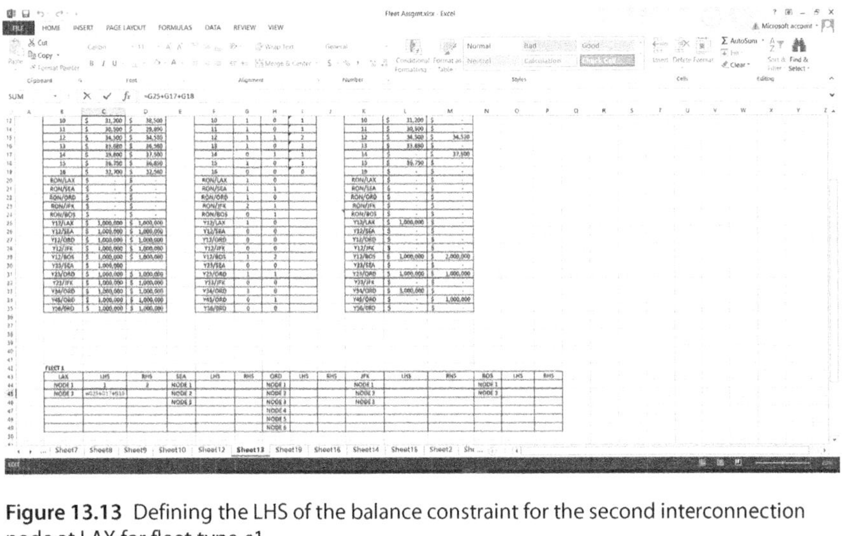

Figure 13.13 Defining the LHS of the balance constraint for the second interconnection node at LAX for fleet type $e1$.

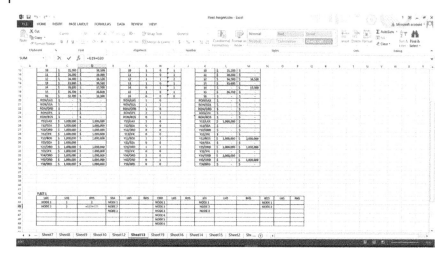

Figure 13.14 Defining the RHS of the balance constraint for the second interconnection node at LAX for fleet type *e*1.

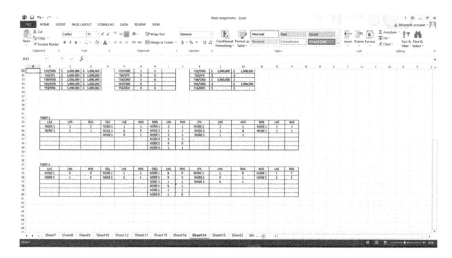

Figure 13.15 The calculated values of the left-hand side and the right-hand side of the balance constraints for the different interconnection nodes for the two fleet types.

number of aircraft remaining overnight at LAX for fleet type *e*1 (cell G20). Thus, cell D45 is set to be equal to the sum of cells G19 and G20.

Figure 13.15 gives the calculated values of LHS and RHS of the balance constraints for all interconnection nodes for the two fleet types. It should be noted that, at this stage, the values of LHS and RHS of the balance constraint at each interconnection node are not necessarily equal. These values depend primarily on the initial values assigned by the analyst to the different decision variables of the problem. As shown later, RHS and LHS of all balance constraints are forced to be equal in order to obtain a feasible fleet assignment solution.

13.3   Setting the Problem in Excel Solver

The next step is to set up the mathematical program formulated for the fleet assignment problem in the Excel Solver in order to determine the problem's optimal solution. Setting up the mathematical program requires identifying which cell is the objective function and whether this objective function is to be minimized or maximized. It also requires defining the decision variables and constraints.

In Microsoft Excel 2013, the Solver function can be accessed under the DATA tab, as shown in Figure 13.16. In case the Solver function is not found, the Solver plug-in should be installed. We highly encourage the readers to familiarize themselves with the Excel Solver before reading this section.

As shown in Figure 13.17, when clicking on the Solver button, the Solver Parameters window appears. Through this window, the analyst enters the information on the objective function, the decision variable, and the constraints. The objective function is set by selecting cell O3, which corresponds to the total incurred cost. This objective function is to be minimized by selecting the "Min" check box. Next, the decision variables are defined. These are the variables that the Solver is permitted to change to determine the

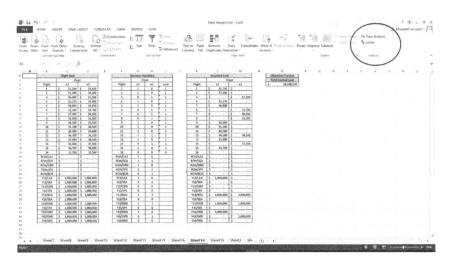

Figure 13.16 The Solver function in Microsoft Excel 2013.

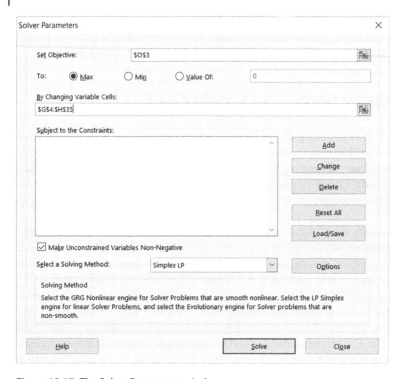

Figure 13.17 The Solver Parameters window.

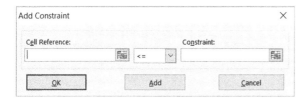

Figure 13.18 The Add Constraint window in the Excel Solver.

optimal solution. As explained above, the decision variables are the variables in columns G and H (row 4 to row 35).

Next, the set of constraints are added. By clicking on the Add button, the Add Constraint window appears, as shown in Figure 13.18. The analyst uses this window to define the problem constraints one by one.

The first set of constraints is the flight coverage constraints. As mentioned above, the coverage constraints ensure that each of the 16 flights is covered by one fleet type. Thus, the problem has 16 coverage constraints – one for each flight. The coverage constraints are entered, as shown in Figure 13.19, by setting cells I4 through I19 to be equal to 1. The 16 constraints can be input at

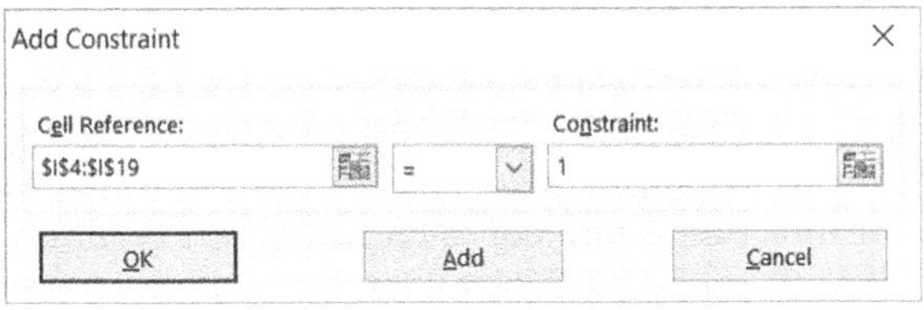

Figure 13.19 Entering the coverage constraints.

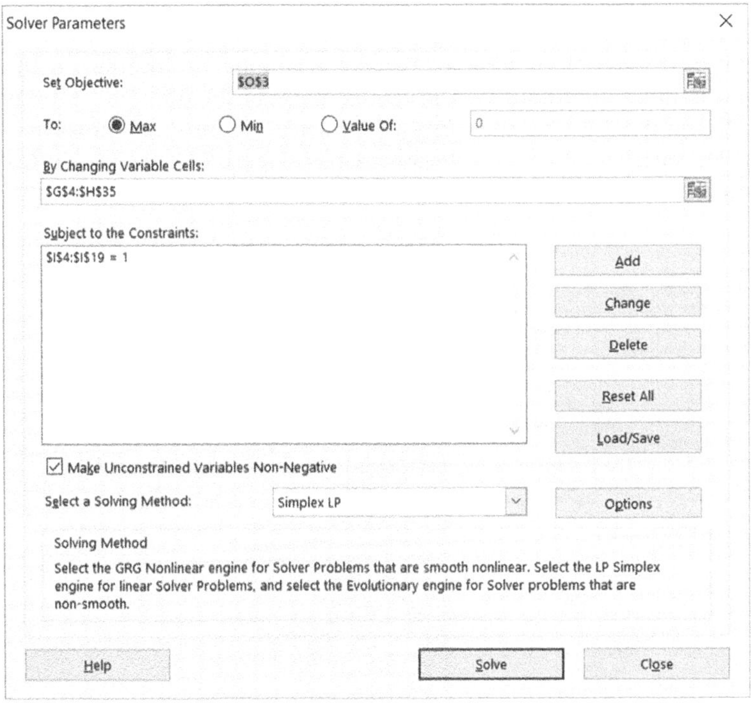

Figure 13.20 The coverage constraints added to the Solver Parameters window.

once in one line in the "Add Constraint" window by selecting the cells I4 to I19 in the "Cell Reference," then selecting the "=" sign, and setting the value of the constraint to be equal to 1. Then, click "Add" to add these 16 constraints to the list of constraints. Figure 13.20 shows the Solver Parameters window with the flight coverage constraints. As mentioned above, cells I4 to I19 include the sum of the decision variables of each flight. When this sum is set to one, it guarantees that each flight is covered by one and only one fleet type.

Next, the balance constraints are added for each interconnection node. As explained above, the problem has 31 interconnection nodes for the two fleet types that need to be balanced. Thus, the problem requires 31 balance constraints. Figure 13.21 shows how the balance constraints are defined for the two interconnection nodes at LAX for fleet type e1. As shown in the figure, the two balance constraints are given in one line. The constraints entail setting RHS and LHS to be equal. Similarly, Figures 13.22 and 13.23 show the input of the balance constraints for the interconnection nodes for fleet type e1 for SEA and ORD, respectively. Figure 13.24 shows the Solver Parameters window after adding all the balance constraints.

Next, the decision variables related to whether the flight is assigned to fleet type e1 or fleet type e2 have to be set as binary variables (i.e. one or zero). Figure 13.25 shows how these decision variables are defined in the Solver. In addition, other decision variables related to number of aircraft on the ground and number of aircraft remaining overnight at the different stations for the two fleet types are set to be integer variables. Figure 13.26 shows how these decision variables are set as integer variables. Figure 13.27 shows the list of constraints after the binary and integer constraints are added.

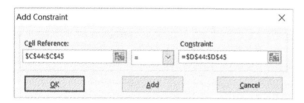

Figure 13.21 Entering the balance constraints for the interconnection nodes at LAX for fleet type e1.

Figure 13.22 Entering the balance constraints for the interconnection nodes at SEA for fleet type e1.

Figure 13.23 Entering the balance constraints for the interconnection nodes at ORD for fleet type e1.

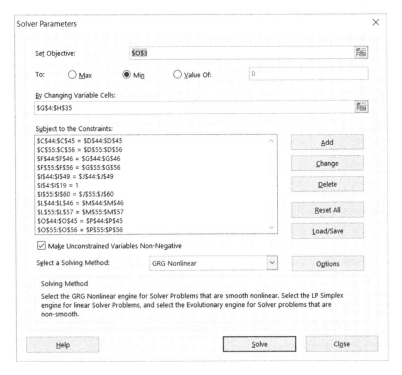

Figure 13.24 The balance constraints added to the Solver Parameters window.

Figure 13.25 Entering the binary constraints.

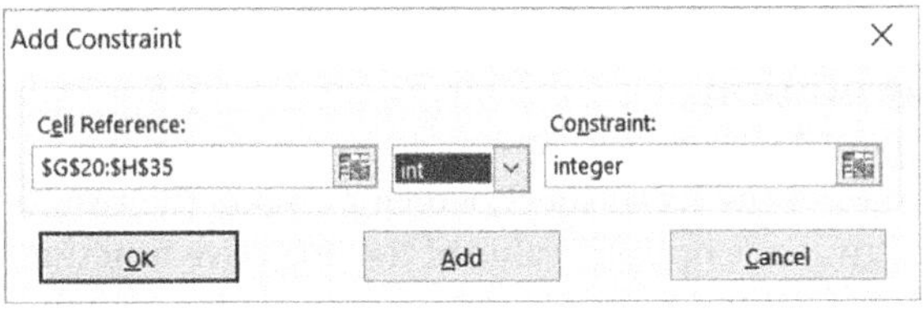

Figure 13.26 Entering the integer constraints.

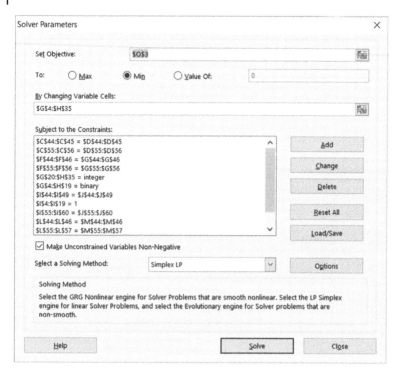

Figure 13.27 The binary and integer constraints added to the Solver Parameters window.

At this stage, the basic fleet assignment problem is still missing the resources constraints. However, solving the problem without these constraints provides a benchmark solution in which the airline is assumed to have all required fleet resources. To obtain this benchmark solution, click on the "Solve" button in the Solver Parameters window to get the Solver Results window, which is shown in Figure 13.28. One should notice the message that Solver found a solution and all constraints and optimality conditions are met. In this case, a solution is found. Click "OK" while noticing the change in the value of objective function and the values of decision variables in the Excel sheet to their optimal values.

13.4   Solution Interpretation

Figure 13.29 shows a snapshot of the Excel sheet, which presents the results of the Excel Solver. Cell O3 gives the optimal value of the objective function, which is $8,548,940. Cells G4–G19 and cells H4–H19 give the flights assigned to fleet type $e1$ and fleet type $e2$, respectively. The flight is assigned to a fleet

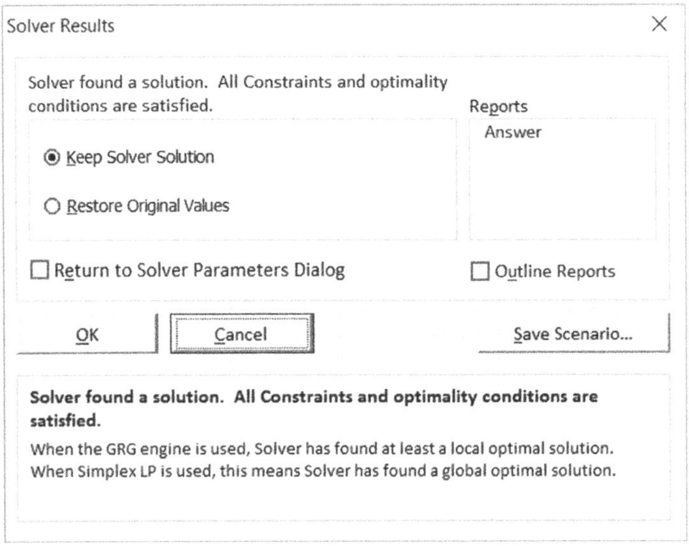

Figure 13.28 The Solver Results window.

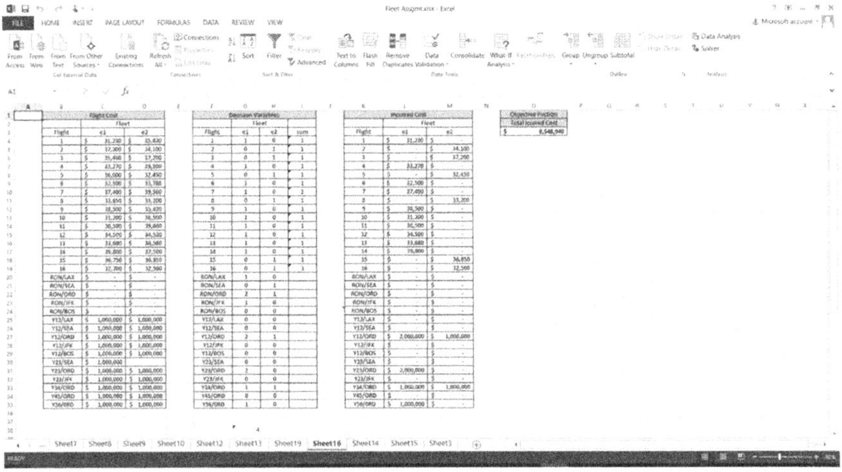

Figure 13.29 The fleet assignment solution without resources constraints.

if the decision variable corresponding to that fleet is equal to 1. As given in the Excel sheet, flights 1, 4, 6, 7, 9, 10, 11, 12, 13, and 14 are assigned to fleet type $e1$, and flights 2, 3, 5, 8, 15, and 16 are assigned to fleet type $e2$. The optimal fleet assignment solution requires four aircraft of fleet type $e1$ to remain overnight at the different stations (one aircraft at LAX, two aircraft at ORD,

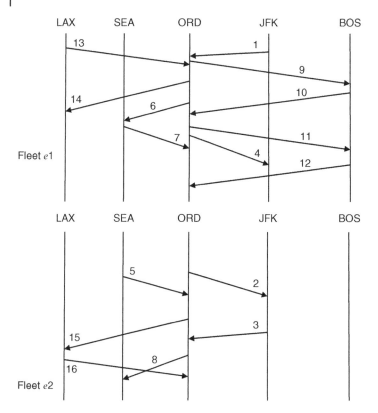

Figure 13.30 Graphical representation of the solution of the fleet assignment problem without resources constraints.

and one aircraft at JFK). The solution also requires two aircraft of fleet type e2 to remain overnight at the different stations (one aircraft at SEA and one at ORD). Figure 13.30 gives a graphical representation of the fleet assignment solution.

13.5 Resources Constraints

As illustrated above, the optimal solution of the fleet assignment problem requires having four aircraft of fleet type e1 and two aircraft of fleet type e2. The objective of this section is to describe how the resources availability constraints can be added to the solution. For example, if there are only three aircraft available of fleet type e1, how the benchmark solution given above will change. A resources constraint can be added to the fleet assignment problem. Figure 13.31 gives a snapshot of the Excel sheet in which cell G38

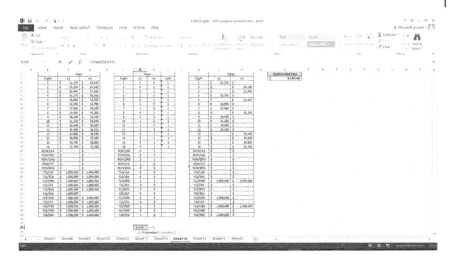

Figure 13.31 Counting the number of aircraft to remain overnight at the different stations for the fleet assignment problem.

Figure 13.32 Adding resources constraints on fleet type e1.

includes the sum of cells G20–G24. This sum represents the number of aircraft of fleet type e1 that remain overnight at the different stations.

Then, in the Excel Solver, an additional constraint is added to force this sum to be less than or equal to 3, as shown in Figure 13.32. The problem is again solved after adding this resources availability constraint.

Figure 13.33 shows a snapshot of the Excel sheet that has the fleet assignment solution after the resources availability constraint for fleet type e1 is added. As shown in this sheet, adding this constraint modifies the benchmark solution obtained for the fleet assignment problem. The objective function, which represents the incurred cost, increased from \$8,548,940 to \$8,549,540. As shown in the figure, flights 1, 4, 6, 7, 9, 10, 11, and 12 are assigned to fleet type e1, and flights 2, 3, 5, 8, 13, 14, 15, and 16 are assigned to fleet type e2. The flights assigned to each fleet type are shown in Figure 13.34. Comparing this solution with the benchmark solution in which no resources constraints are considered, flights 13 and 14 are now assigned to fleet type e2 due to the resources limitation of fleet type e1.

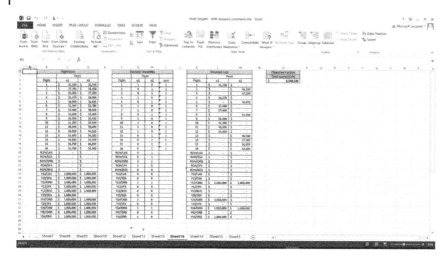

Figure 13.33 The fleet assignment solution with a resources constraint.

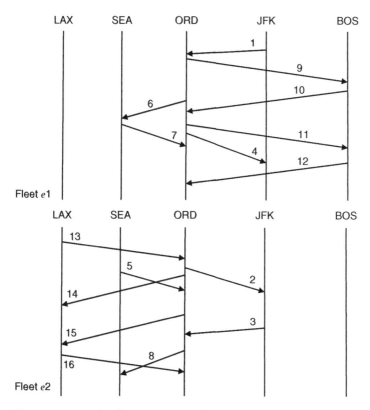

Figure 13.34 Graphical representation of the fleet assignment problem with a resources constraint.

13.6 Additional Constraints

In real applications, several additional constraints are considered. These constraints are typically related to fleet maintenance requirements, crew assignment, and through flights (Clarke et al. 1996). For example, maintenance constraints are added to ensure that aircraft of certain fleet type visit stations where the maintenance activities of this fleet are performed. Other maintenance constraints are added to ensure that the number of aircraft of a given fleet type that visit a maintenance station is within the operation capacity of this maintenance facility. In addition, constraints are added to guarantee that each aircraft spends adequate time at the maintenance station such that the maintenance activity is completed. Crew constraints are added to ensure the number of flights and flight time assigned to each fleet type are proportional to the number of crew members qualified to fly each fleet type. In addition, constraints are added to assign flights originating from crew domicile (home cities) to fleet types that this crew is qualified to fly. Also, constraints are added to guarantee that all crew legalities rules (e.g. minimum rest periods) are satisfied. Finally, as explained earlier, through flights constraints are added to ensure that the two connected flights are assigned to the same fleet type.

Section 4

14

The Schedule Adjustment Problem

14.1 Introduction

Most major airlines publish multiple schedules throughout the year to respond to changes in demand levels associated with seasonality, competition, and changes in available resources (e.g. new aircraft, slots, gates, etc.). These schedules are usually published ahead of time (9–12 months) such that prospective travelers and travel agents can access these schedule and reserve seats on the scheduled flights. A common practice is to publish a schedule for every month in the year. Airlines develop these monthly schedules while maintaining their network structure and preserving adequate service frequency in the different city-pairs to avoid losing loyal customers. Accordingly, most major airlines develop their schedules by tweaking and adjusting previous schedules to minimize schedule changes. This approach is known as the warm start approach (or incremental approach), in which an initial schedule(s) is adjusted in order to obtain the target schedule. There are several reasons for airlines to rely on the warm start approach (Lohatepanont and Barnhart 2004). First, building an entirely new schedule requires data that might not be available to the airline. Second, developing a new schedule from scratch is computationally intractable. Third, significant investment at airport stations will be needed if significant changes in network structure are introduced. Last, airlines prefer schedule consistency between seasons, especially in business markets in which reliability and consistency are highly valued by business travels.

The schedule planning for airlines consists of two sequential main steps, which are route planning and schedule development. In the route planning step, airlines decide on which routes to serve based on demand, competition, and availability of resources. The schedule development step is composed of two main tasks including developing the flight schedules (schedule design) and

Airline Network Planning and Scheduling, First Edition. Ahmed Abdelghany and Khaled Abdelghany.

developing the schedule of the resources (aircraft, crew, gates, etc.). Developing the flight schedules also consists of two sequential steps, which are frequency planning and timetable development. The objective of frequency planning is to determine service frequency in each city-pair. The timetable development determines the departure time of each proposed flight (i.e. distribute the proposed service frequency throughout the day). The schedule of resources entails determining the fleet assignment for each flight, aircraft rotations, crew schedule, gate schedules, etc.

In this chapter, we present the current practice of the warm start approach. In particular, the schedule design problem including the frequency planning and the timetable development is discussed. The approach depends mainly on the fleet assignment formulation presented in Chapter 11.

14.2   Schedule Adjustment Decisions

Several adjustments to an existing schedule can be considered. These adjustments include flight deletion (removal), flight addition, and flight departure time adjustment. Flight deletion is considered in case an airline decides to cut service frequency in a given city-pair. Airlines cut frequency usually when demand in the city-pair declines due to seasonality, competition, or decline in the economic activities. For example, an airline might have been flying five flights daily in a city-pair. If the airline decided to reduce the frequency to four, the least important flight among the five flights is eliminated from the schedule. The process of flight deletion might result in terminating service in one or more of the city-pairs served by the airline (i.e. frequency is equal to 0). Flight deletion requires identifying city-pair targets for service frequency reduction, number of flights to be eliminated, and flights to be eliminated from the schedule. On the other hand, flight addition is adopted when the airline decides to expand its service in one of its currently served markets or start service in a new market. Flight addition is usually considered when an airline aims at strengthening its competition advantage in a city-pair to increase its market share. Flight addition requires identifying city-pairs eligible for flight addition, number of flights to be added, and specifying the departure time of these flights. Flight addition is constrained by the airline's resources availability including aircraft, crew, slots, gates, etc. Schedule adjustment might also require changing the departure time of some existing flights. For this purpose, new departure times for these flights have to be identified. This new departure time should make the flights attractive for perspective travelers and also result in efficient schedule for the different resources (aircraft, crew, gates, etc.). The formulation of the schedule adjustment problem is presented in the next section.

14.3 Problem Formulation

The formulation of the schedule adjustment problem is derived from the fleet assignment formulation presented in Chapter 11. The basic fleet assignment problem minimizes the total cost (which includes the flight operation cost and the demand spill cost) associated with assigning flights to the available fleet type. The formulation ensures that all flights are covered, fleets are balanced at the different stations, and available resources are not exceeded. The basic fleet assignment problem is formulated as follows:

Minimize

$$C = \sum_{f \in F} \sum_{e \in E} c_{fe} \cdot x_{fe}$$

or, maximize

$$P = \sum_{f \in F} \sum_{e \in E} p_{fe} \cdot x_{fe}$$

Subject to:
The coverage constraints:

$$\sum_{e \in E} x_{fe} = 1, \qquad \forall f \in F$$

The balance constraints:

$$\sum_{f \in IN_s} x_{fe} = \sum_{f \in OUT_s} x_{fe}, \qquad \forall e \in E, \forall s \in S$$

The resources availability constraints:

$$\sum_{s \in S} y_{se} \le N_e, \qquad \forall e \in E$$

where

f: Index of flights in the schedule, $f \in F$
e: Index of fleet types (equipment), $e \in E$
s: Index of stations, $s \in S$
x_{fe}: A decision variable that is equal to 1 if flight $f \in F$ is assigned to equipment $e \in E$, and zero otherwise, $x_{fe} \in X$
y_{se}: A decision variable, represents the number of aircraft of type $e \in E$ that remain overnight at station $s \in S$

p_{fe}: The net revenue (profit) associated with assigning the flight $f \in F$ to the fleet type $e \in E$

c_{fe}: The cost c_{fe} associated with assigning the flight $f \in F$ to the fleet type $e \in E$

IN_s and OUT_s: The set of inbound and outbound flights for the airline at station $s \in S$

N_e: The number of available aircraft of type $e \in E$

The schedule adjustment problem is derived from the fleet assignment problem by (i) using the maximization of the net revenue as an objective function and (ii) relaxing the coverage constraints to inequality constraints instead of equality constraints. Thus, the schedule adjustment problem can be presented as follows:

Maximize

$$P = \sum_{f \in F} \sum_{e \in E} p_{fe} \cdot x_{fe}$$

Subject to:
The coverage constraints:

$$\sum_{e \in E} x_{fe} \leq 1, \qquad \forall f \in F$$

The balance constraints:

$$\sum_{f \in IN_s} x_{fe} = \sum_{f \in OUT_s} x_{fe}, \qquad \forall e \in E, \forall s \in S$$

The resources availability constraints:

$$\sum_{s \in S} y_{se} \leq N_e, \qquad \forall e \in E$$

As the net revenue is maximized, relaxing the coverage constraints for a flight implies that this flight is not necessarily to be covered by any fleet type. If the solution recommends that a flight is not covered by any fleet, it explicitly suggests that this flight can be removed from the schedule. In the next chapter, the fleet assignment problem example presented in Chapters 12 and 13 is used to illustrate the schedule adjustment problem.

15

Examples on the Schedule Adjustment Problem

15.1   Flight Deletion

Consider the hypothetical airline described in Chapters 12 and 13. Assume this airline is cutting part of its service to respond to a network-wide demand reduction and/or to restructure its cost. Adjusting the schedule requires deciding which flights to eliminate from the schedule. To solve this problem, we use the formulation presented for the schedule adjustment problem in Chapter 14. The net revenue (profit) of each flight is calculated and assumed given as an input to the problem. Figure 15.1 shows a snapshot of the Excel sheet developed for the problem. The table in columns C and D gives the net revenue associated with assigning each of the 16 flights to each of the two fleet types.

Next, the decision variables are defined, and the incurred net revenue is calculated by multiplying the net revenue for each flight by its corresponding decision variable. Figure 15.2 gives another snapshot of the Excel sheet, in which the decision variables for the fleet assignments are initialized in columns G and H. Columns L and M give the incurred net revenue associated with the initial values assigned to the decision variables. The decision variables associated with the number of aircraft that remain overnight at the different stations and the number of aircraft that remain on the ground between the different interconnection nodes are defined. In addition, a negative value of $1000 is assumed to represent the net revenue (i.e. revenue loss) associated with each aircraft remaining on the ground between interconnection nodes. The total net revenue is calculated by adding all net revenue elements in columns L and M. The total net revenue represents the objective function of the problem ($28,535) that is to be maximized. Figure 15.3 shows a snapshot of the Excel sheet of the schedule adjustment problem with the calculation of the objective function in cell O3.

The total number of aircraft remaining overnight for each fleet type is calculated. Figure 15.4 shows the calculation of the total number of aircraft remaining overnight for fleet type $e1$ (cell G38), which is the sum of the

Airline Network Planning and Scheduling, First Edition. Ahmed Abdelghany and Khaled Abdelghany.
© 2019 John Wiley & Sons, Inc. Published 2019 by John Wiley & Sons, Inc.

Flight Net Revenue (profit)		
	Fleet	
Flight	e1	e2
1	$ 2,655	$ 2,548
2	$ 4,587	$ 3,782
3	$ 1,589	$ 2,250
4	$ 3,200	$ 2,980
5	$ 2,650	$ 3,005
6	$ 4,587	$ 3,904
7	$ 3,325	$ 2,289
8	$ 3,478	$ 3,578
9	$ 1,588	$ 2,689
10	$ 1,202	$ 2,200
11	$ 1,782	$ 1,590
12	$ 2,248	$ 1,800
13	$ 6,345	$ 1,780
14	$ 2,478	$ 2,504
15	$ 3,525	$ 2,298
16	$ 1,560	$ 2,580

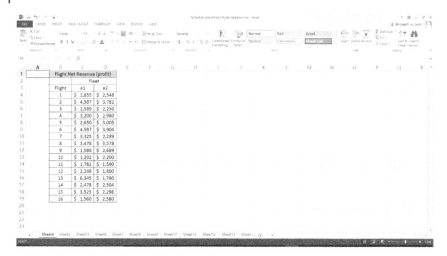

Figure 15.1 Net revenue for the 16 flights when assigned to each fleet type.

=C4*G4

Flight Net Revenue (profit)		
	Fleet	
Flight	e1	e2
1	$ 2,655	$ 2,548
2	$ 4,587	$ 3,782
3	$ 1,589	$ 2,250
4	$ 3,200	$ 2,980
5	$ 2,650	$ 3,005
6	$ 4,587	$ 3,904
7	$ 3,325	$ 2,289
8	$ 3,478	$ 3,578
9	$ 1,588	$ 2,689
10	$ 1,202	$ 2,200
11	$ 1,782	$ 1,590
12	$ 2,248	$ 1,800
13	$ 6,345	$ 1,780
14	$ 2,478	$ 2,504
15	$ 3,525	$ 2,298
16	$ 1,560	$ 2,580

Decision Variables			
	Fleet		
Flight	e1	e2	sum
1	0	1	1
2	0	0	0
3	1	0	1
4	1	0	1
5	1	0	1
6	1	0	1
7	0	1	1
8	0	0	0
9	1	0	1
10	1	0	1
11	1	0	1
12	1	0	1
13	1	0	1
14	1	1	2
15	1	0	1
16	0	0	0

Incurred Net Revenue		
	Fleet	
Flight	e1	e2
1	=C4*G4	$ 2,548
2	$ -	$ -
3	$ 1,589	$ -
4	$ 3,200	$ -
5	$ 2,650	$ -
6	$ 4,587	$ -
7	$ -	$ 2,289
8	$ -	$ -
9	$ 1,588	$ -
10	$ 1,202	$ -
11	$ 1,782	$ -
12	$ 2,248	$ -
13	$ 6,345	$ -
14	$ 2,478	$ 2,504
15	$ 3,525	$ -
16	$ -	$ -

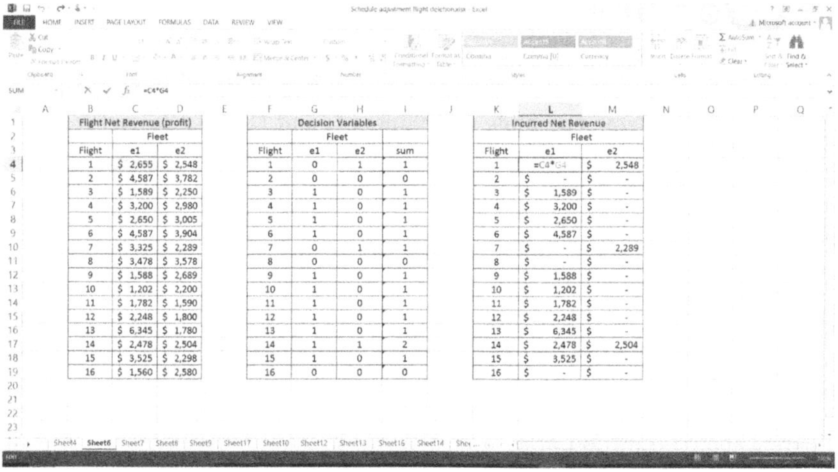

Figure 15.2 The initial decision variables and the associated net revenues for the schedule adjustment problem for the flight deletion example.

number of aircraft at each of the five stations: LAX, SEA, ORD, JFK, and BOS. The step is repeated to determine the total number of aircraft remaining overnight for fleet type $e2$ (cell H38).

Next, the balance constraints at each station are defined for each interconnection node and for each fleet type. For more details on defining these constraints, the reader is advised to review the example given for the fleet

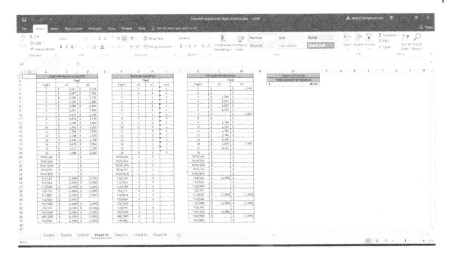

Figure 15.3 The calculation of the total incurred net revenue for the schedule adjustment problem for the flight deletion example.

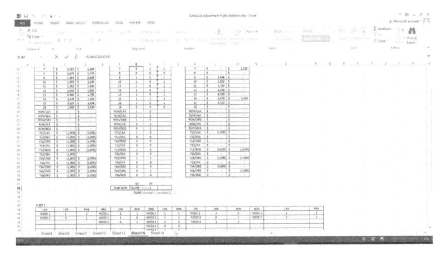

Figure 15.4 Counting the number of aircraft to remain overnight at the different stations for the schedule adjustment problem for the flight deletion example.

assignment problem in Chapter 13. Figure 15.5 shows a snapshot of the Excel sheet with the definition of the balance constraints.

Figure 15.6 shows the Solver Parameters window. First, the objective function is maximized, which represents the total net revenue as given in cell O3. The decision variables are similar to the one defined in the fleet assignment

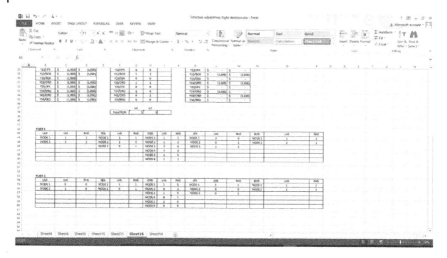

Figure 15.5 The balance constraints for the schedule adjustment problem for the flight deletion example.

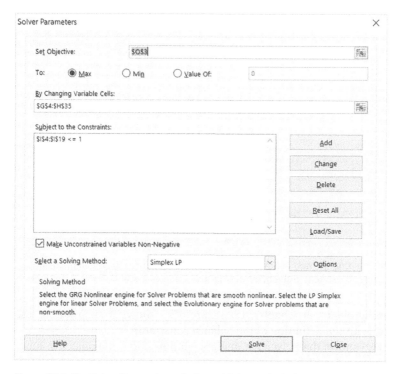

Figure 15.6 The Solver Parameters window with inputs including the objective function, the decision variables, and the coverage constraints for the schedule adjustment problem for the flight deletion example.

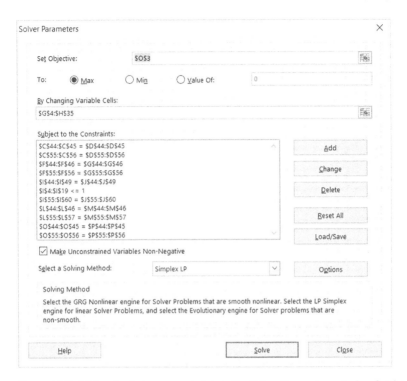

Figure 15.7 Adding the balance constraints in the Solver Parameters window for the schedule adjustment problem for the flight deletion example.

formulation, which are given in columns G and H. The coverage constraints are relaxed as described in the previous chapter as inequality constraints.

Figure 15.7 shows another snapshot of the Solver Parameters window after adding the balance constraints. Next, the resources constraints are added. In this example, it is assumed that there are only two aircraft of fleet type $e1$ and two aircraft of fleet type $e2$. Thus, the values in cell G38 and H38 should be less than or equal to 2. Figure 15.8 shows a snapshot of the Solver Parameters window with the addition of the resources constraints.

Finally, constraints are added to define the binary and integer decision variables. In addition, nonnegativity constraints are added to prevent assigning number of aircraft remaining on the ground a negative value. Figure 15.9 shows a snapshot of the Solver Parameters window after adding all constraints.

Figure 15.10 shows a snapshot of the Excel sheet of the schedule adjustment problem after obtaining the optimal solution. As shown in the figure, the value of the optimal objective function is $37,312. The solution eliminates flights 11, 12, 15, and 16 from the schedule. The right-hand side of the coverage constraint is equal to 0 for these four flights. Finally, Figure 15.11 shows the schedule after eliminating these flights.

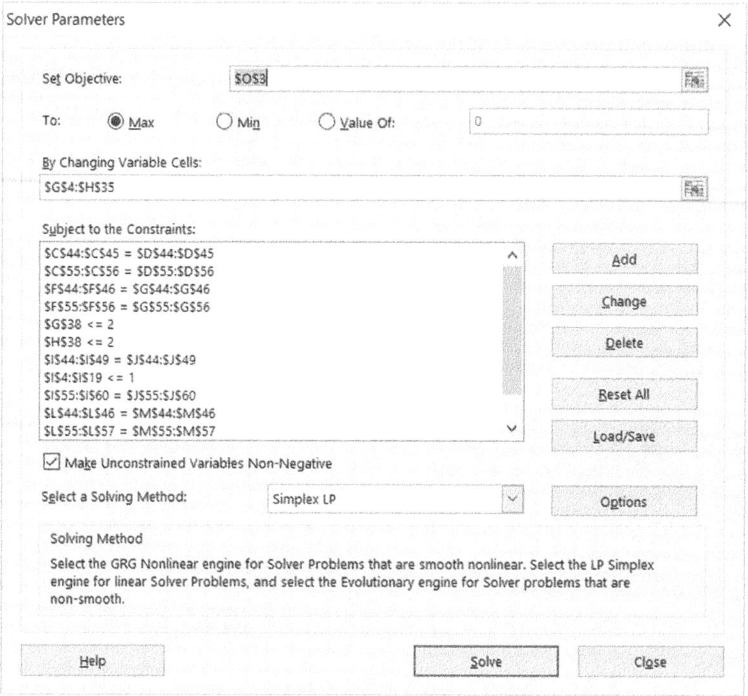

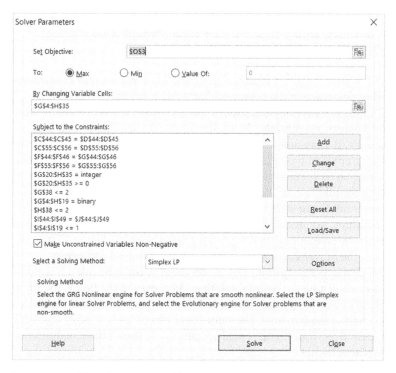

Figure 15.8 Adding the resources constraints in the Solver Parameters window for the schedule adjustment problem for the flight deletion example.

Figure 15.9 Adding the integer and binary constraints in the Solver Parameters window for the schedule adjustment problem for the flight deletion example.

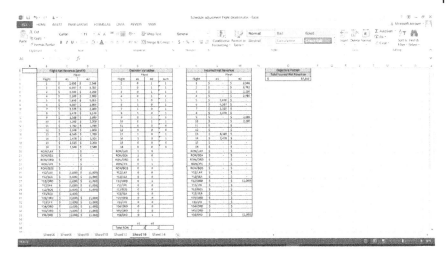

Figure 15.10 The schedule adjustment problem for the flight deletion example.

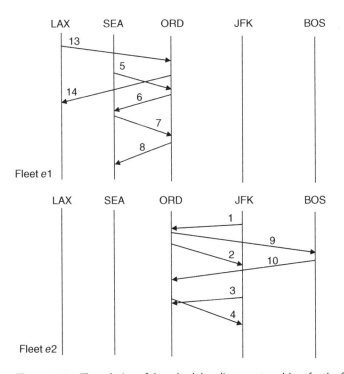

Figure 15.11 The solution of the schedule adjustment problem for the flight deletion example.

15.2 Flight Addition

To investigate the feasibility of adding new flights to the schedule, a methodology similar to the flight deletion methodology presented above is adopted. The solution methodology consists of two main steps. All newly proposed flights are first included in the schedule (initial set). Then, the flight deletion methodology is applied to check whether these proposed flights should be kept or eliminated from the schedule. The only difference is that the coverage constraints are relaxed (i.e. with inequality) only for the proposed flights. All flights that will certainly remain in the new schedule are assigned equality coverage constraints. For example, consider the hypothetical airline presented in Chapters 12 and 13, which operates 16 flights with a hub at ORD. Assume this airline is investigating on adding new service to another spoke such as Washington Dulles International Airport (IAD), as shown in Figure 15.12. The new proposed two-way route is shown with dashed lines connecting ORD to IAD. Assume that this hypothetical airline is proposing two inbound flight and two outbound flights between ORD and IAD. Thus, four flights numbered 17, 18, 19, and 20 are added to the schedule. The proposed schedule of these flights is given in Figure 15.13. The four flights are shown by dashed lines, where flights 17 and 19 originate at ORD and flights 18 and 20 originate at IAD. Assume the airline will continue to schedule flights 1 through 16 as they reported to be highly profitable flights. Thus, the question is how feasible are flights 17, 18, 19, and 20 to include in the schedule.

The problem is formulated as presented in Chapter 14. First, the interconnection nodes are defined considering the four proposed flights. Figures 15.14 and 15.15 show the different interconnection nodes at the different stations for fleet type $e1$ and fleet type $e2$, respectively. The addition of the four proposed flights resulted in one additional interconnection node at ORD and two interconnection nodes at IAD. The Excel sheet is prepared by adding the decision

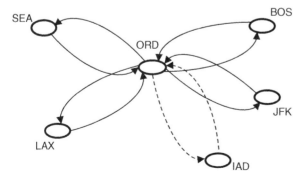

Figure 15.12 The proposed routes for the airline with additional destination at IAD.

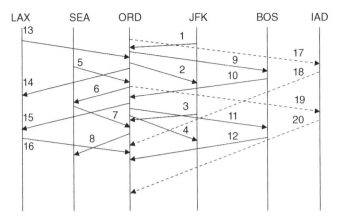

Figure 15.13 The proposed schedule considering all 20 flights.

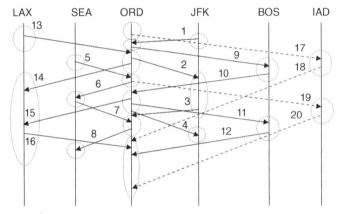

Fleet type *e*1

Figure 15.14 The interconnection nodes for the schedule when operated by fleet type *e*1 for the flight addition problem.

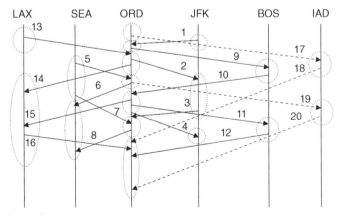

Fleet type *e*2

Figure 15.15 The interconnection nodes for the schedule when operated by fleet type *e*2 for the flight addition problem.

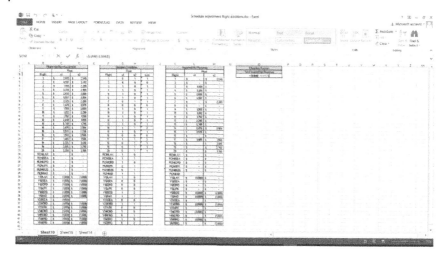

Figure 15.16 A snapshot of the Excel sheet representing the decision variables and the net revenue of the flight addition problem.

variables for flights 17, 18, 19, and 20. In addition, additional variables that represent number of aircraft on the ground between the different interconnection nodes are added. Finally, additional decision variables that represent aircraft remaining overnight at IAD are considered. A snapshot of the Excel sheet that shows all decision variables is given in Figure 15.16. Based on the assumed initial values of the decision variables, the value of the objective function is $34,410 as given in cell O3. Figure 15.17 gives a snapshot of the Excel sheet showing the values of the left-hand and right-hand sides of the balance constraints considering the initial values given for the decision variables.

Figure 15.18 shows a snapshot of the Solver Parameters window. As shown in this figure, the objective function in cell O3 is maximized. In addition, the decision variables are given in columns G and H. The Solver Parameters window also shows the coverage constraints. The flights are divided into two sets. The first set (flights 1 through 16) includes flights that the airline already decided to consider in the new schedule. The second set (flights 17 through 20) includes the newly proposed flights. The coverage constraints of the first set of flights are in the form of equality constraints, while the coverage constraints of the second set of flights are in the form of inequality constraints.

Figure 15.19 shows another snapshot of the Solver Parameters window after adding all constraints except for the resources availability constraints. Not including the resources availability constraints replicates a scenario in which the airline is assumed to have no limitations on resources. Figure 15.20 gives a snapshot of the solution. The model recommends serving all 20 flights yielding a net revenue value of $45,518. Serving all flights in the new schedule requires

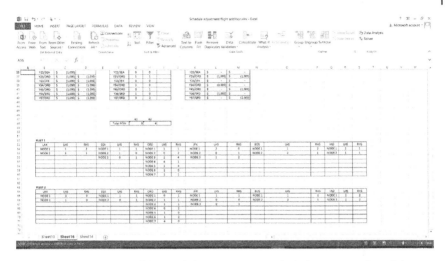

Figure 15.17 A snapshot of the Excel sheet representing the balance constraints of the flight addition problem.

Figure 15.18 The Solver Parameters window for the flight addition problem with the objective function, decision variables, and coverage constraints.

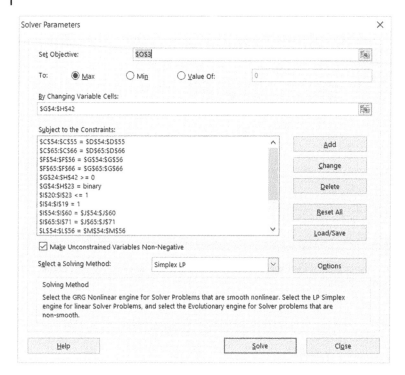

Figure 15.19 The Solver Parameters window for the flight addition problem with all constraints except resources constraints.

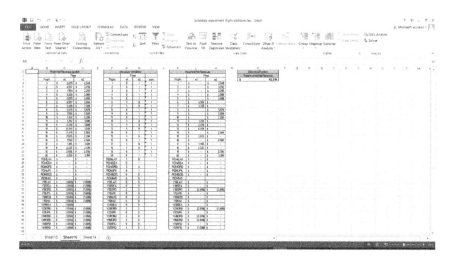

Figure 15.20 A snapshot of the optimal solution of the flight addition problem with no resources constraints.

four aircraft of fleet type $e1$ (one aircraft at LAX and three aircraft at ORD) and another four aircraft of fleet type $e2$ (one aircraft at SEA, two aircraft at ORD, and one at JFK).

Now, consider this airline has limited number of aircraft. The airline owns only three aircraft of fleet type $e1$ and four aircraft of fleet type $e2$. The resources constraints are added in the Solver Parameters window as shown in Figure 15.21. As shown in the figure, cell G48 and cell H48, which define the total number of aircraft to remain overnight at the different stations for fleet type $e1$ and fleet type $e2$, are constrained to be less than three and four, respectively.

Figure 15.22 shows the balance and resources availability constraints of the solution. Figure 15.23 gives a snapshot of the Excel sheet of the solution after adding the resources availability constraints. As shown in the figure, the optimal solution suggests adding only flights 17 and 20 to the schedule. Both flights are operated using fleet type $e2$. The corresponding total net revenue is reduced to $42,153. The solution suggests that flights 5, 6, 7, 8, 11, 12, 13, and 15 are assigned to fleet type $e1$, while flights 1, 2, 3, 4, 9, 10, 14, 16, 17, and 20 are assigned to fleet type $e2$. Figures 15.24 and 15.25 show a graphical

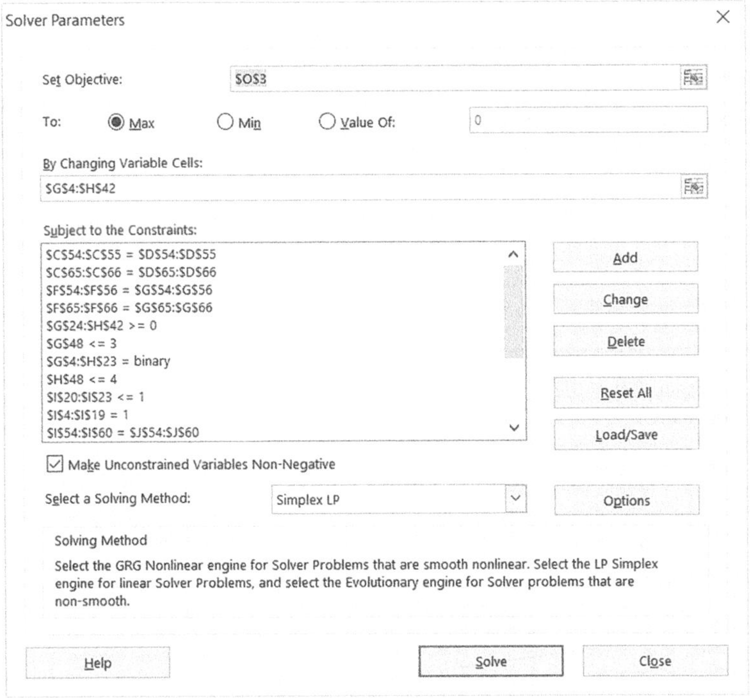

Figure 15.21 The Solver Parameters window for the flight addition problem with all constraints including resources constraints.

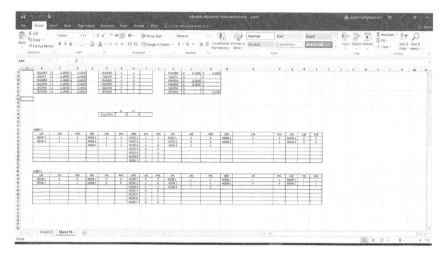

Figure 15.22 The balance and resources constraints for the flight addition problem.

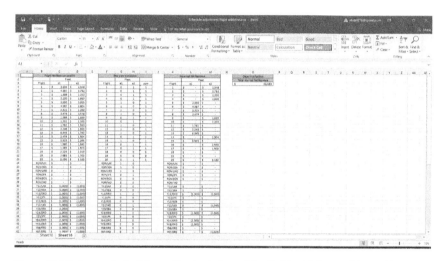

Figure 15.23 A snapshot of the optimal solution of the flight addition problem with resources constraints.

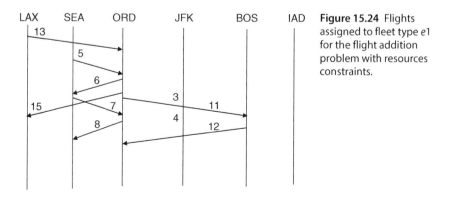

Figure 15.24 Flights assigned to fleet type e1 for the flight addition problem with resources constraints.

Figure 15.25 Flights assigned to fleet type *e*2 for the flight addition problem with resources constraints.

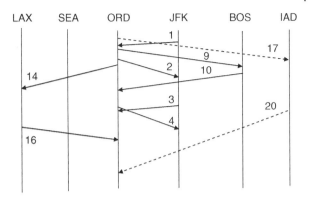

representation of the fleet assignment solution for fleet type *e*1 and fleet type *e*2, respectively. As illustrated in Figure 15.25, flights 17 and 20 are included in the schedule assigned to fleet type *e*2.

Several limitations are identified for adopting this solution approach for the schedule adjustment problem. For example, the approach requires the availability of an initial solution, and this initial solution has to be optimal on its own. In addition, the problem size typically increases as the number of proposed flights increases, which makes the problem to be computationally intractable. Furthermore, the problem does not guarantee that the initially selected flights are optimal. There could be other flights that yield higher revenue than the revenue associated with the selected flights. However, as these flights are not proposed in the initial set, the solution methodology has not identified any of them.

15.3 Flight Departure Time

As part of the schedule adjustment process, an airline might consider modifying the departure time of one or more of its existing flights. The flight departure time might be adjusted to capture more demand, take advantage of a freed slot, create more efficient aircraft rotation and crew schedules, etc. For any flight in the schedule, several departure time options might be suggested, and only one of them is selected. To elaborate on the departure time adjustment, consider the example of the hypothetical airline presented in Chapter 12. Assume that the airline is proposing to adjust the departure time of flight 16 from LAX to ORD. Four departure times are proposed for this flight in addition to the original one. Figure 15.26 shows the proposed five departure times for this flight, which are shown by dashed lines. As shown in the figure, five different copies are considered for flight 16. The objective is to decide which of these five flights to remain in the schedule.

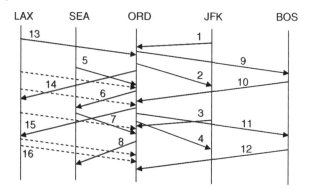

Figure 15.26 The proposed five departure times of flight 16 between LAX and ORD.

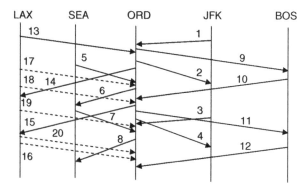

Figure 15.27 Flight numbers for the proposed copies of flight 16 between LAX and ORD.

The problem can be solved based on the schedule adjustment formulation presented in Chapter 14. For this purpose, the different copies of flight 16 are assigned unique identification numbers. As shown in Figure 15.27, the four proposed copies of flight 16 are numbered as 17, 18, 19, and 20. Then, the schedule adjustment program is used to solve the problem with the appropriate flight coverage constraints. The coverage constraints of flights 1 through 15 are in the form of equality constraints. The coverage constraint of flight 16 is adjusted to ensure that only one of the flights 16 through 20 (i.e. one flight out of all copies of flight 16) is included in the schedule. This constraint is as follows:

$$x_{16e1} + x_{17e1} + x_{18e1} + x_{19e1} + x_{20e1} + x_{16e2} + x_{17e2} + x_{18e2} + x_{19e2} + x_{20e2} = 1$$

To solve the problem, the interconnection nodes are defined after adding the new proposed flights. Figures 15.28 and 15.29 show the different interconnection nodes at the different stations for fleet type $e1$ and fleet type $e2$, respectively.

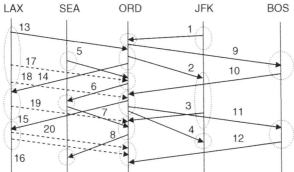

Figure 15.28 The interconnection nodes for the schedule if operated by fleet type e1 for the flight departure time adjustment problem.

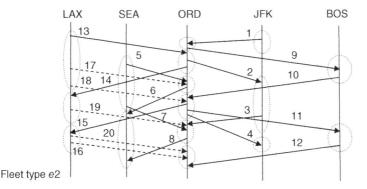

Figure 15.29 The interconnection nodes for the schedule if operated by fleet type e2 for the flight departure time adjustment problem.

Next, the Excel sheet that defines the decision variables is created. A snapshot of the Excel sheet that shows all decision variables is given in Figure 15.30. The Excel sheet is prepared by adding the decision variables for flight 17, 18, 19, and 20. In addition, the additional variables that represent the number of aircraft on the ground between the different interconnection nodes are added. Based on the assumed initial values of the decision variables, the value of the objective function is $43,799. Figure 15.31 gives a snapshot of the Excel sheet showing the values of the left-hand side and the right-hand side of the balance constraints for all interconnection nodes. Cell I49 gives the sum of the decision variables $x_{16\,e1}$, $x_{17\,e1}$, $x_{18\,e1}$, $x_{19\,e1}$, $x_{20\,e1}$, $x_{16\,e2}$, $x_{17\,e2}$, $x_{18\,e2}$, $x_{19\,e2}$, and $x_{20\,e2}$. It is used to represent the coverage constraint for flight 16 and its copies.

Figure 15.32 shows the Solver Parameters window. The objective function representing the total net revenue given in cell O3 is maximized. In addition, the decision variables are given in columns G and H. The Solver Parameters window is also showing the coverage constraints. As shown in the figure, the

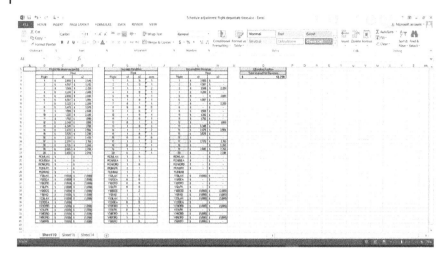

Figure 15.30 A snapshot of Excel sheet that shows the initial values of the decision variables for the flight departure time adjustment problem.

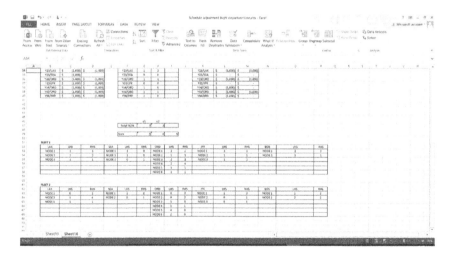

Figure 15.31 A snapshot of the Excel sheet that shows the balance constraints for the flight departure time adjustment problem.

flights are divided into two groups. The first group includes flights (flights 1–15) that the airline is certain about their departure time. The coverage constraints of these flights are in the form of equality constraints. The second group includes flight(s) that their departure time is not confirmed yet (i.e. flight 16). The coverage constraint of this flight is in the form of equality constraint that includes all copies of the flight considering the two fleet types (cell I49).

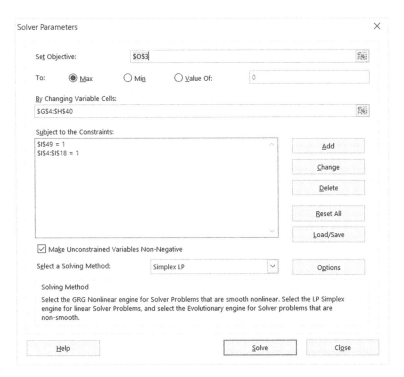

Figure 15.32 The Solver Parameters window for the flight departure time adjustment problem with the objective function and the coverage constraints.

Figure 15.33 shows another snapshot of the Solver Parameters window, when all problem constraints are added, except the resources availability constraints. These model settings assume a scenario in which the airline has no limitation on resources. Figure 15.34 shows the optimal solution in the Excel sheet. The solution gives the best departure time for flight 16, which is the one represented by flight 18. The solution indicates that this flight is assigned to fleet type $e1$. The optimal net revenue value is $43,236. The solution requires five aircraft of fleet type $e1$ (two aircraft at LAX, one aircraft at SEA, and two aircraft at ORD) and one aircraft of fleet type $e2$ (one aircraft at JFK). The solution assigns flights 2, 3, 5, 6, 7, 8, 11, 12, 13, 14, 15, and 16 to fleet type $e1$, while flights 1, 4, 9, and 10 are assigned to fleet type $e2$. Figures 15.35 and 15.36 show a graphical representation of the fleet assignment solution for fleet type $e1$ and fleet type $e2$, respectively. The new departure time for flight 16 can be noted in Figure 15.35.

Similar to the flight addition problem presented above, the main limitations of this version of the schedule adjustment problem is that an initial solution has to be available and this initial solution has to be optimal on its own. Also,

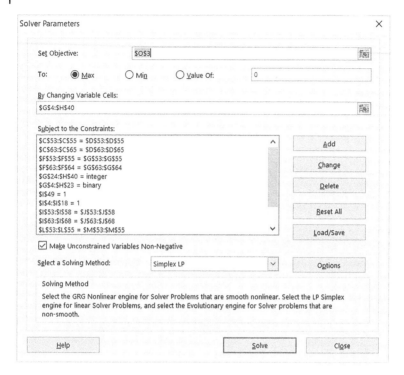

Figure 15.33 The Solver Parameters window for the flight departure time adjustment problem with all constraints.

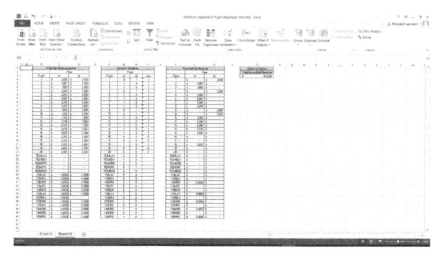

Figure 15.34 A snapshot of the Excel sheet for the flight departure time adjustment problem with the values of the optimal solution.

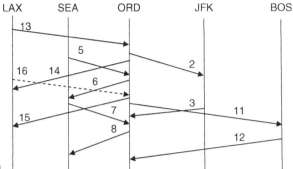

Figure 15.35 Flights assigned to fleet type e1 for the flight departure time adjustment problem.

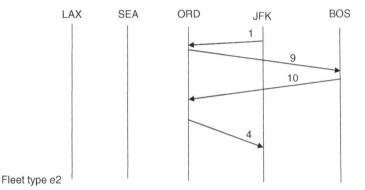

Figure 15.36 Flights assigned to fleet type e2 for the flight departure time adjustment problem.

the problem size rapidly increases with the increase in number of flights to be adjusted and number of departure times (copies) suggested for each flight. The large problem size increases the problem computation intractability. Furthermore, the problem does not guarantee that the new departure times assigned to the flights are optimal as there could be other departure times that give a better solution and not considered.

Section 5

16

Itinerary-based Fleet Assignment Model (IFAM)

16.1 Introduction

As explained in Chapter 11, the main objective of the fleet assignment model (FAM) is to maximize the net revenue of the airline. The net revenue is defined as the sum over all flights of the difference between the unconstrained revenue and the cost of the flight. The unconstrained revenue is assumed to be constant, which is the maximum possible revenue for a particular flight regardless of the assigned fleet capacity. The cost depends on the assigned fleet type. It includes the flight operating cost and the spill cost. The operating cost incorporates fuel cost, crew cost, landing fees, etc. The spill cost of a flight is the revenue lost in case the assigned aircraft (fleet type) cannot accommodate all passengers. The spill cost is incurred only when the passenger demand of the flight is more than the seat capacity of the assigned fleet.

The basic FAM is a flight-based (or leg-based) model. It assumes that both the demand and the revenue of a flight are independent of all other flights in the schedule. As explained earlier, for airlines that adopt the hub-and-spoke network structure, the demand of a flight is a mix of local and connecting passengers. Local passengers are traveling from the origin city of the flight to its destination city. Connecting passengers use the origin or the destination airport (or both) of the flight as a transfer point(s) to continue their travel to their final destinations. Passenger demand is usually predicted at the origin–destination (OD) level (i.e. the itinerary level). Thus, the basic FAM requires converting the itinerary-based demand (i.e. connecting demand) into its corresponding flight-based demand. Similarly, fares are paid at the itinerary level, and consequently the fare should also be prorated from an itinerary to a flight.

As mentioned earlier, in the FAM, if the capacity of the assigned fleet is less than the flight's estimated demand, the excessive demand is assumed to spill off the flight and lost to other airlines or to other travel modes (driving, rail, etc.). The basic FAM ignores the possibility that this spilled demand could be

Airline Network Planning and Scheduling, First Edition. Ahmed Abdelghany and Khaled Abdelghany.
© 2019 John Wiley & Sons, Inc. Published 2019 by John Wiley & Sons, Inc.

recaptured by other flights operated by the same airline. In this case, the fleet type assigned to a flight could affect the demand of other flights.

Calculating the demand spill cost requires information on which passengers are spilled from each flight and what fare they would have paid. It also requires information on whether this spilled demand (or part of it) is recaptured on other itineraries. Most major airlines adopt advanced revenue management (RM) systems that control seat availability on the different flights. RM allocates the seats of a flight among all itinerary–fare classes that this flight is serving (de Boer et al. 2002). An itinerary–fare class is a booking class defined in terms of an itinerary, a set of ticket restrictions, and a fare. When demand exceeds the capacity of the flight, RM systems give airlines control on deciding which passengers to accommodate and which passengers to spill.

Without accurate representation of the spill-and-recapture mechanism, the solution quality of the fleet assignment problem could be seriously impacted. In this chapter, the itinerary-based fleet assignment model (IFAM) that considers itinerary-based demand with spill-and-recapture mechanism is presented. The presentation of this chapter is based on the work of Barnhart et al. (2002) and earlier work by Farkas (1996). The next two sections explain in more details the spill-and-recapture mechanism and its impact on the demand of the flights.

16.2 Spill Cost Estimates and Network Effect

To illustrate the impact of flight dependencies on spill cost estimation, we follow an example similar to the one presented by Barnhart et al. (2002). In this example, a hypothetical airline with only two flights (F1 and F2) is assumed as shown in Figure 16.1. Flight F1 serves in city-pair X–Y and flight F2 serves city-pair Y–Z. Accordingly, three city-pairs (markets) are served by this airline, which are X–Y, Y–Z, and X–Z through Y. Table 16.1 gives the itineraries, demand, and fares of this hypothetical airline. The demand of flight F1 is equal to 150 passengers, which includes local passengers of market X–Y (75 passengers) and connecting passengers in market X–Z (75 passengers). Similarly, the demand of flight F2 is equal to 225 passengers, which includes local passengers in market Y–Z (150 passengers) and connecting passengers in market X–Z (75 passengers).

Assume these two flights (F1 and F2) are assigned a fleet type (aircraft) with a capacity of 100 seats. Clearly, this seating capacity is less than the passenger demand for both flights. Thus, some passengers have to be spilled from both flights. There are 50 passengers to be spilled from flight F1 and 125 passengers to be spilled from flight F2.

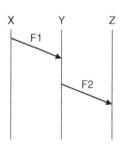

Figure 16.1 A hypothetical airline with two flights.

Table 16.1 Markets and itineraries served by a hypothetical airline.

Market	Itinerary (sequence of flights)	Demand (number of passengers)	Average fare ($)
X–Y	F1	75	200
Y–Z	F2	150	225
X–Z	F1–F2	75	300

The methodology by which passengers are spilled is crucial for determining the spill cost and consequently the fleet assignment solution. To illustrate, consider the following two different demand spilling strategies:

Spilling strategy I: In this strategy, each flight is considered independently while ignoring the effects of its demand spilling to other flights in the network. This strategy is identical to what is adopted in the basic FAM. Available seats are filled with passengers willing to pay high fares, and spilled passengers are the ones searching for low fares. Assume the lowest fare is $200 for the X–Y market. Thus, 50 passengers traveling in the X–Y market are spilled at a fare of $200. The next higher fare is $225 and is charged in the Y–Z market. Thus, another 125 passengers are spilled from that market at a fare of $225. This strategy results in a total spill cost of $38,125 (50 × $200 + 125 × $225 = $38,125).

Spilling strategy II: In terms of used seats, one should notice that spilling one connecting passenger traveling in the market X–Z is equivalent to spilling two nonconnecting passengers (one passenger from each of the two local markets X–Y and Y–Z). In addition, spilling one connecting passenger in the market X–Z has a spill cost that is less than the cost of spilling the two nonconnecting passengers. The cost of spilling the connecting passenger in the market X–Z is $300, while the cost of spilling the two nonconnecting passengers in the market X–Y and Y–Z is $425 (the sum of $200 for market X–Y and $225 for market Y–Z). Accordingly, to minimize the total spilling cost, consider the following strategy. First, spill 50 passengers from the demand traveling in the connecting market X–Z. The remaining demand on flight F1 no longer exceeds capacity. Another 75 passengers (125 minus 50) are still needed to be spilled from flight F2. This extra demand on flight F2 is spilled from nonconnecting passengers traveling in the market Y–Z because it has less spilling cost than that of market X–Z. Thus, the total cost of this strategy is $31,875 (50 × $300 + 75 × $225 = $31,875).

This example illustrates the effect of the strategy used to spill passengers on the spilling cost. The example shows that using a strategy that considers the network effect and interdependencies between connecting and nonconnecting itineraries could reduce the spilling cost.

Next, we demonstrate how the spill cost is represented mathematically. To simplify the presentation, and without loss of generality, we consider an airline that serves one city-pair. As such, an implicit assumption is made here that the demand of a market is independent of the demand of other markets. Consider the set of itineraries P scheduled in this city-pair. This set includes a dummy itinerary to represent all competitors' itineraries. This dummy itinerary recaptures the demand that cannot be served by the airline. The unconstrained demand (number of passengers requesting travel on the itinerary) and the average fare of itinerary p are denoted as D_p and $fare_p$, respectively. Define t_p as the number of passengers to be spilled from itinerary p. The spill cost S (or lost revenue due to spill) is the sum of the spilled revenue from all of its itineraries as follows:

$$S = \sum_{p \in P} t_p . \text{fare}_p$$

It should also be noted that the demand spilled from the itinerary cannot exceed the unconstrained demand of the itinerary. Thus, for each itinerary $p \in P$, the following constraint should be satisfied:

$$0 \leq t_p \leq D_p$$

Also, the total unconstrained airline revenue R can be computed as follows:

$$R = \sum_{p \in P} D_p . \text{fare}_p$$

16.3 Demand Recapture

As mentioned above, when seat capacity allocated to an itinerary is less than its unconstrained demand D_p, some of this demand has to be spilled. This spilled demand t_p represents the difference between the unconstrained demand and the allocated seat capacity of the itinerary. Typically, some of this spilled demand is recaptured by other itineraries offered by the airline in the same city-pair. The remaining of the spilled demand that is not recaptured is assumed to be lost to other airlines or other transportation modes. To further elaborate on demand recapture, consider t_p^r as the number of passengers requesting itinerary p that the airline attempts to redirect to itinerary r due to limited capacity on itinerary p. For each itinerary $p \in P$, the demand that is spilled from itinerary p and recaptured on itinerary r should satisfy the following constraints:

$$\sum_{r \in P} t_p^r = t_p$$

$$\sum_{r \in P} t_p^r \le D_p$$

$$t_p^r \ge 0$$

The first constraint indicates that the total demand spilled from itinerary $p \in P$ is equal to the sum of the demand spilled from itinerary p and recaptured on equivalent itinerary r considering all available equivalent itineraries. The second constraint indicates that the total demand spilled from itinerary $p \in P$ is less than its unconstrained demand D_p. Finally, the third constraint defines the recaptured demand as a nonnegative value.

In some cases, itinerary $r \in P$ cannot accommodate the entire (or a specified portion) demand spilled from itinerary $p \in P$. The variable b_p^r is defined as the recapture rate from itinerary p to itinerary r. This rate is the fraction of passengers spilled from itinerary p and redirected to itinerary r. Thus, the number of passengers that is actually spilled from itinerary p and recaptured on itinerary r is $b_p^r t_p^r$ with revenue $b_p^r t_p^r$ fare$_r$, where fare$_r$ is the fare value of itinerary r. Accordingly, the total recaptured revenue M from recapturing passengers can be calculated as follows:

$$M = \sum_{r \in P} \sum_{p \in P} b_p^r t_p^r \, \text{fare}_r$$

Consider the airline network example given in Figure 16.2. This airline serves one city-pair with four itineraries (itineraries 1–4). A dummy itinerary is considered to represent all other travel options (shown as dotted line). The seat capacity, average fare, and unconstrained demand of each itinerary are as shown in the figure.

The total unconstrained revenue R is calculated as the product of the unconstrained demand and the average fare and summed up for the four itineraries:

$$R = 220 \times \$250 + 150 \times \$270 + 165 \times \$280 + 227 \times \$240 = \$196,180$$

The unconstrained demands for itineraries 1 and 4 are more than their seat capacity, and accordingly itinerary 1 has to spill 40 passengers, while itinerary 4 has to spill 47 passengers. Thus, the spill cost S is calculated as follows:

$$S = 40 \times \$250 + 47 \times \$240 = \$21,280$$

Calculating the recaptured revenue M is less straightforward as it requires determining how the spilled demand is recaptured and distributed among the different itineraries that can accommodate this spilled demand. Itinerary 2 has 30 available seats (180 minus 150), and itinerary 3 has 15 available seats (180 minus 165). Thus, itineraries 2 and 3 can recapture 45 of the spilled demand. In this example, the total demand spilled from itineraries 1 and 4 is 87 passengers, which is greater than the available seats on itineraries 2 and 3

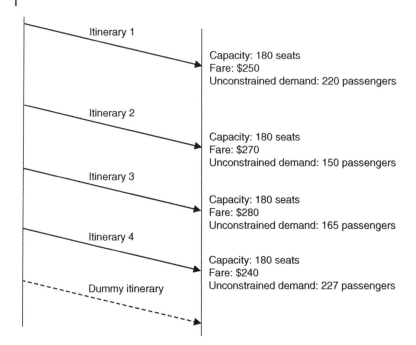

Figure 16.2 Example used to show the calculations of the unconstrained revenue, the spill cost, and the recaptured revenue.

(45 seats). Accordingly, some of the passengers are expected to be accommodated by other airlines (i.e. the dummy itinerary). Up to this point of the discussion, it is not clear yet how the 87 spilled passengers of itineraries 1 and 4 are distributed among itineraries 2, 3, and the dummy itinerary.

A spill-and-recapture mechanism has to be used to reassign the excess demand among the available itineraries. The recaptured revenue can be represented mathematically for this example as follows:

$$M = b_1^2 \times t_1^2 \times \$270 + b_4^2 \times t_4^2 \times \$270 + b_1^3 \times t_1^3 \times \$280 + b_4^3 \times t_4^3 \times \$280$$

where the first and second terms represent the revenue recaptured on itinerary 2 and the third and fourth terms represent the revenue recaptured on itinerary 3.

Now, assume that the airline assigns 21 excess passengers that requested booking on itineraries 1 to itinerary 2 (i.e. $t_1^2 = 21$) and assigns another 19 excess passengers that requested booking on itineraries 1 to itinerary 3 (i.e. $t_1^3 = 19$). Likewise, the airline assigns 24 excess passengers that requested bookings on itineraries 4 to itinerary 2 (i.e. $t_4^2 = 24$) and assigns another 23 excess passengers that requested bookings on itineraries 4 to itinerary 3 (i.e. $t_4^3 = 23$).

The spill and recapture depends on the recapture rate (b_p^r) between itineraries. As mentioned above, the recapture rate represents the fraction of passengers spilled from itinerary p and recaptured on itinerary r. Assume the recapture rates b_1^2, b_4^2, b_1^3, and b_4^3 have the values of 0.809, 0.541, 0.158, and 0.522, respectively. According to these recapture rates, itinerary 2 recaptures 17 passengers (i.e. 0.809×21) from itinerary 1 and 13 passengers (0.541×24) from itinerary 4. Similarly, itinerary 3 recaptures 3 passengers (0.158×19) from the demand spilled from itinerary 1 and 12 passengers (0.522×23) from itinerary 4. The dummy itinerary recaptures the remaining spilled passengers, which are 20 passengers from itinerary 1 and 22 passengers from itinerary 4. As such, the recaptured revenue M can be calculated, as shown below. Out of the spill cost of \$21,280, the airline can recapture \$12,300:

$$M = (0.809) \times 21 \times \$270 + (0.541) \times 24 \times \$270 + (0.158) \times 19 \times \$280$$
$$+ (0.522) \times 23 \times \$280 = \$12,300$$

In this example, it is assumed that the airline is successful in redirecting the spilled passengers to fill all available seats on itineraries that have excess seat capacity (i.e. itineraries 2 and 3). However, in other scenarios, spilled passengers might choose to use other airlines. If more passengers choose other airlines, the airline might fail to fill all available seats on some of its itineraries using the recaptured demand.

16.4 The Flight–Itinerary Interaction

An itinerary serving a city-pair might consist of one or more flights (legs). The origin of the first leg is the origin of the itinerary, and the destination of the final leg is the destination of the itinerary. To capture the interaction between the itinerary demand and the seat capacity of its flights, an itinerary–flight incidence parameter δ_f^p is defined, which is equal to one if flight $f \in F$ is part of itinerary $p \in P$, and zero otherwise. For example, Figure 16.3 shows nine hypothetical itineraries serving the city-pairs A–B, B–C, and A–C. The city-pair A–C is served with several connecting (single-stop) itineraries through station B. Flight $f1$ is part of three different itineraries. The first itinerary $p1$ consists of one flight $f1$. The second itinerary $p2$ consists of flights $f1$ and $f2$. The third itinerary $p3$ consists of flights $f1$ and $f3$. Accordingly, δ_{f1}^{p1}, δ_{f1}^{p2}, and δ_{f1}^{p3} are equal to 1.

Given the unconstrained demand on all itineraries, the unconstrained demand Q_f on a flight $f \in F$ can be determined using the unconstrained demand D_p of all itineraries constructed using this flight as follows:

$$Q_f = \sum_{p \in P} \delta_f^p D_p$$

In other words, the unconstrained demand Q_{f1} of flight $f1$ is equal to the sum of the unconstrained demand of itineraries $p1$, $p2$, and $p3$, which incorporate this flight. For flight $f1$, which is part of itineraries $p1$, $p2$, and $p3$, the unconstrained demand Q_{f1} is calculated as follows:

$$Q_{f1} = D_{p1} + D_{p2} + D_{p3}$$

Now consider the variable SEAT$_f$ to represent the seat capacity of flight $f \in F$. If the flight's unconstrained demand Q_f is greater than its seat capacity SEAT$_f$, the difference between Q_f and SEAT$_f$ is the number of passengers that has to be spilled of flight f.

The spilled demand from any flight is equal to the sum of the spilled demand of all itineraries that include this flight. For flight $f \in F$, the demand spilled of this flight can be computed as follows:

$$\text{Spilled demand of flight } f = \sum_{r \in P} \sum_{p \in P} \delta_f^p \, t_p^r$$

For example, the spilled demand of flight $f1$ is equal to the demand spilled from itineraries $p1$, $p2$, and $p3$. As shown in Figure 16.3, the spilled demand

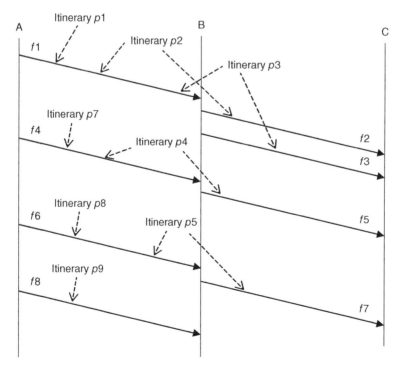

Figure 16.3 A hypothetical airline with several connecting itineraries.

from itinerary $p1$ is directed to itineraries $p7$, $p8$, and $p9$. These recaptured demands are represented as t_{p1}^{p7}, t_{p1}^{p8}, and t_{p1}^{p9}, respectively. Similarly, the spilled demand from itinerary $p2$ is directed to itineraries $p3$, $p4$, and $p5$, which are represented as t_{p2}^{p3}, t_{p2}^{p4}, and t_{p2}^{p5}, respectively. Finally, the spilled demand from itinerary $p3$ is directed to itineraries $p2$, $p4$, and $p5$ and are represented as t_{p3}^{p2}, t_{p3}^{p4}, and t_{p3}^{p5}, respectively. Hence, the demand spilled from flight $f1$ is calculated as follows:

$$\text{Spilled demand of } f1 = t_{p1}^{p7} + t_{p1}^{p8} + t_{p1}^{p9} + t_{p2}^{p3} + t_{p2}^{p4} + t_{p2}^{p5} + t_{p3}^{p2} + t_{p3}^{p4} + t_{p3}^{p5}$$

Similarly, the recaptured demand on a flight $f \in F$ is equal to the recaptured demand on all itineraries that include this flight. The demand recaptured by flight f can be computed as follows:

$$\text{Recaptured demand } on \text{ flight } f = \sum_{r \in P}\sum_{p \in P} \delta_f^p \, b_r^p \, t_r^p$$

For example, the recaptured demand on flight $f1$ is equal to the demand recaptured on itineraries $p1$, $p2$, and $p3$. As shown in Figure 16.3, the demand recaptured on itinerary $p1$ is spilled from itineraries $p7$, $p8$, and $p9$. These recaptured demand portions are represented as $b_{p7}^{p1} t_{p7}^{p1}$, $b_{p8}^{p1} t_{p8}^{p1}$, and $b_{p9}^{p1} t_{p9}^{p1}$, respectively. Similarly, the demand recaptured on itinerary $p2$ is spilled from itineraries $p3$, $p4$, and $p5$ and are represented as $b_{p3}^{p2} t_{p3}^{p2}$, $b_{p4}^{p2} t_{p4}^{p2}$, and $b_{p5}^{p2} t_{p5}^{p2}$, respectively. Finally, the demand recaptured on itinerary $p3$ is spilled from itineraries $p2$, $p4$, and $p5$ and are represented as $b_{p2}^{p3} t_{p2}^{p3}$, $b_{p4}^{p3} t_{p4}^{p3}$, and $b_{p5}^{p3} t_{p5}^{p3}$, respectively. Thus, the demand recaptured on flight $f1$ is calculated as follows:

$$\text{Recaptured demand } on \text{ flight } f1 = b_{p7}^{p1} t_{p7}^{p1} + b_{p8}^{p1} t_{p8}^{p1} + b_{p9}^{p1} t_{p9}^{p1} + b_{p3}^{p2} t_{p3}^{p2} + b_{p4}^{p2} t_{p4}^{p2}$$
$$+ b_{p5}^{p2} t_{p5}^{p2} + t_{p2}^{p3} t_{p2}^{p3} + b_{p4}^{p3} t_{p4}^{p3} + b_{p5}^{p3} t_{p5}^{p3}$$

For any flight $f \in F$, the difference between the spilled demand of the flight and its recaptured demand should always exceed the number of passengers that must be spilled from the flight ($Q_f - \text{SEAT}_f$). In other words, for any flight, the demand attracted (unconstrained demand and recaptured demand) minus the spilled demand must be less than the seat capacity of the flight. This constraint can be written for flight f as follows:

$$\sum_{r \in P}\sum_{p \in P} \delta_f^p \, t_p^r - \sum_{r \in P}\sum_{p \in P} \delta_f^p \, b_r^p \, t_r^p \geq Q_f - SEAT_f$$

16.5   The Itinerary-based Fleet Assignment Problem

The objective function of the IFAM extends the objection function defined for the basic FAM to include the spill-and-recapture components. In IFAM, the objective of the airline is to maximize $R - S + M - C$, where C is the operating cost of the schedule. Please note the difference between the definition of the term C for the IFAM and its definition for the basic FAM as described in Chapter 11. In FAM, the term C represents the assignment cost, which includes the operating cost and the spill cost. However, in IFAM, the term C represents only the operating cost of the flight and computed as follows:

$$C = \sum_{f \in F} \sum_{e \in E} c_{fe} \cdot x_{fe}$$

where c_{fe} denotes the operating cost associated with assigning flight $f \in F$ to fleet type $e \in E$. The decision variable $x_{fe} \in X$ is equal to one if flight $f \in F$ is assigned to equipment $e \in E$, and zero otherwise. Since R is a fixed number, the objective of the airline is alternatively to minimize $C + (S - M)$.

One should note that the decision variables of the IFAM include the fleet assignment decisions X (as defined in FAM) as well as the associated amount of demand spill and recapture on the different itineraries/flights.

The constraints of IFAM are similar to those of the basic fleet assignment, which include the coverage, the balance, and the resources constraints. Additional constraints are added to ensure that the spilled and recaptured demands are feasible. The next chapter provides a detained numerical example on the IFAM.

17

Example on IFAM

17.1　Problem Definition

To further explain the different components of the itinerary-based fleet assignment model (IFAM) described in the previous chapter, this chapter provides an application of the IFAM considering a small hypothetical airline network. Figure 17.1 shows the flight schedule of this airline, which includes six flights. As shown in the figure, the airline serves three stations: LAX, ORD, and BOS. The six flights are identified as LAX-ORD ($f1$), ORD-BOS ($f2$), LAX-BOS ($f3$), BOS-ORD ($f4$), ORD-LAX ($f5$), and BOS-LAX ($f6$). Thus, city-pairs LAX-ORD, ORD-LAX, ORD-BOS, and BOS-ORD are served by nonstop itineraries, and city-pairs LAX-BOS and BOS-LAX are served by nonstop and connecting itineraries.

Accordingly, the network includes eight different itineraries, which are LAX-ORD, ORD-BOS, LAX-BOS, BOS-ORD, ORD-LAX, BOS-LAX, LAX-ORD-BOS, and BOS-ORD-LAX. The demand and average fare value for the eight itineraries are given in Table 17.1. Assume that each of these six flights can be assigned to fleet types $e1$ or $e2$ with seat capacity of 150 and 250, respectively. These two fleet types are assumed to have the same speed. Thus, the flight's arrival time is independent of its assigned fleet type (i.e. identical time-staggered diagrams and interconnection nodes for the schedule when operated by both fleet types). The operations cost associated with assigning each of the six flights to the two fleet types is given in Table 17.2.

The problem has 12 assignment decision variables, where each of the six flights is possibly to be assigned to one of the two fleet types. Table 17.3 gives the notations of the decision variables. For example, the variable $x_{5,1}$ is equal to 1 if flight 5 is assigned to fleet $e1$, and zero otherwise.

To construct the balance constraints of the problem, the interconnection nodes are defined. Considering the similarities in aircraft speed of the two fleet types, the interconnection nodes are identical for the two fleet types. Figure 17.2 shows the interconnection nodes at the different stations. As shown in the

Airline Network Planning and Scheduling, First Edition. Ahmed Abdelghany and Khaled Abdelghany.
© 2019 John Wiley & Sons, Inc. Published 2019 by John Wiley & Sons, Inc.

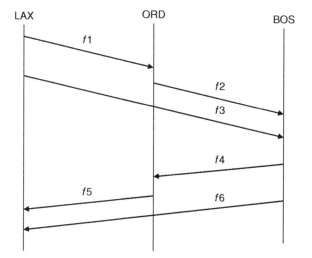

Figure 17.1 An example of a hypothetical airline with six flights for the IFAM problem.

Table 17.1 The passenger demand and fares of each of the eight itineraries for the IFAM problem.

Itinerary	LAX-ORD	ORD-BOS	LAX-BOS	BOS-ORD	ORD-LAX	BOS-LAX	LAX-ORD-BOS	BOS-ORD-LAX
Passenger demand	50	50	300	50	50	300	50	50
Fare ($)	200	225	350	225	200	350	300	300

Table 17.2 The operating cost of each flight for each fleet type for the IFAM problem.

Flight		Fleet	
		e1 ($)	e2 ($)
LAX-ORD	f1	30,000	16,600
ORD-BOS	f2	15,100	13,500
LAX-BOS	f3	15,500	45,000
BOS-ORD	f4	57,000	13,500
ORD-LAX	f5	42,000	16,600
BOS-LAX	f6	15,500	45,000

Table 17.3 The decision variables of fleet assignment for the IFAM example.

Flight	e1	e2
LAX-ORD	$x_{1,1}$	$x_{1,2}$
ORD-BOS	$x_{2,1}$	$x_{2,2}$
LAX-BOS	$x_{3,1}$	$x_{3,2}$
BOS-ORD	$x_{4,1}$	$x_{4,2}$
ORD-LAX	$x_{5,1}$	$x_{5,2}$
BOS-LAX	$x_{6,1}$	$x_{6,2}$

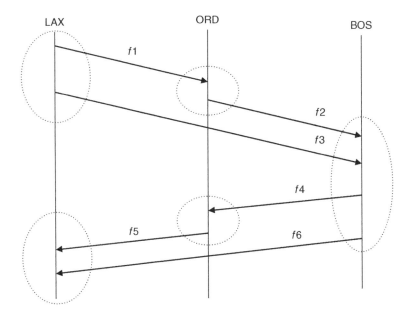

Figure 17.2 The interconnection nodes for the IFAM example.

figure, there are two interconnection nodes at LAX, two interconnection nodes at ORD, and one interconnection node at BOS.

Given these interconnection nodes, Table 17.4 defines the decision variables related to the number of aircraft to remain overnight at the different stations and the number of aircraft to remain on the ground between the different interconnection nodes.

Table 17.4 The decision variables for the ground arcs and the overnight arcs for the IFAM example.

$RON_{LAX\,e1}$	Number of aircraft of type $e1$ that remain overnight at LAX
$RON_{ORD\,e1}$	Number of aircraft of type $e1$ that remain overnight at ORD
$RON_{BOS\,e1}$	Number of aircraft of type $e1$ that remain overnight at BOS
$y_{(1-2)\,LAX\,e1}$	Number of aircraft of type $e1$ that remain on the ground link that connects interconnection nodes 1 and 2 at LAX
$y_{(1-2)\,ORD\,e1}$	Number of aircraft of type $e1$ that remain on the ground link that connects interconnection nodes 1 and 2 at ORD
$RON_{LAX\,e2}$	Number of aircraft of type $e2$ that remain overnight at LAX
$RON_{ORD\,e2}$	Number of aircraft of type $e2$ that remain overnight at ORD
$RON_{BOS\,e2}$	Number of aircraft of type $e2$ that remain overnight at BOS
$y_{(1-2)\,LAX\,e2}$	Number of aircraft of type $e2$ that remain on the ground link that connects interconnection nodes 1 and 2 at LAX
$y_{(1-2)\,ORD\,e2}$	Number of aircraft of type $e2$ that remain on the ground link that connects interconnection nodes 1 and 2 at ORD

17.2 The Constraints of the IFAM Example

The coverage constraints of the problem entail that each flight is assigned to one fleet type. The problem has six coverage constraints as follows:

$$x_{1,1} + x_{1,2} = 1$$

$$x_{2,1} + x_{2,2} = 1$$

$$x_{3,1} + x_{3,2} = 1$$

$$x_{4,1} + x_{4,2} = 1$$

$$x_{5,1} + x_{5,2} = 1$$

$$x_{6,1} + x_{6,2} = 1$$

The problem has ten balance constraints that are defined as follows:

$$RON_{LAX\,e1} = y_{(1-2)LAX\,e1} + x_{1,1} + x_{3,1}$$

$$y_{(1-2)LAX\,e1} + x_{5,1} + x_{6,1} = RON_{LAX\,e1}$$

$$RON_{LAX\,e2} = y_{(1-2)LAX\,e2} + x_{1,2} + x_{3,2}$$

$$y_{(1-2)LAX\,e2} + x_{5,2} + x_{6,2} = RON_{LAX\,e2}$$

$$RON_{ORD\,e1} + x_{1,1} = y_{(1-2)ORD\,e1} + x_{2,1}$$

$$y_{(1-2)ORD\,e1} + x_{4,1} = RON_{ORD\,e1} + x_{5,1}$$

$$RON_{ORD\,e2} + x_{1,2} = y_{(1-2)ORD\,e2} + x_{2,2}$$

$$y_{(1-2)ORD\,e2} + x_{4,2} = RON_{ORD\,e2} + x_{5,2}$$

$$RON_{BOS\,e1} + x_{2,1} + x_{3,1} = RON_{BOS\,e1} + x_{4,1} + x_{6,1}$$

$$RON_{BOS\,e2} + x_{2,2} + x_{3,2} = RON_{BOS\,e2} + x_{4,2} + x_{6,2}$$

For simplicity, we assume that there is no limitation on the available fleet resources. In addition, maintenance and crew constraints are ignored.

17.3 The Objective Function

As mentioned earlier, the objective of the problem is to minimize the sum of operating cost and the spill cost minus the recapture revenue (i.e. $C + S - M$). To explain the different components of the objective function, consider the snapshots of the Excel sheet shown in Figures 17.3 and 17.4. This Excel sheet gives the solution of the itinerary-based fleet assignment problem for the hypothetical airline presented above. The spreadsheet in Figure 17.3 gives the formula representation of the cells, while the spreadsheet in Figure 17.4 gives the corresponding values. Column B (rows 4–9) gives the list of flights in the network. Each flight is defined by its origin–destination pair.

Columns C and D give the operating cost associated with assigning each of the six flights to fleet types $e1$ and $e2$, respectively. For example, it costs the airline \$15,500 to operate flight LAX-BOS using fleet type $e1$ and \$45,000 to operate the same flight using fleet type $e2$. Column B (rows 10–12) gives the remaining-overnight (RON) arcs for the three stations: LAX, ORD, and BOS. Finally, the ground arcs between the interconnection nodes are given in rows 13 and 14. Columns C and D give the cost associated with an aircraft that remains overnight at each of the three stations for each fleet type, which are assumed to be zero in this example. A cost of \$3000 is assumed for each aircraft to remain on the ground between the interconnection nodes for both fleet types.

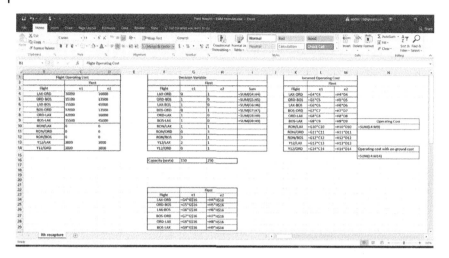

Figure 17.3 The decision variables and formulas for cost calculations for the IFAM example.

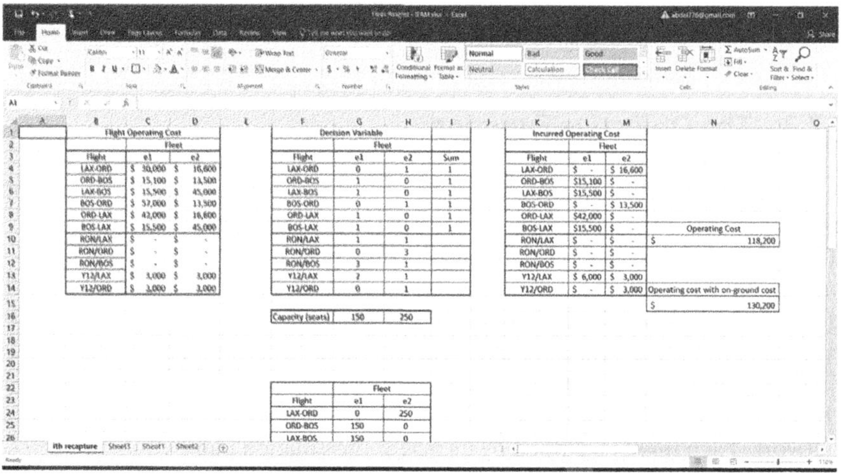

Figure 17.4 The decision variables and initial cost values for the IFAM example.

Columns G and H give the initial values of the assignment decision variables. For example, when the value of cell H4 is equal to 1, it implies that the LAX-ORD flight is assigned to fleet type $e2$. The value of cell G12 implies that the number of aircraft of type $e1$ to remain overnight at BOS has an initial value of three aircraft. Also, as given in cell H14, the number of aircraft of fleet type $e2$ to remain on ground during the time between the two defined interconnection nodes at ORD is initialized to 1.

The table in columns K, L, and M gives the incurred operating cost based on the given initial values of the decision variables. The total operations cost is given in cell N15, which is the sum of the cost elements given in columns L and M (rows 4–14). For example, given the initial values of the decision assignment variables, the operating cost is $130,200. Finally, column I (rows 4–9) gives the sum of the flight assignment decision variables for all fleet types. These values are used to set up the coverage constraints for the flights (i.e. each flight is assigned one fleet type).

In Figures 17.5 and 17.6, cells G16 and H16 give the seat capacity of fleet types $e1$ and $e2$, respectively. Columns G and H (rows 24–29) give the seat capacity of each flight based on the initial values of the decision variables.

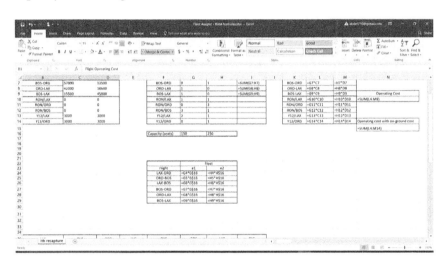

Figure 17.5 The formulas representing the seat capacity of each flight for the IFAM example.

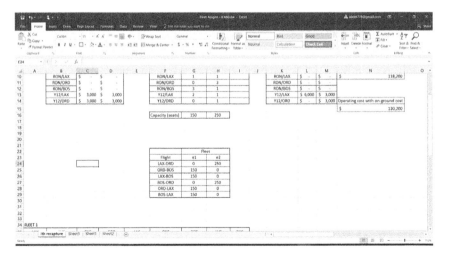

Figure 17.6 The initial values of the seat capacity of each flight for the IFAM example.

The spreadsheet in Figure 17.5 gives the formula used to determine the value of each cell, while the spreadsheet in Figure 17.6 gives the computed values.

Figures 17.7 and 17.8 give the balance constraints (starting at row 35) of the fleet assignment problem. As described above, there are five interconnection nodes for each fleet type. Thus, 10 balance constraints are defined for the problem. The spreadsheet in Figure 17.7 gives the formula representation of the cells, while the spreadsheet in Figure 17.8 gives the corresponding

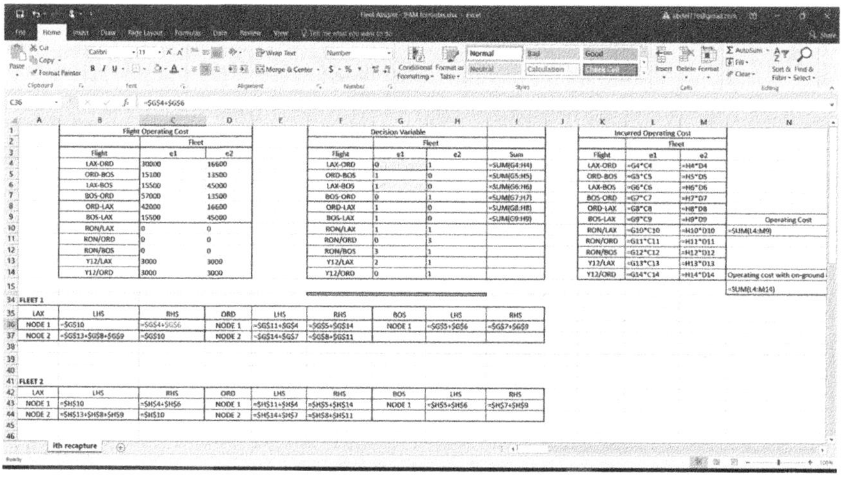

Figure 17.7 The formulas for the balance constraints for the IFAM example.

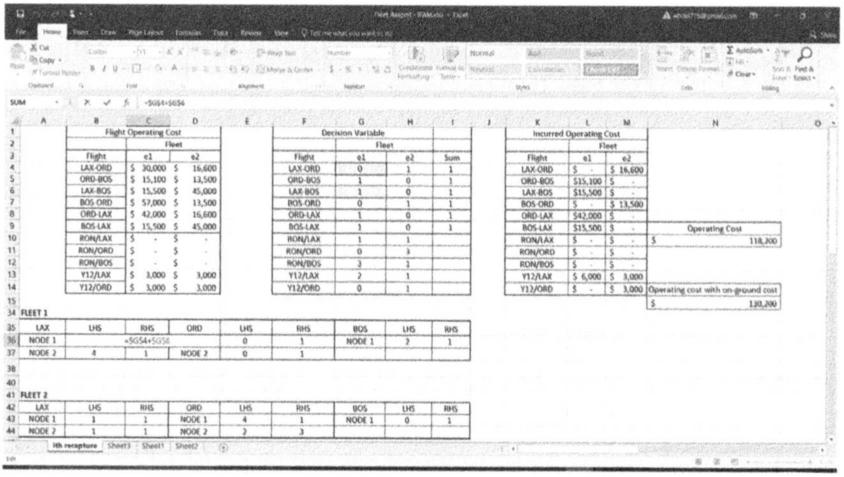

Figure 17.8 The initial values of the balance constraints for the IFAM example.

computed values. For each constraint, the values of the right-hand side (RHS) and the left-hand side (LHS) are given. It should be noted that the balance constraints are not yet satisfied, as initial values are just assigned to the decision variables. Using these initial values, the RHS and the LHS of the constraints might not be equal. As shown hereafter, constraints are added to force the equality of the RHS and the LHS for each balance constraint.

In Figures 17.9 and 17.10, the table in columns Q–X (rows 4–9) gives the itinerary–flight incidence matrix. The incidence matrix defines the relation

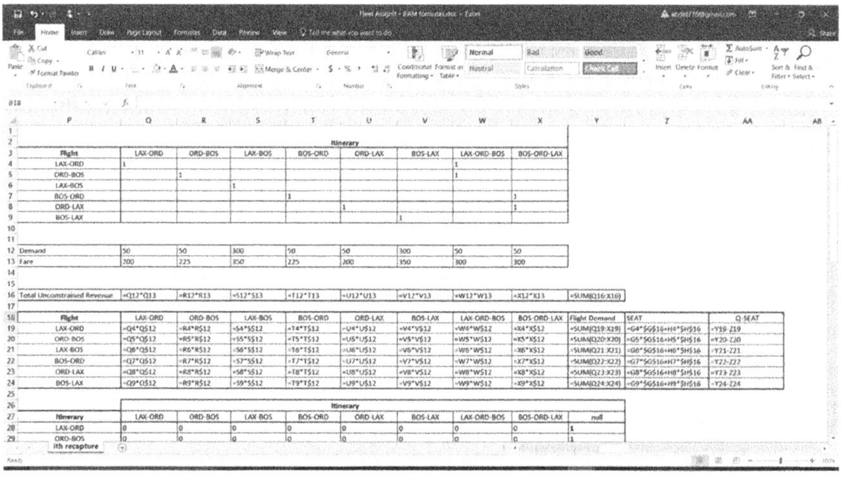

Figure 17.9 The formulas for the unconstrained demand for each flight for the IFAM example.

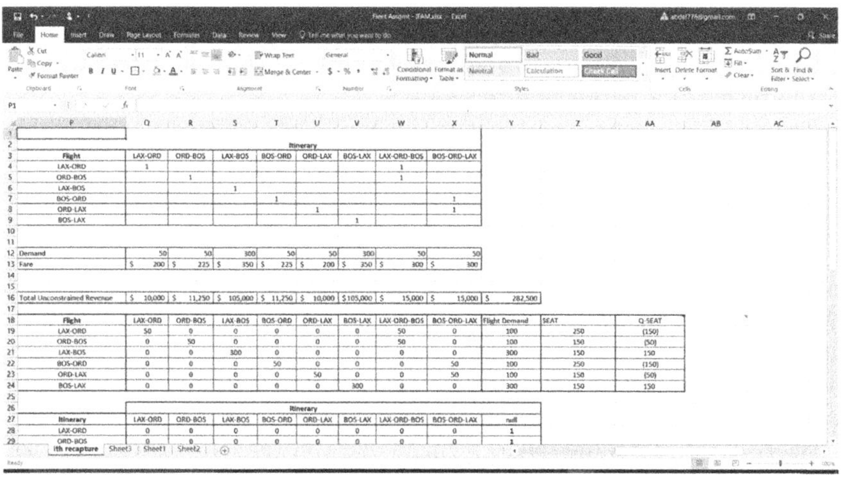

Figure 17.10 The values for the unconstrained demand for each flight for the IFAM example.

between the flights and the itineraries in the network. In this matrix, rows represent flights and columns represent itineraries. A cell representing a flight–itinerary combination is equal to 1, if the itinerary includes the flight as part of its structure, and zero otherwise. For example, the value in cell W5 is equal to 1 as flight ORD-BOS is part of the itinerary LAX-ORD-BOS. Rows 12 and 13 give the unconstrained passenger demand and the fare for each of the eight itineraries. As defined earlier, the unconstrained demand of an itinerary is defined as number of passengers traveling on this itinerary assuming unlimited seat capacity. The unconstrained revenue for each itinerary is calculated in row 16. The unconstrained revenue of each itinerary is calculated by multiplying its unconstrained demand by its fare. The total unconstrained revenue is calculated in cell Y16, which is the sum of unconstrained revenue for all itineraries.

Figures 17.9 and 17.10 also give the calculations of the unconstrained demand at the flight level. First, the unconstrained demand of each itinerary is associated with the flight(s) constituting the itinerary. Then, the unconstrained demand of a flight is calculated as the sum of the unconstrained demand of all itineraries that include this flight. For example, the BOS-ORD flight is included in two itineraries, which are BOS-ORD and BOS-ORD-LAX. Each of these itineraries has unconstrained demand of 50 passengers. Hence, the unconstrained demand of this flight is equal to 100. The total unconstrained demand (Q) of each flight is given in column Y (rows 19–24). Column Z (rows 19–24) gives the seat capacity of each flight. The seat capacity of each flight depends on the fleet assignment solution of each flight, which is given in columns G and H (rows 4–9). For example, the initial value of the assignment decision variable of flight LAX-BOS indicates that this flight is assigned to fleet type $e1$ (i.e. cell G6 = 1). As such, the seat capacity of this flight is 150. Finally, column AA (rows 19–24) gives the difference between the unconstrained demand of the flight and its seat capacity. For example, the LAX-BOS flight (row 21) has a total demand of 300 passengers and a seat capacity of 150. Thus, the excess demand of this flight is 150 passengers (cell AA21).

Figure 17.11 shows the itinerary–itinerary spill–recapture relationship matrix. This matrix is given in columns Q–Y and rows 28–35. The rows in the matrix represent the itineraries from which demand is spilled, while the columns represent the itineraries that possibly can recapture the spilled demand. The cells in the matrix indicate whether there is a spill–recapture relationship between the two itineraries represented by the cell. For example, as cell W30 is equal to 1, it implies that itinerary LAX-BOS can spill demand to itinerary LAX-ORD-BOS. Each itinerary can spill demand to an additional imaginary dummy itinerary, which represents the case in which the airline cannot recapture the spilled demand on any of its itineraries.

Figure 17.12 shows the spilled demand decision variables. These variables are given in columns Q–Y and rows 39–46. The given values represent the number of passengers to be spilled and recaptured among itineraries.

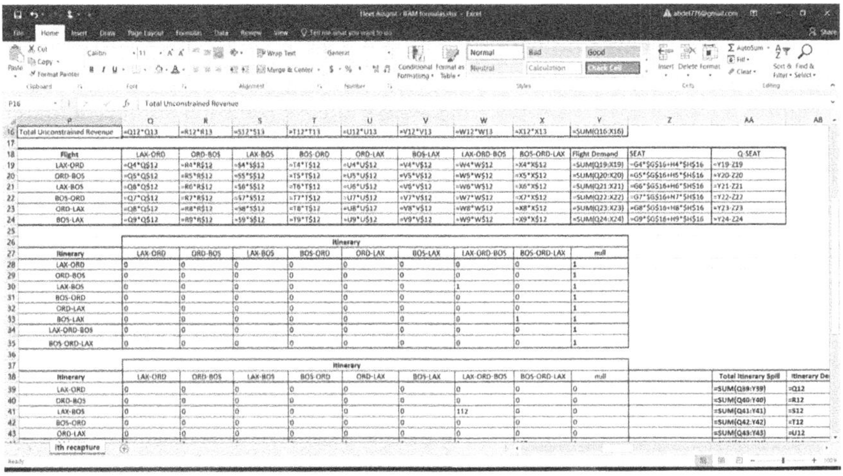

Figure 17.11 The itinerary–itinerary spill–recapture relationship matrix for the IFAM example.

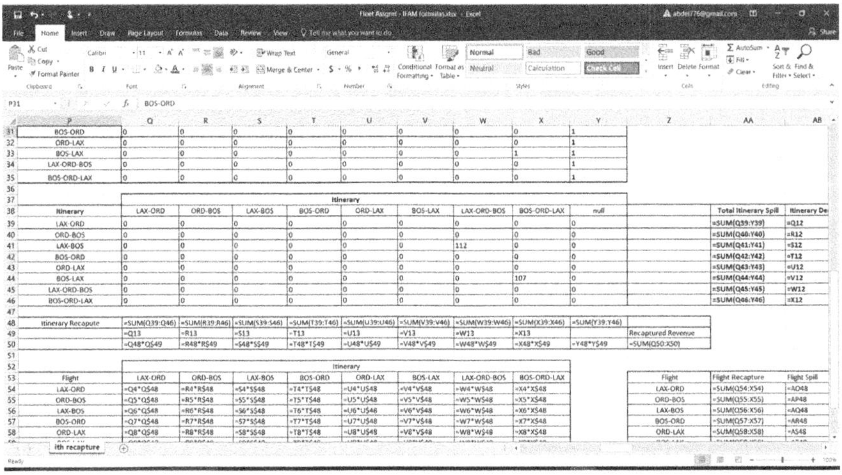

Figure 17.12 The initial values of the spilled demand decision variables for the IFAM example.

For example, it is assumed that 107 passengers are to be spilled from itinerary BOS-LAX to BOS-ORD-LAX (cell X44). These numbers are associated with the itinerary–itinerary spill–recapture relationship matrix defined above. In other words, the value of the demand to be spilled and recaptured between any two itineraries is valid only if the corresponding value in the itinerary–itinerary spill–recapture relationship matrix is equal to 1. At this stage, initial values are given in the table. When the fleet assignment problem is solved, the optimal values of the spilled and recaptured demand are to be determined in conjunction with the optimal fleet assignment of the different flights.

Figures 17.13 and 17.14 show the calculations of the total spilled demand for each itinerary. The total itinerary's spilled demand is the sum of the demand spilled on all recommended (substitute) itineraries including the

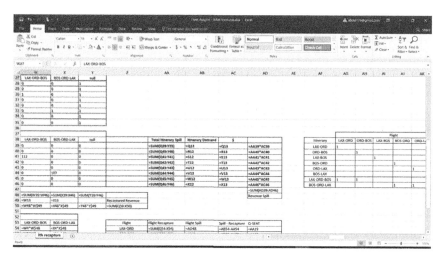

Figure 17.13 The formulas for the calculations of the total spilled demand for each itinerary for the IFAM example.

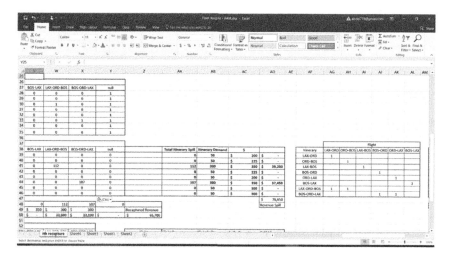

Figure 17.14 The initial values of the total spilled demand for each itinerary for the IFAM example.

dummy itinerary. Column AA in Figures 17.13 and 17.14 gives the calculations and the corresponding values of the total spilled demand for each itinerary (rows 39–46), respectively. Columns AB and AC (rows 39–46) give the itinerary demand and the itinerary fares, respectively, which are previously defined. Column AD (rows 39–46) gives the spilled revenue of each itinerary. The spilled revenue is calculated as the product of the spilled demand of the itinerary and its fare. Finally, cell AD47 gives the total spilled revenue, which is the sum of spilled revenue of all itineraries in the network. For example, given the initial values of the decision variables for the fleet assignment and the decision variables for the spilled demand, the spilled revenue is $76,650. The spilled revenue is corresponding to the value S in the objective function defined above.

Figure 17.15 shows the transposed format of itinerary–flight incidence matrix, which is previously defined in Figures 17.9 and 17.10. The transposed matrix is given in columns AG–AL (rows 39–46), where the columns define the flights and the rows define the itineraries.

Figures 17.16 and 17.17 give the calculations (formulas) and the values of the spilled demand at the flight level as shown in the table given in columns AO–AT (rows 39–46). The logic here is that when a demand is spilled from an itinerary, it is spilled from the flights of the itinerary. A flight could be part of more than one itinerary. Therefore, the demand spilled of a flight is calculated as the sum of the demand spilled of all itineraries that include the flight. The flight's spilled demand is given in row 48 (columns AO–AT).

Figures 17.18 and 17.19 show the formulas and the values of the recaptured demand on the different itineraries. As mentioned above, when demand is spilled from any itinerary, the airline suggests another itinerary (itineraries) to

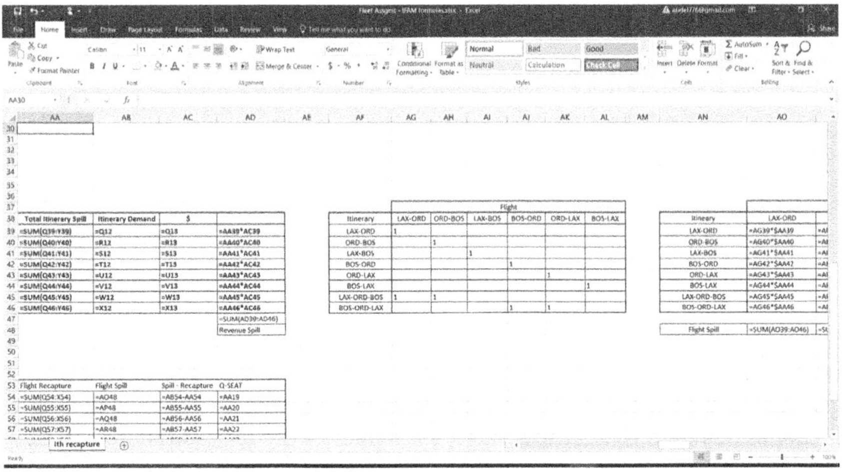

Figure 17.15 The transposed form of the itinerary–flight incidence matrix for the IFAM example.

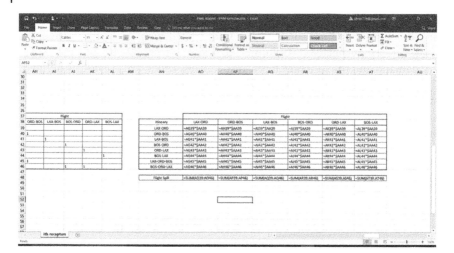

Figure 17.16 The formulas for the calculations of the spilled demand at the flight level for the IFAM example.

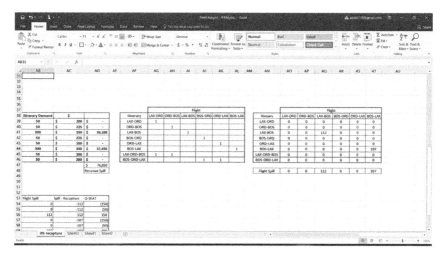

Figure 17.17 The initial values of the spilled demand at the flight level for the IFAM example.

recapture this demand. An itinerary can recapture demand spilled from more than one itinerary. Thus, the demand recaptured on any itinerary is the sum of the demand spilled of other itineraries to this itinerary. Row 48 (columns Q–Y) gives the demand recaptured on each itinerary including the dummy itinerary. It should be noted that, at this stage, the values of the recaptured demand for each itinerary depends on the initial values of the different decision variables (i.e. the fleet assignment variables and the demand spill variables). These initial values are not yet optimal nor satisfying the problem constraints. Consequently,

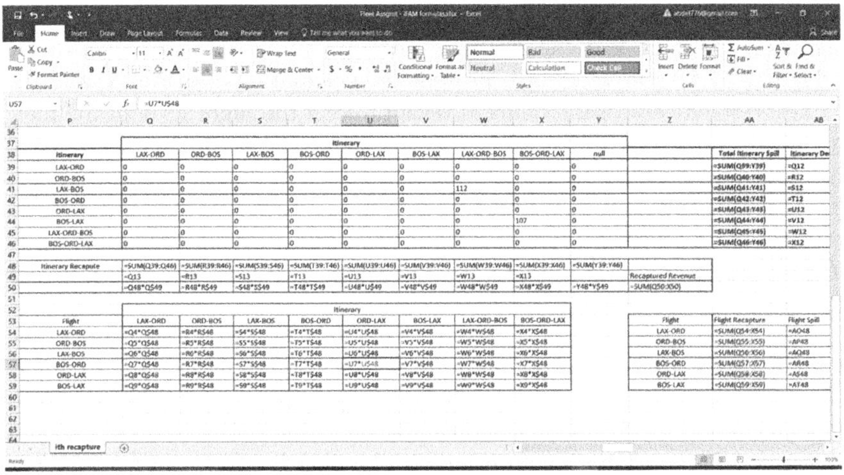

Figure 17.18 The calculations of the recaptured demand and recaptured revenue for the IFAM example.

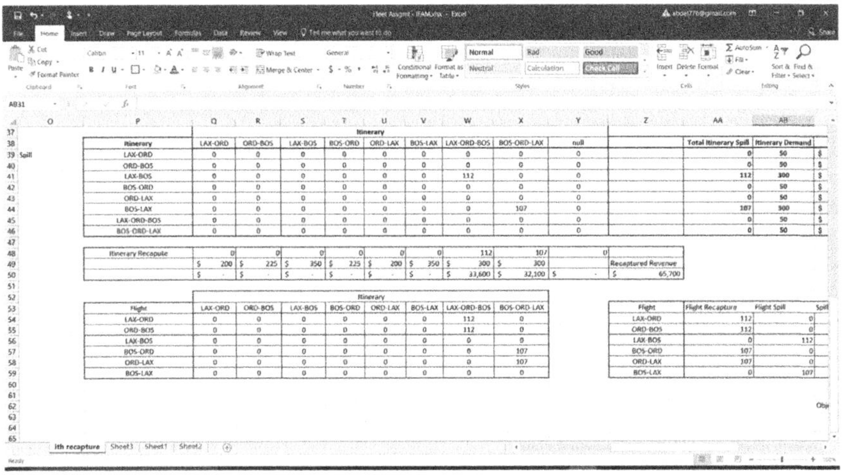

Figure 17.19 The initial values of the recaptured demand and recaptured revenue for the IFAM example.

the demand recaptured for each itinerary is not necessarily satisfying the seat capacity constraints of the itinerary. Row 49 (columns Q–Y) gives the itinerary fare as defined previously in row 13. Finally, the recaptured revenue of each itinerary is estimated by multiplying the recaptured demand of the itinerary by its fare. The total recaptured revenue by the airline is given in cell Z50, which is the sum of the recaptured revenue of all itineraries except the dummy itinerary. For example, given the initial values of the assignment and spill decision variables, the recaptured revenue is $65,700. The recaptured revenue corresponds to the value M in the objective function.

Figures 17.18 and 17.19 also present the calculations (formulas) and the values, respectively, of the demand recaptured at the flight level in the table given in columns Q–X (rows 54–59). The logic here is that when a demand is recaptured on an itinerary, it is recaptured on all flights constituting this itinerary. A flight could be part of more than one itinerary. Thus, the demand recaptured on a flight is calculated as the sum of the demand recaptured on all itineraries that include the flight. The recaptured demand for each of the six flights is given in column AA (rows 54–59).

Figures 17.20 and 17.21 present the calculations and the initial values of the flight demand characteristics and the objective function of the IFAM example. The table in columns AA–AD (rows 54–59) gives the main statistics of each flight including recaptured demand (column AA) and spilled demand (column AB). The difference between recaptured and spilled demand is given in column AC. Finally, the difference between the unconstrained demand and the seat capacity of the flight is given for each flight in column AD. Finally, cell AD62 gives the objective function value $C + S - M$ that required to be minimized. As shown in Figure 17.21, the initial value of the objective function is $141,150.

17.4 Problem Solution

The Excel Solver is used to obtain the optimal solution of the fleet assignment problem presented above. Figures 17.22 and 17.23 give snapshots of the Solver

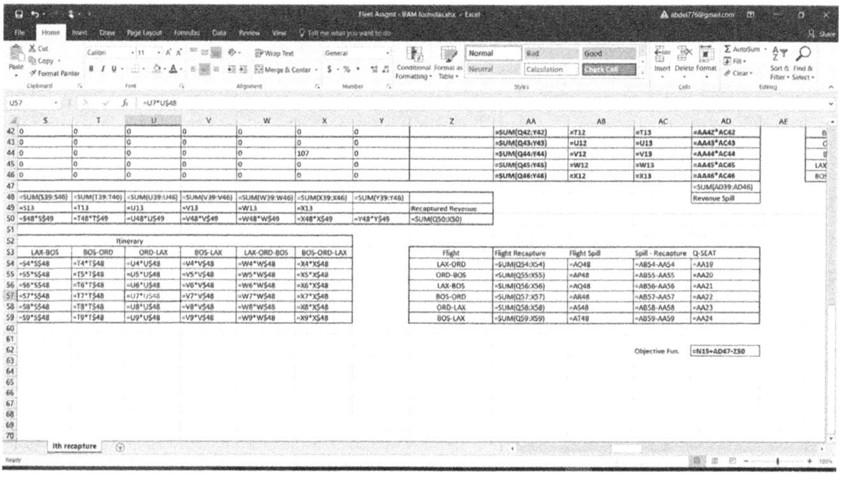

Figure 17.20 The calculations of the flight demand characteristics and the objective function of the IFAM example.

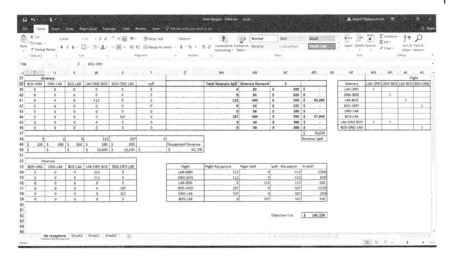

Figure 17.21 The initial values of the flight demand characteristics and the objective function of the IFAM example.

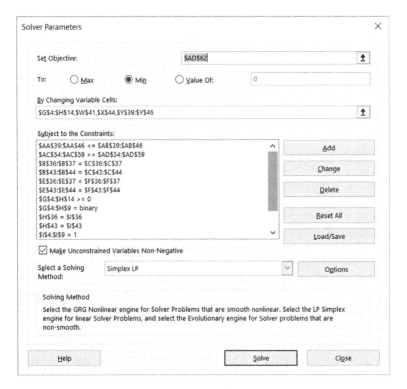

Figure 17.22 The Solver Parameters window for the IFAM example (part 1).

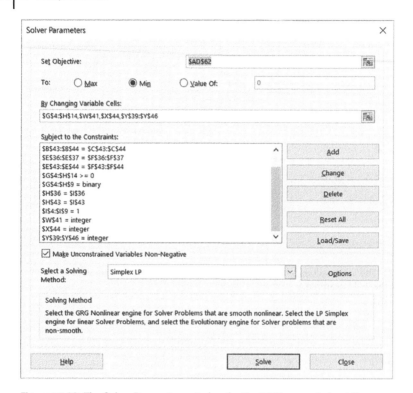

Figure 17.23 The Solver Parameters window for the IFAM example (part 2).

Parameters window that shows the corresponding values of the objective function, the decision variables, and the constraints. Two figures are given to show the complete list of the constraints. As mentioned earlier, the objective of the problem is to minimize the objective function defined in cell AD62.

The decision variables include the fleet assignment variables, the number of aircraft remaining on the ground between the interconnection nodes, and the number of aircraft remaining overnight at the three destinations. These variables are given in columns G and H (rows 4–14). The decision variables also include the number of passengers to be spilled between itineraries. The variables are given in cells W41 and X44. In addition, itinerary demand spilled to the dummy itineraries is included in the decision variables given in cells Y39–Y46.

The set of constraints are similar to the ones presented in the basic fleet assignment problem including the coverage constraints, the balance constraints, and the resources availability constraints. As mentioned earlier, the resources availability constraints are ignored in this example assuming

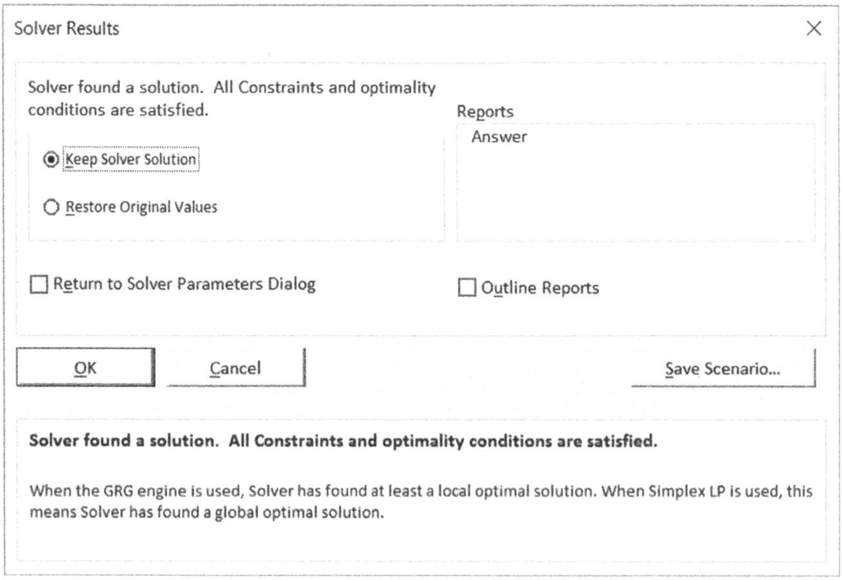

Figure 17.24 The Solver Results window for the IFAM example.

availability of resources of both fleets. In addition, other constraints are added to make sure that flight seat capacity is not exceeded and the spilled demand does not exceed the unconstrained demand. The coverage constraints require the values of cells I4–I9 to be equal to 1. The balance constraints force the equality of the values of the RHS cell and the LHS cell for each interconnection node and for the two fleet types. The seat capacity constraints guarantee that, for each flight, the difference between the spilled and recaptured demand of the flight is greater than or equal to the number of passengers that must be spilled from the flight ($Qf-$ SEATf). The seat capacity constraints are the first set of constraints in the constraints list in Figure 17.22. Other constraints are given for binary, integer, and nonnegative variables. When the problem is solved, the "Solver Results" window indicates that a solution is found as shown in Figure 17.24.

Figures 17.25 through 17.32 show different snapshots of the Excel sheet presenting the solution of the problem. For example, as shown in Figure 17.25, the LAX-BOS flight and BOS-LAX flight are assigned to fleet type $e1$ (cells G6 and G9 are equal to 1). The other four flights are assigned to fleet type $e2$ (cells H4, H5, H7, and H8 are equal to 1). There is one aircraft of each fleet type to remain overnight at LAX and no aircraft to remain on the ground between the interconnection nodes. As shown in Figure 17.32, the value of the objective

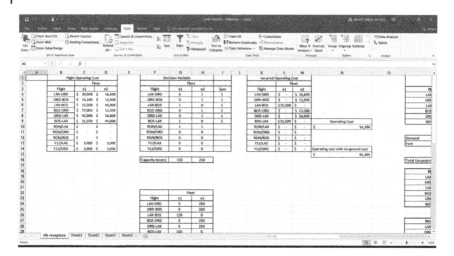

Figure 17.25 The solution of the IFAM example (part 1).

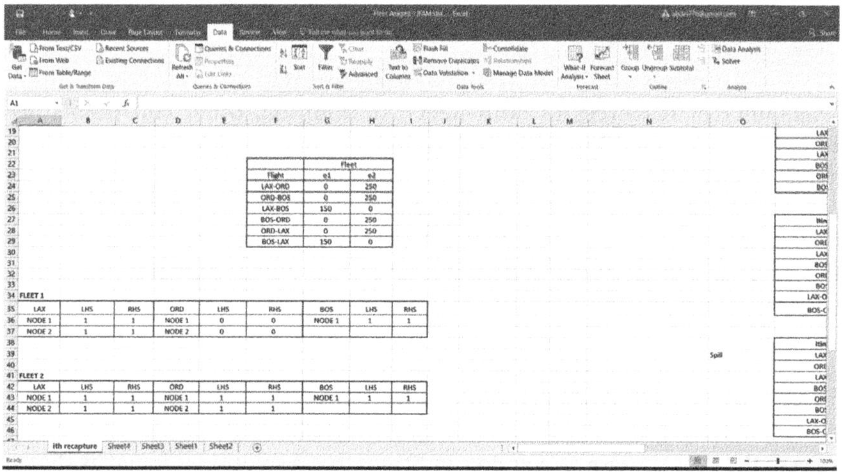

Figure 17.26 The solution of the IFAM example (part 2).

function (cell AD62) is \$106,200 (reduced from an initial value of \$141,150). The cost of assigning the six flights (i.e. flight operating cost) is \$91,200 (cell N10), as shown in Figure 17.25. The model suggests that 150 passengers are spilled from itinerary LAX-BOS to itinerary LAX-ORD-BOS (cell W41). In

	Itinerary							
Flight	LAX-ORD	ORD-BOS	LAX-BOS	BOS-ORD	ORD-LAX	BOS-LAX	LAX-ORD-BOS	BOS-ORD-LAX
LAX-ORD	1						1	
ORD-BOS		1					1	
LAX-BOS			1					
BOS-ORD				1				1
ORD-LAX					1			1
BOS-LAX						1		

Demand	50	50	300	50	50	300	50	50	
Fare	$ 200	$ 225	$ 350	$ 225	$ 200	$ 350	$ 300	$ 300	
Total Unconstrained Revenue	$ 10,000	$ 11,250	$ 105,000	$ 11,250	$ 10,000	$105,000	$ 15,000	$ 15,000	$ 282,500

Flight	LAX-ORD	ORD-BOS	LAX-BOS	BOS-ORD	ORD-LAX	BOS-LAX	LAX-ORD-BOS	BOS-ORD-LAX	Flight Demand	SEAT	Q-SEAT
LAX-ORD	50	0	0	0	0	0	50	0	100	250	(150)
ORD-BOS	0	50	0	0	0	0	50	0	100	250	(150)
LAX-BOS	0	0	300	0	0	0	0	0	300	150	150
BOS-ORD	0	0	0	50	0	0	0	50	100	250	(150)
ORD-LAX	0	0	0	0	50	0	0	50	100	250	(150)
BOS-LAX	0	0	0	0	0	300	0	0	300	150	150

	Itinerary								
Itinerary	LAX-ORD	ORD-BOS	LAX-BOS	BOS-ORD	ORD-LAX	BOS-LAX	LAX-ORD-BOS	BOS-ORD-LAX	null
LAX-ORD	0	0	0	0	0	0	0	0	1
ORD-BOS	0	0	0	0	0	0	0	0	1

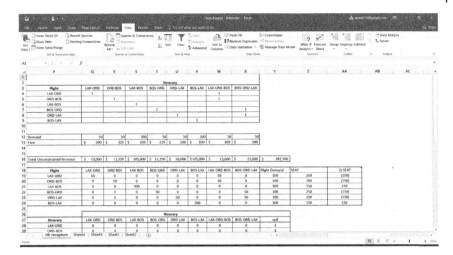

Figure 17.27 The solution of the IFAM example (part 3).

	Itinerary								
Itinerary	LAX-ORD	ORD-BOS	LAX-BOS	BOS-ORD	ORD-LAX	BOS-LAX	LAX-ORD-BOS	BOS-ORD-LAX	null
LAX-ORD	0	0	0	0	0	0	0	0	1
ORD-BOS	0	0	0	0	0	0	0	0	1
LAX-BOS	0	0	0	0	0	0	1	0	1
BOS-ORD	0	0	0	0	0	0	0	0	1
ORD-LAX	0	0	0	0	0	0	0	0	1
BOS-LAX	0	0	0	0	0	0	0	1	1
LAX-ORD-BOS	0	0	0	0	0	0	0	0	1
BOS-ORD-LAX	0	0	0	0	0	0	0	0	1

	Itinerary												
Itinerary	LAX-ORD	ORD-BOS	LAX-BOS	BOS-ORD	ORD-LAX	BOS-LAX	LAX-ORD-BOS	BOS-ORD-LAX	null		Total Itinerary Spill	Itinerary Demand	$
LAX-ORD	0	0	0	0	0	0	0	0	0		0	50	$ 200
ORD-BOS	0	0	0	0	0	0	0	0	0		0	50	$ 225
LAX-BOS	0	0	0	0	0	0	150	0	0		150	300	$ 350
BOS-ORD	0	0	0	0	0	0	0	0	0		0	50	$ 225
ORD-LAX	0	0	0	0	0	0	0	0	0		0	50	$ 200
BOS-LAX	0	0	0	0	0	0	0	150	0		150	300	$ 350
LAX-ORD-BOS	0	0	0	0	0	0	0	0	0		0	50	$ 300
BOS-ORD-LAX	0	0	0	0	0	0	0	0	0		0	50	$ 300

Itinerary Recapute	0	0	0	0	0	0	150	150	0	
	$ 200	$ 225	$ 350	$ 225	$ 200	$ 350	$ 300	$ 300		Recaptured Revenue
	$ -	$ -	$ -	$ -	$ -	$ -	$ 45,000	$ 45,000	$ -	$ 90,000

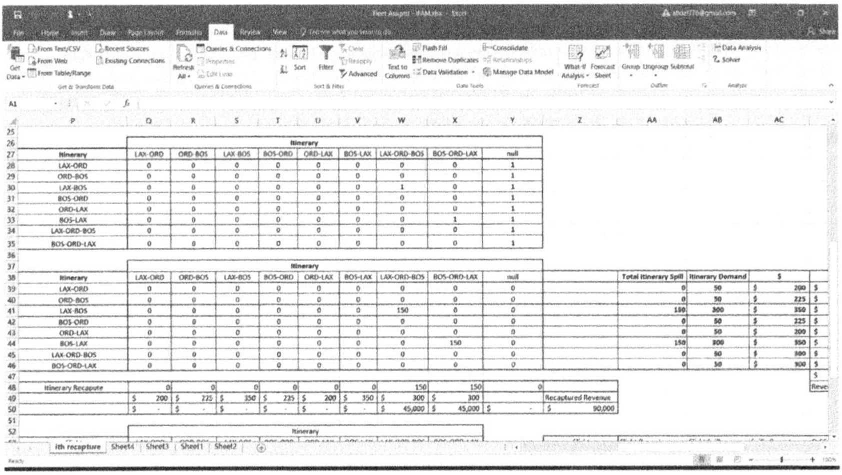

Figure 17.28 The solution of the IFAM example (part 4).

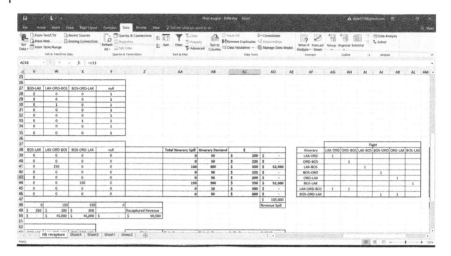

Figure 17.29 The solution of the IFAM example (part 5).

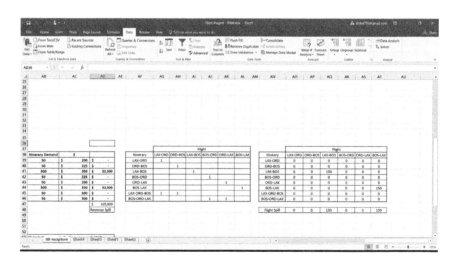

Figure 17.30 The solution of the IFAM example (part 6).

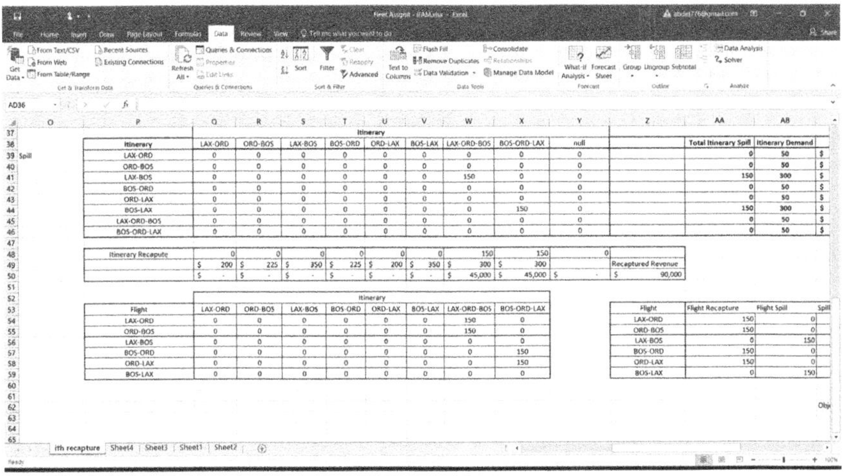

Figure 17.31 The solution of the IFAM example (part 7).

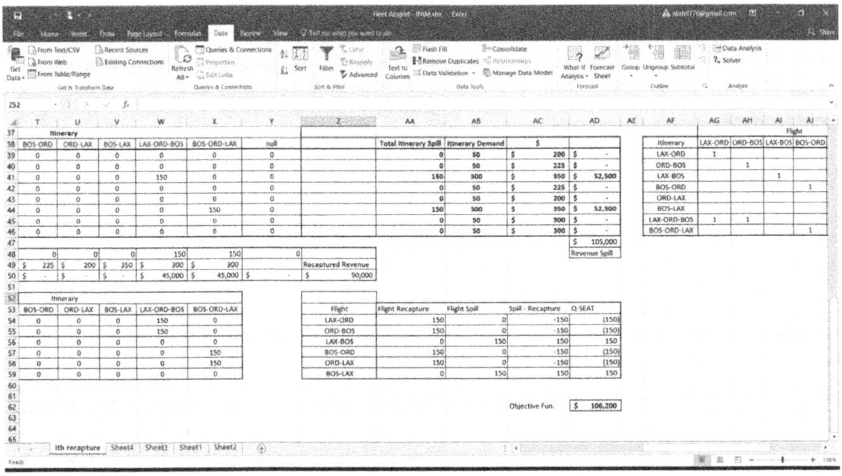

Figure 17.32 The solution of the IFAM example (part 8).

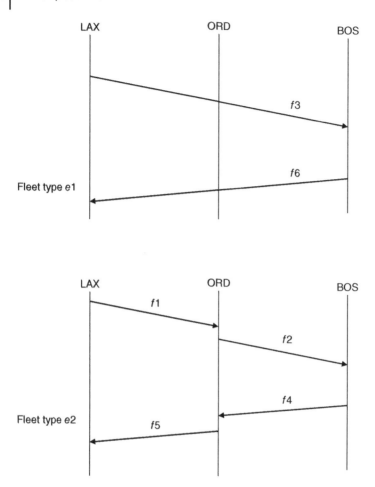

Figure 17.33 Graphical representation of the solution for the IFAM example.

addition, another 150 passengers are spilled from itinerary BOS-LAX to itinerary BOS-ORD-LAX (cell X44). No demand is spilled to the dummy itineraries (the value of cells Y39–Y46 is zero). Finally, Figure 17.33 shows the graphical representation of the solution, where two flights are assigned to fleet type $e1$ and four flights are assigned to fleet type $e2$.

18

Comparing FAM and IFAM

18.1 Problem Definition

To illustrate the benefits of using the itinerary-based fleet assignment model (IFAM) over the basic fleet assignment model (FAM), the same example given in Chapter 17 is solved using the basic FAM. As illustrated earlier, the main difference between the two models is that the FAM model is formulated at the flight level, while the IFAM model is itinerary based. As given in Chapter 11, the objective function of the basic FAM minimizes the sum of the flights operating cost and the spilled revenue (i.e. $C + S'$), where C represents the operating cost and S' represents the spilled revenue at the flight level. The problem constraints include the flight coverage constraints and the balance constraints. To match the settings of the IFAM problem, the resources availability constraints are ignored. Similar to the IFAM example given in Chapter 17, the basic FAM problem is formulated using the Excel Solver. Figures 18.1 through 18.10 present the problem solution steps in formula and value formats.

As shown in Figures 18.1 and 18.2, column B (rows 4–9) gives the list of flights in the network. Each flight is defined by its origin–destination pair. Columns C and D give the operating cost associated with assigning each of the six flights to fleet types $e1$ and $e2$, respectively. These cost elements are identical to the ones given in the IFAM example in Chapter 17. Column B (rows 10–12) also represents the remaining-overnight (RON) arcs for the three stations LAX, ORD, and BOS. Finally, the representation of the ground arcs between the interconnection nodes is given in rows 13 and 14. Columns C and D give the cost associated with an aircraft if it remains overnight at each of the three stations for each fleet type. In addition, the cost associated with each aircraft if it remains idle on the ground between the interconnection nodes for each fleet type is also given. These cost elements are identical to the ones given in the IFAM example in Chapter 17.

Airline Network Planning and Scheduling, First Edition. Ahmed Abdelghany and Khaled Abdelghany.
© 2019 John Wiley & Sons, Inc. Published 2019 by John Wiley & Sons, Inc.

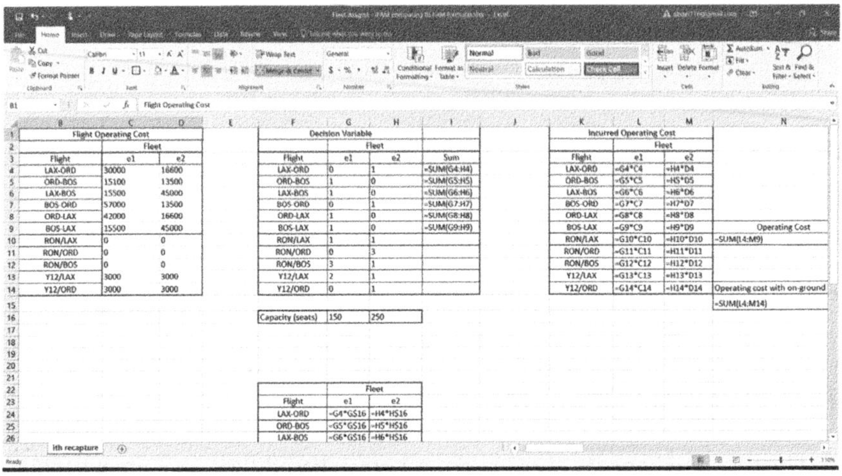

Figure 18.1 The decision variables and formulas for cost calculation for the comparable FAM example.

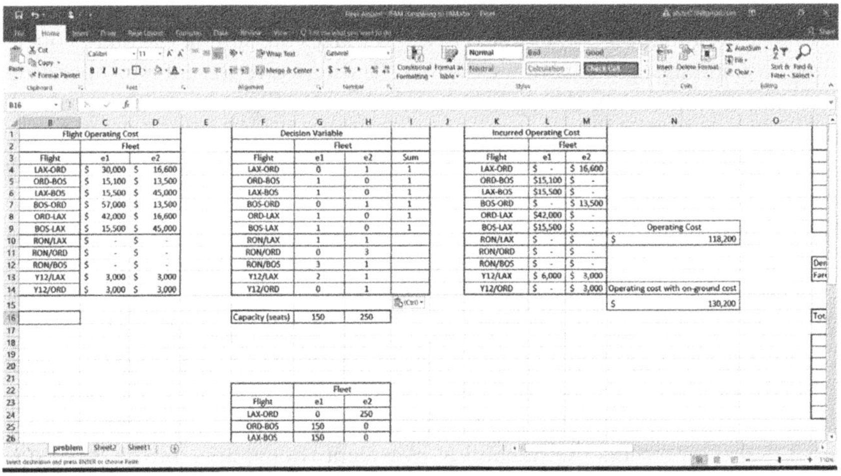

Figure 18.2 The decision variables and values for cost calculation for the comparable FAM example.

Columns G and H give the initial values of the assignment decision variables. These initial values are also identical to the ones given in the IFAM example in Chapter 17. The table in columns K, L, and M gives the incurred operating cost based on the initial values given to the decision variables. The spreadsheet in Figure 18.1 gives the formula representation of the cells, while the spreadsheet in Figure 18.2 gives the corresponding values. The operation cost is given in cell N15, which is the sum of the cost elements given in columns L and M

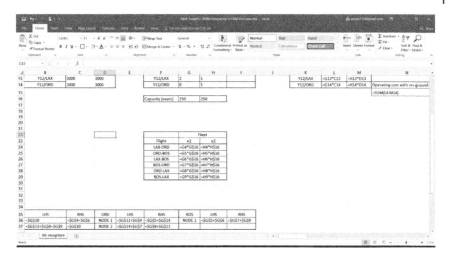

Figure 18.3 The formulas representing the seat capacity of each flight for the comparable FAM example.

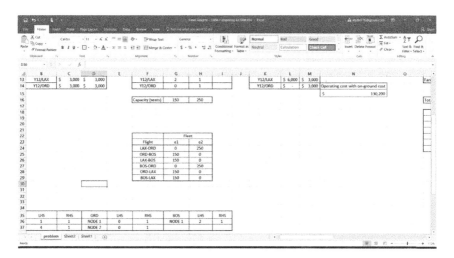

Figure 18.4 The initial values of the seat capacity of each flight for the comparable FAM example.

(rows 4–14). For example, given the initial values of the decision assignment variables, the operating cost is $130,200 (cell N15). This cost corresponds to the value C in the objective function of the IFAM model. Finally, column I (rows 4–9) gives the sum of the flight assignment decision variables for the two fleet types. This value is used to set up the coverage constraints of the problem.

In Figures 18.3 and 18.4, cells G16 and H16 give the seat capacity of the aircraft of fleet types $e1$ and $e2$, respectively. Columns G and H (rows 24–29) give

	Fleet	
Flight	e1	e2
LAX-ORD	=G4*G$16	=H4*H$16
ORD-BOS	=G5*G$16	=H5*H$16
LAX-BOS	=G6*G$16	=H6*H$16
BOS-ORD	=G7*G$16	=H7*H$16
ORD-LAX	=G8*G$16	=H8*H$16
BOS-LAX	=G9*G$16	=H9*H$16

FLEET 1

LAX	LHS	RHS	ORD	LHS	RHS	BOS	LHS	RHS
NODE 1	=$G$10	=$G$4+$G$6	NODE 1	=$G$11+$G$4	=$G$5+$G$14	NODE 1	=$G$5+$G$6	=$G$7+$G$9
NODE 2	=$G$13+$G$8+$G$9	=$G$10	NODE 2	=$G$14+$G$7	=$G$8+$G$11			

FLEET 2

LAX	LHS	RHS	ORD	LHS	RHS	BOS	LHS	RHS
NODE 1	=$H$10	=$H$4+$H$6	NODE 1	=$H$11+$H$4	=$H$5+$H$14	NODE 1	=$H$5+$H$6	=$H$7+$H$9
NODE 2	=$H$13+$H$8+$H$9	=$H$10	NODE 2	=$H$14+$H$7	=$H$8+$H$11			

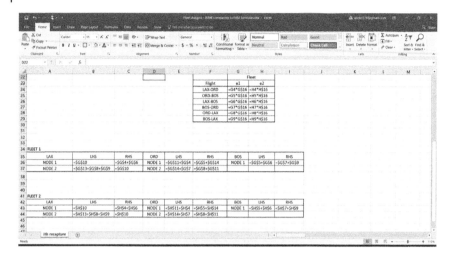

Figure 18.5 The formulas for the balance constraints for the comparable FAM example.

	Fleet	
Flight	e1	e2
LAX-ORD	0	250
ORD-BOS	150	0
LAX-BOS	150	0
BOS-ORD	0	250
ORD-LAX	150	0
BOS-LAX	150	0

FLEET 1

LAX	LHS	RHS	ORD	LHS	RHS	BOS	LHS	RHS
NODE 1	1	1	NODE 1	0	1	NODE 1	2	1
NODE 2	4	1	NODE 2	0	1			

FLEET 2

LAX	LHS	RHS	ORD	LHS	RHS	BOS	LHS	RHS
NODE 1	1	1	NODE 1	4	1	NODE 1	0	1
NODE 2	1	1	NODE 2	2	3			

Figure 18.6 The initial values of the balance constraints for the comparable FAM example.

the seat capacity of each flight based on the initial values of the decision variables. The spreadsheet in Figure 18.3 gives the formula representation of the cells, while the spreadsheet in Figure 18.4 gives the corresponding values.

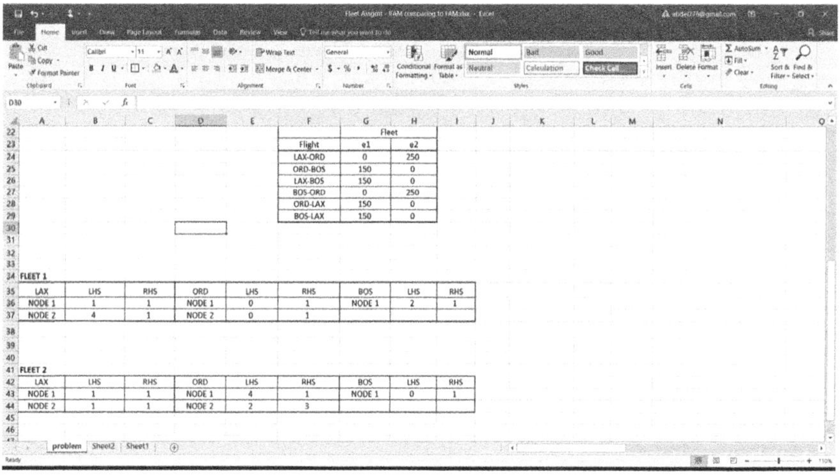

Figures 18.5 and 18.6 show snapshots of the Excel sheet that gives the balance constraints (starting at row 35). Similar to the IFAM example, there are five interconnection nodes for each fleet type, and accordingly ten balance constraints are defined for the problem. The spreadsheet in Figure 18.5 shows the formula representation of the cells, while the spreadsheet in Figure 18.6

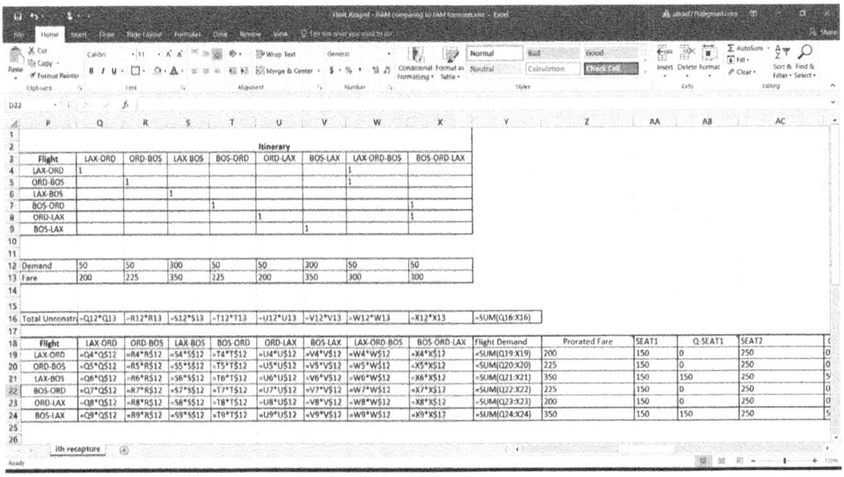

Flight	LAX-ORD	ORD-BOS	LAX-BOS	BOS-ORD	ORD-LAX	BOS-LAX	LAX-ORD-BOS	BOS-ORD-LAX
LAX-ORD	1						1	
ORD-BOS		1					1	
LAX-BOS			1					
BOS-ORD				1				1
ORD-LAX					1			1
BOS-LAX						1		
Demand	50	50	300	50	50	300	50	50
Fare	200	225	350	225	200	350	300	300
Total Unconstr	=Q12*Q13	=R12*R13	=S12*S13	=T12*T13	=U12*U13	=V12*V13	=W12*W13	=X12*X13

Total: =SUM(Q16:X16)

Flight	LAX-ORD	ORD-BOS	LAX-BOS	BOS-ORD	ORD-LAX	BOS-LAX	LAX-ORD-BOS	BOS-ORD-LAX	Flight Demand	Prorated Fare	SEAT1	Q-SEAT1	SEAT2
LAX-ORD	=Q4*Q$12	=R4*R$12	=S4*S$12	=T4*T$12	=U4*U$12	=V4*V$12	=W4*W$12	=X4*X$12	=SUM(Q19:X19)	200	150	0	250
ORD-BOS	=Q5*Q$12	=R5*R$12	=S5*S$12	=T5*T$12	=U5*U$12	=V5*V$12	=W5*W$12	=X5*X$12	=SUM(Q20:X20)	225	150	0	250
LAX-BOS	=Q6*Q$12	=R6*R$12	=S6*S$12	=T6*T$12	=U6*U$12	=V6*V$12	=W6*W$12	=X6*X$12	=SUM(Q21:X21)	350	150	150	250
BOS-ORD	=Q7*Q$12	=R7*R$12	=S7*S$12	=T7*T$12	=U7*U$12	=V7*V$12	=W7*W$12	=X7*X$12	=SUM(Q22:X22)	225	150	0	250
ORD-LAX	=Q8*Q$12	=R8*R$12	=S8*S$12	=T8*T$12	=U8*U$12	=V8*V$12	=W8*W$12	=X8*X$12	=SUM(Q23:X23)	200	150	0	250
BOS-LAX	=Q9*Q$12	=R9*R$12	=S9*S$12	=T9*T$12	=U9*U$12	=V9*V$12	=W9*W$12	=X9*X$12	=SUM(Q24:X24)	350	150	150	250

Figure 18.7 The formulas for the unconstrained demand for each flight for the comparable FAM example.

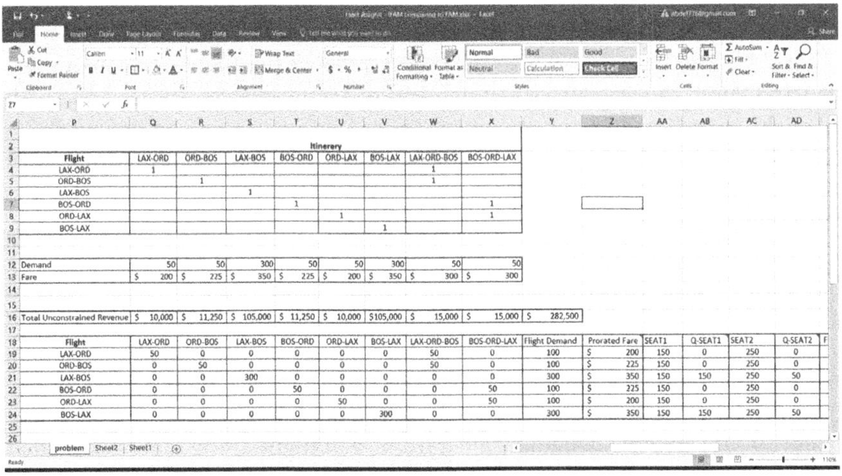

Flight	LAX-ORD	ORD-BOS	LAX-BOS	BOS-ORD	ORD-LAX	BOS-LAX	LAX-ORD-BOS	BOS-ORD-LAX
LAX-ORD	1						1	
ORD-BOS		1					1	
LAX-BOS			1					
BOS-ORD				1				1
ORD-LAX					1			1
BOS-LAX						1		
Demand	50	50	300	50	50	300	50	50
Fare	$ 200	$ 225	$ 350	$ 225	$ 200	$ 350	$ 300	$ 300
Total Unconstrained Revenue	$ 10,000	$ 11,250	$ 105,000	$ 11,250	$ 10,000	$105,000	$ 15,000	$ 15,000

Total: $ 282,500

Flight	LAX-ORD	ORD-BOS	LAX-BOS	BOS-ORD	ORD-LAX	BOS-LAX	LAX-ORD-BOS	BOS-ORD-LAX	Flight Demand	Prorated Fare	SEAT1	Q-SEAT1	SEAT2	Q-SEAT2
LAX-ORD	50	0	0	0	0	0	50	0	100	$ 200	150	0	250	0
ORD-BOS	0	50	0	0	0	0	50	0	100	$ 225	150	0	250	0
LAX-BOS	0	0	300	0	0	0	0	0	300	$ 350	150	150	250	50
BOS-ORD	0	0	0	50	0	0	0	50	100	$ 225	150	0	250	0
ORD-LAX	0	0	0	0	50	0	0	50	100	$ 200	150	0	250	0
BOS-LAX	0	0	0	0	0	300	0	0	300	$ 350	150	150	250	50

Figure 18.8 The values for the unconstrained demand for each flight for the comparable FAM example.

shows the corresponding values. For each constraint, the values of the right-hand side (RHS) and the left-hand side (LHS) are given. The balance constraints are not satisfied based on the initial values of the decision variables. Accordingly, at this stage, the RHS and the LHS of each constraint may not be equal.

Figures 18.7 and 18.8 show the itinerary–flight incidence matrix in columns Q–X (rows 4–9). The incidence matrix defines the relation between the flights

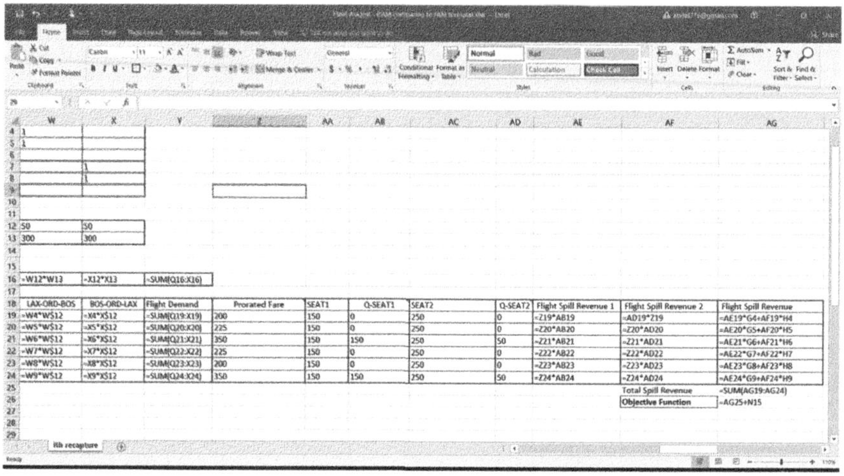

Figure 18.9 The calculation of the flight spilled revenue for the comparable FAM example.

and the itineraries in the network. As explained above, a cell in the matrix is equal to 1, if the itinerary corresponding to the cell includes the flight, and zero otherwise. The figures also give the calculations of the flights' unconstrained demand. First, the unconstrained demand of each itinerary is associated with all flight(s) included in the itinerary. Then, the unconstrained demand of a flight is calculated as the sum of the unconstrained demand of all itineraries that include this flight. The total unconstrained demand (Q) of each flight is given in column Y (rows 19–24).

Figures 18.9 and 18.10 give the formulas and values for the calculation of the flights' spilled revenue S'. First, a prorated fare is assumed for each flight, as given in column Z (rows 19–24). For this example, the fare of each flight is assumed to be equal to the fare of its corresponding nonstop itinerary. Columns AA (rows 19–24) and AC (rows 19–24) give the seat capacity of each flight for fleet types $e1$ and $e2$, respectively. Columns AB (rows 19–24) and AD (rows 19–24) give the demand spilled from each flight, if the flight is assigned to fleet types $e1$ and $e2$, respectively. The spilled demand is calculated as the difference between the flight's demand and seat capacity. Columns AE and AF (rows 19–24) give the spilled revenue corresponding to assigning the flights to each of the two fleet types $e1$ and $e2$, respectively. The spilled revenue is calculated by multiplying the fare by the spilled demand. Column AG (rows 19–24) gives the spilled revenue for each flight. The spilled revenue depends on which fleet type is assigned to the flight. The total spilled revenue S' is given in cell AG25. Finally, the objective function (i.e. $C + S'$) of the problem is calculated as given in cell AG26.

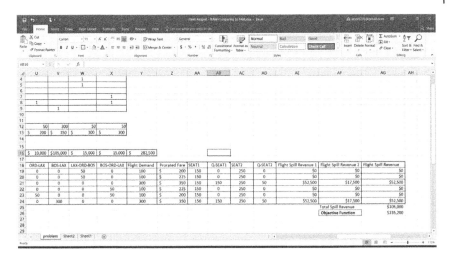

Figure 18.10 The flight spilled revenue for the comparable FAM example.

18.2 Problem Solution

Figure 18.11 shows a snapshot of the Solver Parameters window. It gives the corresponding values of the objective function, the decision variables, and the constraints. As mentioned earlier, the objective function minimizes the value defined in cell AG26. The decision variables include the fleet assignment variables, the number of aircraft that remain on the ground between the interconnection nodes, and the number of aircraft that remain on the ground at the three destinations. These variables are given in columns G and H (rows 4–14). The set of constraints includes the coverage constraints, the balance constraints, and the resources availability constraints (ignored in this example assuming availability of adequate resources of both fleets). The coverage constraints require each of the cells I4–I9 to be one. The balance constraints require the RHS cell to be equal to its corresponding LHS for each interconnection node and for the two fleet types. Other constraints are given to ensure that the decision variables are either binary or nonnegative integers. The Solver Results window shows that a solution is found as shown in Figure 18.12.

Figures 18.13 through 18.16 show different snapshots of the solution. As shown in Figure 18.13, all flights are assigned to fleet type $e2$ (cells H4–H9 are equal to 1). The solution requires two aircraft of fleet type $e2$ to remain overnight at LAX. As shown in Figure 18.16, the value of the objective function (cell AG26) is reduced from $235,200 to $185,200. The cost of assigning the six flights (i.e. flight operating cost) is $150,200 (cell N10).

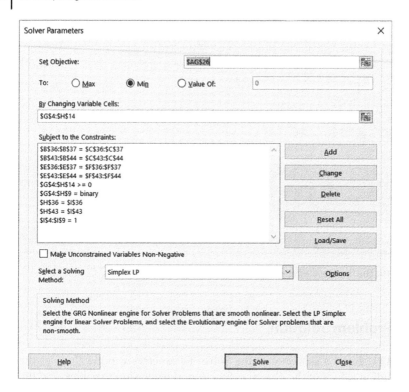

Figure 18.11 The Solver Parameters window for the comparable FAM example.

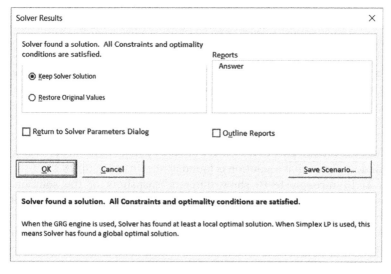

Figure 18.12 The Solver Results window for the comparable FAM example.

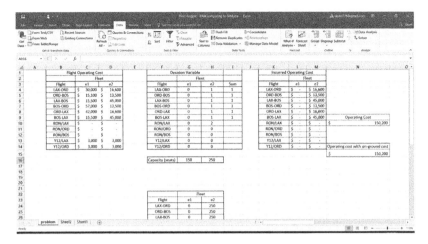

Figure 18.13 The solution of the comparable FAM example (part 1).

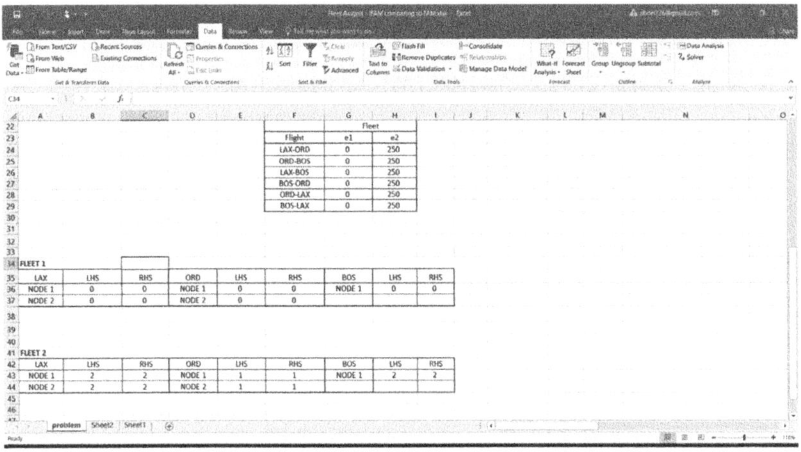

Figure 18.14 The solution of the comparable FAM example (part 2).

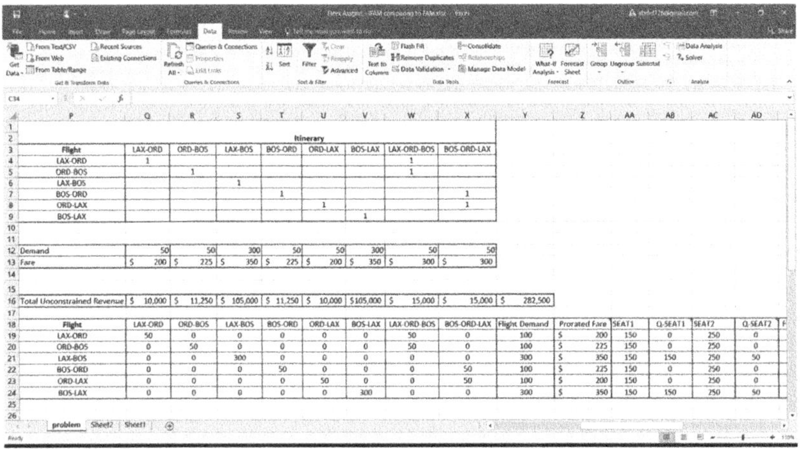

Figure 18.15 The solution of the comparable FAM example (part 3).

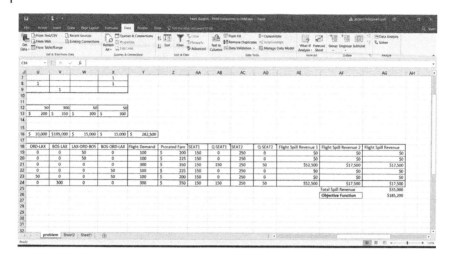

Figure 18.16 The solution of the comparable FAM example (part 4).

It is obvious that this solution is different from the IFAM model. The solution of the basic FAM tends to minimize the spilled revenue at the flight level, and hence large aircraft (fleet type $e2$) are used for all flights. On the other hand, the IFAM model considers that the spilled demand can be recaptured on other itineraries. Thus, large aircraft (fleet type $e2$) are not necessarily needed for all flights.

Section 6

19

Integrated Schedule Design with the Itinerary-based Fleet Assignment Model (ISD-IFAM)

19.1 Introduction

In Chapter 14, the schedule design problem is described as an extension of the basic fleet assignment model (FAM). The problem allows flight schedule modification in terms of flight additions, flight deletions, and departure time adjustment for a limited number of flights. One main limitation of this approach is the lack of demand–supply interactions. In other words, it is assumed that the demand is constant and independent of any suggested schedule changes. However, if a flight is eliminated in a city-pair, the passenger demand is expected to shift to other existing flights/itineraries serving that city-pair. Similarly, adding a flight to the schedule might induce new demand and/or shift existing demand from other flights/itineraries to the new flight. A change in the demand pattern is also expected with any change in the departure time of the flights. Of course, one should expect that if a significant change is introduced to the schedule, the impact on the demand estimated for each flight would also be significant. As such, assuming fixed flight demand simplifies the integrated schedule design and fleet assignment problem. Disregarding this assumption would require the capability to predict how the demand is redistributed among the different itineraries/flights as a result of any proposed schedule modification.

In this chapter, the itinerary-based fleet assignment model (IFAM) presented in Chapter 16 is extended to solve the integrated schedule design and fleet assignment problem. The main contribution over the methodology presented in Chapter 14 is that the interaction between flight demand and supply is considered. The methodology simultaneously seeks to optimize the selection of the flights and the assignment of fleet to the selected flights. The discussion of this chapter is based on the work presented by Lohatepanont and Barnhart (2004), which is also adopted as basis for schedule design and FAM developed by several major airlines.

Airline Network Planning and Scheduling, First Edition. Ahmed Abdelghany and Khaled Abdelghany.
© 2019 John Wiley & Sons, Inc. Published 2019 by John Wiley & Sons, Inc.

Following this approach, the set of flights proposed by the airline in a future schedule is categorized in two groups, namely, (i) mandatory flights and (ii) optional flights. The mandatory flights are flights that the airline is certain about including them in its future schedule as they are historically shown to be profitable flights. On the other hand, the profitability of the optional flights is uncertain, and hence they might or might not be included in the schedule. The mandatory flights and the optional flights are joined together to form a master flight list. The objective of the problem is to find which of the optional flights are worth including in the future schedule and to determine the optimal fleet assignment decisions for the mandatory flights and the selected optional flights.

19.2 Example of Demand and Supply Interactions

In Chapter 3, we presented several examples to illustrate the interaction among the different planning decisions including determining the flight frequency in the market, specifying the departure time for each flight, and assigning the optimal fleet for each flight. These examples illustrate the effect of supply-related decisions such as flight frequency (i.e. flight addition/deletion) and flight departure times on the demand of a flight/itinerary. This section extends the discussion in Chapter 3 to illustrate the supply and demand interaction in an airline network.

As explained earlier, on the demand side, the unconstrained demand in any city-pair is defined as the number of passengers traveling between the city-pair, assuming unlimited seat capacity. Passengers choose among nonstop and connecting itineraries scheduled in the city-pair. Given the share of each itinerary, the demand is then determined at the flight level, in order for the airlines to match this demand with the appropriate fleet type (i.e. number of seats). If the assigned capacity is less than the unconstrained demand, the actual demand is set to be equal to the flight capacity. In this case, some demand is spilled, and the airline attempts to recapture this spilled demand by other itineraries. On the supply side, airlines serve the different city-pairs either by nonstop or connecting itineraries. Major airlines adopt a hub-and-spoke network structure to serve connecting traffic through the hubs. Airlines design time banks at the hub to create efficient connecting itineraries. A time bank is defined as a set of flight arrivals to the hub followed by a set of flight departures from the hub.

Figure 19.1 shows an example of a hypothetical airline with a major hub at ATL. The airline serves ten spoke cities in a single time bank. For example, passengers can select to fly between MCO and PHL by connecting at ATL using itinerary MCO-ATL-PHL. As explained earlier in Chapter 2, the demand on each flight consists of local and connecting traffic. For example, the demand of the ATL-PHL flight consists of passengers traveling in the ATL-PHL market

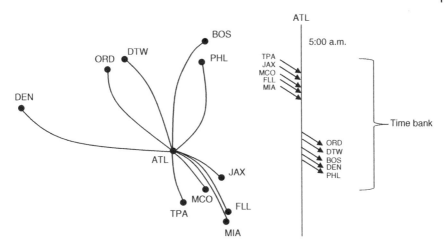

Figure 19.1 An example of a hypothetical airline with a hub and a time bank at ATL.

and connecting passengers traveling in the markets TPA-PHL, JAX-PHL, MCO-PHL, FLL-PHL, and MIA-PHL. When a flight is added or removed from the time bank (i.e. change in supply), the demand of some other flights is expected to change. For example, if the FLL-ATL flight is removed from the schedule, the demand on the ATL-PHL flight is reduced by the number of passengers traveling from FLL to PHL. Similar effect is expected to all other outbound flights from ATL.

19.3 Aspects of Demand–Supply Interactions: Demand Correction Factors

Two aspects should be considered while studying the demand–supply interactions. The first aspect is related to the relationship between the unconstrained demand and changes in the service supply. As discussed in Chapter 3, the demand is not linearly proportional to the service frequency provided in a market. Adding a flight in a market attracts demand to that flight and also induces additional demand for the existing itineraries. Adding a flight usually creates more attractive itineraries that are likely to be selected by customers and hence increases their demand. The opposite impact occurs when a flight is eliminated.

The second aspect is pertinent to understanding the effect of adding/deleting a flight on the demand of other available itineraries/flights. As discussed in Chapter 8, the unconstrained market share of an itinerary depends on the quality of this itinerary as well as all other itineraries in the market. Thus, adding or removing a flight (itinerary) is expected to change the market share of all itineraries in the city-pair. The impact of demand change might extend to other city-pairs as explained in the example given above.

To further explain these two aspects of the problem, consider the hypothetical flight schedule given in Figure 19.2. This schedule serves four stations including MIA, ATL, ORD, and DTW. The city-pairs are served by nonstop and connecting itineraries. The demand of each itinerary is as given in the figure. In *case a*, the city-pair MIA-ORD is served by three itineraries: the nonstop itinerary F1 (itinerary 1), the connecting itinerary F2–F3 (itinerary 2), and the connecting itinerary F4–F5 (itinerary 3). The total demand in this city-pair is 267 passengers. In this market, 188 passengers are selecting itinerary 1, 37 passengers are selecting itinerary 2, and 42 passengers are selecting itinerary 3. As shown in *case b*, the airline eliminated flight F4 (MIA-ATL) from the schedule, which results in the elimination of itinerary 3 (F4–F5) in the city-pair MIA-ORD. Thus, itinerary 3 (F4–F5) lost 42 passengers in the city-pair MIA-ORD.

In addition, since the airline now has less presence in the MIA-ORD market, one might expect shifting of some passengers to other airlines. Assume that

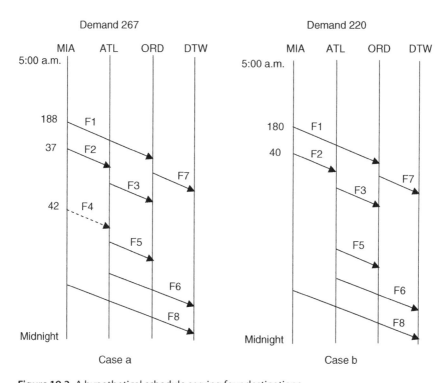

Figure 19.2 A hypothetical schedule serving four destinations.

eight passengers are lost from itinerary 1 and seven passengers are lost from itinerary 2. Meanwhile, assume 10 passengers (out of the 42 passengers of itinerary 3) are shifted from the removed itinerary (itinerary 3) to itinerary 2. As a result of the flight deletion, itinerary 1 has 180 passengers (188 – 8). Itinerary 2 has 40 passengers (37 – 7 + 10). Accordingly, the total demand in the market becomes 220 passengers (267 – 42 – 15 + 10) or (180 + 40).

Define P_o as the set of itineraries containing the optional flight legs, which is indexed by q. Define also P as the set of itineraries serving a given market including a dummy itinerary representing demand served by other airlines in this market. Define ΔD_q^p as the change in the demand of itinerary p as a result of the deletion of itinerary q. This variable is also known as the demand correction for itinerary p, when itinerary q is deleted. We also define the variable D_q as the unconstrained demand of itinerary q. Considering the example above, $D_3 = 42$, $\Delta D_3^1 = -8$, and $\Delta D_3^2 = +3$. The total change in the unconstrained demand in this market due to the deletion of itinerary q can be computed as follows:

$$D_q - \sum_{p:p\neq q} \Delta D_q^p$$

For generalization, we define the decision variable z_q that is equal to 1, if itinerary q is included in the proposed schedule, and zero otherwise. Thus, the total change in the unconstrained demand of this market due to the deletion of itinerary q can be determined as follows:

$$\left(D_q - \sum_{p:p\neq q} \Delta D_q^p\right)\cdot\left(1 - z_q\right)$$

Applying to the example above, the total change (reduction) in the unconstrained demand in city-pair MIA-ORD due to the deletion of itinerary 3 is as follows:

$$D_3 - \left[\Delta D_3^1 + \Delta D_3^2\right]$$

The total change in the unconstrained demand in city-pair MIA-ORD due to the deletion of itinerary 3 is 47 (i.e. 42 – (–8 + 3) = 47).

Consider the variable $fare_q$ to represent the fare of itinerary q. The corresponding loss of revenue due to the deletion of itinerary q city-pair can be calculated as the revenue loss due to the deletion of the itinerary minus the revenue recaptured on other itineraries as follows:

$$D_q\, fare_q - \sum_{p:p\neq q} \Delta D_q^p\, fare_p$$

Assume that the fares of itineraries 1, 2, and 3 are $270, $165, and $178, respectively. The loss of the unconstrained revenue in the city-pair MIA-ORD due to the deletion of itinerary 3 can be calculated as follows:

$$42 \cdot \$178 + 8 \cdot \$270 - 3 \cdot \$165 = \$9141$$

In the discussion above, the impact of the deletion of itinerary F4–F5 is calculated. However, it should be noted that the impact of the deletion of flight F4 extends to all city-pairs that are served by itineraries that include flight F4. When flight F4 is removed from the schedule, the nonstop itinerary F4 in city-pair MIA-ATL, the single-stop itinerary F4–F6 in the city-pair MIA-DTW, and the single-stop itinerary F4–F5 in the city-pair MIA-ORD no longer exist. Thus, the impact of the deletion of flight F4 should be considered in all three city-pairs that are served by itineraries that include flight F4.

Define the set P_o as the set of itineraries containing the removed flight F4 (this set of itineraries is indexed by q). In the above example, the set P_o contains three itineraries F4, F4–F5, and F4–F6. Thus, the impact of the deletion of a flight, in terms of the change of the unconstrained revenue ΔR, can be represented mathematically as follows:

$$\Delta R = \sum_{q \in P^o} \left(D_q \, \text{fare}_q - \sum_{p:p \neq q} \Delta D_q^p \, \text{fare}_p \right) \cdot \left(1 - z_q \right)$$

Assume in the example above that flight F3 is also removed. This deletion implies that itinerary 2 in the city-pair MIA-ORD is removed. As a result of this deletion, we assume $\Delta D_2^1 = 25$, that is, 25 of the 40 passengers of itinerary 2 are shifted to itinerary 1 (the other 15 passengers are assumed to be assigned to the dummy itinerary). Thus, the impact of the deletion of itinerary 3 and itinerary 2 on the demand of itinerary 1 can be calculated as follows:

$$\Delta D_3^1 + \Delta D_2^1 = -8 + 25 = 17$$

The demand of itinerary 1 after the deletion of itineraries 2 and 3 is $188 + 17 = 205$.

If several itineraries are deleted in a given city-pair (each itinerary is denoted by q), the impact of the deletion of these itineraries on the demand of an itinerary $p \in P$ can be calculated as follows:

$$\sum_{q \in P^o} \Delta D_q^p \cdot \left(1 - z_q \right)$$

In the above example, the city-pair MIA-DTW has two itineraries, namely, itinerary 7 and itinerary 8. Itinerary 7 is a single-stop itinerary composed of flights F1 and F7, while itinerary 8 is a nonstop itinerary using flight F8. Assume that the unconstrained demand of itinerary 7 is 35 passengers and that of itinerary 8 is 70 passengers. Assume that itinerary 8 is an optional itinerary and is removed from the network. We further assume that 47 passengers are shifted from itinerary 8 to itinerary 7 (i.e. $\Delta D_8^7 = 47$).

One might be interested to estimate the impact of the deletion of itineraries at the flight level rather than at the itinerary level. For example, flight F1 is part of itineraries 1 and 7. The change in the demand of flight 1 due to the deletion of itineraries 2 and 3 in city-pair MIA-ORD and the deletion of itinerary 8 in city-pair MIA-DTW is computed as follows:

$$\left(\Delta D_3^1 + \Delta D_2^1\right) + \left(\Delta D_8^7\right) = (-8 + 25) + 47 = 64$$

Thus, the demand correction for a flight f that is part of itinerary $p \in P$ resulting from the deletion of optional itineraries $q \in P_o$ can be expressed mathematically as follows:

$$\sum_{p \in P} \sum_{q \in P^o} \delta_f^p \Delta D_q^p \left(1 - z_q\right)$$

where δ_f^p is equal to 1, if itinerary p includes flight leg f, and zero otherwise. In the above example, $\delta_{F1}^1 = 1$ and $\delta_{F1}^7 = 1$.

Finally, it should be noted that the estimation of ΔR requires knowledge of the value D_q, which is the unconstrained demand of each itinerary $q \in P_o$. In addition, there should be a mechanism to estimate the impact of the deletion of an itinerary $q \in P_o$ on the unconstrained demand D_q^p of other itineraries p in the same city-pair. The values D_q and D_q^p are expected to get different values for each schedule proposed by the airline.

It is important at this stage to distinguish between the demand correction terms defined above and the recapture rate. Both the demand correction term and the recapture rate are used for demand adjustment (i.e. updating the demand of the itineraries). Both factors estimate the demand of alternative itineraries, if a preferred itinerary is longer available (deleted or become full). The main difference is that the demand correction term deals with the unconstrained demand of the city-pair, while the recapture rate deals with the constrained demand. The demand correction term estimates the change in the unconstrained demand of an itinerary, when an alternative itinerary is eliminated. The recapture rate, on the other hand, attempts to relocate passengers on alternative itineraries, when the desired itinerary is full. The alternative itineraries should have enough seats to accommodate these passengers. The recapture rate does not impact the total unconstrained demand of the itinerary or the city-pair.

19.4 The Schedule Design and Adjustment Problem

19.4.1 The Objective Function of ISD-IFAM

The objective function of the ISD-IFAM problem is similar to the objective function of the IFAM problem presented in Chapter 16, with an additional term added to consider minimizing the unconstrained revenue loss due to flight deletion. Accordingly, the objective function can be represented mathematically as follows:

$$\text{Maximize } R - S + M - C - \Delta R$$

where R is the total unconstrained revenue, S is the revenue spilled due to limited seat capacity on some itineraries, M is the portion of the spilled revenue that could be recaptured on other alternative itineraries, C is the operation cost, and ΔR is the unconstrained revenue loss due to flight deletion. Since the total unconstrained revenue is constant, an equivalent cost-minimizing objective function can be written as follows:

$$\text{Minimize } C + S - M + \Delta R$$

Following the notations presented in Chapter 16, the objective function can be written in a detailed format as follow:

$$\text{Minimize } \sum_{f \in F} \sum_{e \in E} c_{fe} \cdot x_{fe} + \sum_{p \in P} t_p \cdot \text{fare}_p - \sum_{r \in P} \sum_{p \in P} b_p^r \, t_p^r \, \text{fare}_r +$$
$$\sum_{q \in P^o} \left(D_q \, \text{fare}_q - \sum_{p: p \neq q} \Delta D_q^p \, \text{fare}_p \right) \cdot \left(1 - z_q \right)$$

19.4.2 The Constraints of the ISD-IFAM

The constraints of the ISD-IFAM problem are similar to those considered for the FAM and the IFAM problems. They include the coverage, balance, and aircraft availability constraints. One difference is that the coverage constraints are different between mandatory and optional flights. Define L^F and L^O as the sets of mandatory and optional flights, respectively. Each of these sets is indexed by f. The coverage constraints of the mandatory flights, which ensure that each flight is assigned to one fleet type, are written as follows:

$$\sum_{e \in E} x_{fe} = 1, \qquad \forall f \in L^F$$

The coverage constraints of the optional flights are written as shown below. If an option flight f is selected as part of the schedule, a fleet should be assigned to this flight:

$$\sum_{e\in E} x_{fe} \leq 1, \qquad \forall f \in L^O$$

The demand constraints presented in the IFAM problem are extended to consider the demand correction when an optional itinerary is removed from the schedule. The constraints below ensure that for each itinerary $p\in P$, the total number of passengers requesting itinerary $p\in P$ and redirected to other itineraries $r\in P$ should be less than or equal to the demand of itinerary p plus the demand correction for itinerary $p\in P$ due to the deletion of any optional itineraries:

$$\sum_{r\in P} t_p^r \leq D_p + \sum_{q\in P^O} \Delta D_q^p \left(1-Z_q\right) \qquad \forall p \in P$$

Similarly, the capacity constraints presented in the IFAM problem are extended to consider the demand correction when an optional itinerary is removed from the schedule. The constraints below ensure that the number of passengers of a flight does not exceed the capacity assigned to this flight. For any flight $f\in F$, the difference between the spilled and recaptured demands should exceed the number of passengers that must be spilled from the flight:

$$\sum_{r\in P}\sum_{p\in P} \delta_f^p t_p^r - \sum_{r\in P}\sum_{p\in P} \delta_f^p b_r^p t_r^p \geq Q_f + \sum_{p\in P}\sum_{q\in P^O} \delta_f^p \Delta D_q^p \left(1-z_q\right) - \text{SEAT}_f$$

where Q_f is the unconstrained demand of flight $f\in F$, in case all itineraries are included in the schedule.

Another two constraints are added, which are pertinent to the decision variable Z_q. As mentioned above, the decision variable z_q is equal to 1, if itinerary q is included in the proposed schedule, and zero otherwise. Accordingly, constraints are added to ensure that Z_q is set to zero if at least one flight in itinerary q is eliminated. These constraints can be written as follows:

$$Z_q \leq \sum_{e\in E} x_{fe} \qquad \forall f \in L\left(q\right)$$

where $L(q)$ is the set of flight legs in itinerary q.

In the above example, itinerary 3 is composed of two flights, F4 and F5. Thus, two constraints need to be considered for itinerary 3:

$$Z_{\text{itinerary 3}} \leq \sum_{e\in E} x_{F4e}$$

$$Z_{\text{itinerary 3}} \leq \sum_{e \in E} x_{F5e}$$

Since flight F4 is removed and not assigned to any fleet (i.e. $\sum_{e \in E} x_{F4e} = 0$), it implies that $Z_{\text{itinerary 3}} = 0$, that is, itinerary 3 is also eliminated.

Other constraints are added to ensure that Z_q is equal to 1 in case all flights in itinerary q are considered to be part of the schedule:

$$Z_q - \sum_{f \in L(q)} \sum_{e \in E} x_{fe} \geq 1 - N_q \qquad \forall q \in P^O$$

where N_q is the number of flight legs in itinerary q.

In the above example, itinerary 7 has two flights, which are F1 and F7 (i.e. $N_q = 2$). Thus, the right-hand side is equal to -1. If flights F1 and F7 of itinerary 7 are considered in the schedule, the sum $\sum_{f \in L(q)} \sum_{e \in E} x_{fe}$, which is $\sum_{e \in E} x_{F1e} + \sum_{e \in E} x_{F7e}$, is equal to 2. Thus, Z_7 should be equal to 1. On the other hand, if $\sum_{e \in E} x_{F1e} = 0$ and/or $\sum_{e \in E} x_{F7e} = 0$, Z_7 should be set to zero to satisfy the constraint above for itinerary 7.

20

Example on ISD-IFAM

20.1   Problem Definition

This chapter provides an example that illustrates the application of the integrated schedule design problem based on the itinerary-based fleet assignment model (ISD-IFAM). The example presented here extends the example presented in Chapter 17 to consider the case in which the schedule can be adjusted by possibly eliminating some of its optional flights. Figure 20.1 shows the flight schedule of a hypothetical airline that consists of six flights serving between LAX, ORD, and BOS. These six flights are identified as LAX-ORD ($f1$), ORD-BOS ($f2$), LAX-BOS ($f3$), BOS-ORD ($f4$), ORD-LAX ($f5$), and BOS-LAX ($f6$). They represent the set of master flights. Flights $f1$, $f3$, $f5$, and $f6$ are assumed to be mandatory, while flights $f2$ and $f4$ are optional.

As shown in the figure, each of the city-pairs LAX-ORD, ORD-LAX, ORD-BOS, and BOS-ORD is served by nonstop flights. The city-pairs LAX-BOS and BOS-LAX are served by both nonstop and connecting itineraries. For example, the city-pair LAX-BOS is served by the nonstop itinerary $f3$ and the connecting itinerary $f1 - f2$. Similarly, the city-pair BOS-LAX is served by the nonstop itinerary $f6$ and the connecting itinerary $f4 - f5$. Accordingly, the network includes a total of eight itineraries, which are LAX-ORD, ORD-BOS, LAX-BOS, BOS-ORD, ORD-LAX, BOS-LAX, LAX-ORD-BOS, and BOS-ORD-LAX. The passenger demand and average fare of each itinerary are given in Table 20.1.

Assume that each of these six flights can be assigned to aircraft of fleet types $e1$ or $e2$ with seat capacity of 150 and 250 seats, respectively. The two aircraft types are assumed to have the same speed. Thus, each flight has the same arrival time regardless of its assigned fleet. The operating cost of each of the six flights considering the two fleet types are given in Table 20.2.

Airline Network Planning and Scheduling, First Edition. Ahmed Abdelghany and Khaled Abdelghany.
© 2019 John Wiley & Sons, Inc. Published 2019 by John Wiley & Sons, Inc.

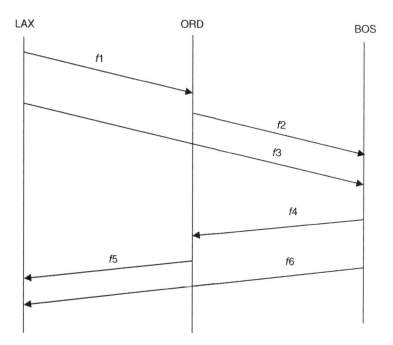

Figure 20.1 An example of a hypothetical airline with six flights for the ISD-IFAM problem.

Table 20.1 The passenger demand and fares on each of the eight itineraries for the ISD-IFAM problem.

Itinerary	LAX-ORD	ORD-BOS	LAX-BOS	BOS-ORD	ORD-LAX	BOS-LAX	LAX-ORD-BOS	BOS-ORD-LAX
Passenger demand	50	50	300	50	50	300	50	50
Fare ($)	200	225	350	225	200	350	300	300

Table 20.2 The operating cost of each flight for each fleet type for the ISD-IFAM problem.

		Fleet	
Flight		e1 ($)	e2 ($)
LAX-ORD	f1	30,000	16,600
ORD-BOS	f2	15,100	13,500
LAX-BOS	f3	15,500	45,000
BOS-ORD	f4	57,000	13,500
ORD-LAX	f5	42,000	16,600
BOS-LAX	f6	15,500	45,000

With six flights and two fleet types, the problem has 12 assignment decision variables, where each of the six flights is assigned to one of the two fleet types. Table 20.3 gives the notation of the decision variables of the problem. For example, $x_{5,1}$ is equal to 1 if flight 5 is assigned to fleet $e1$, and zero otherwise.

To consider the balance constraints of the problem, the interconnection nodes are defined at all three airports. As both fleet types have the same speed, their interconnection nodes are identical as shown in Figure 20.2. As illustrated, there are two interconnection nodes at LAX, two interconnection nodes at ORD, and one interconnection node at BOS.

Table 20.3 The decision variables of fleet assignment for the ISD-IFAM example.

Flight	e1	e2
LAX-ORD	$x_{1,1}$	$x_{1,2}$
ORD-BOS	$x_{2,1}$	$x_{2,2}$
LAX-BOS	$x_{3,1}$	$x_{3,2}$
BOS-ORD	$x_{4,1}$	$x_{4,2}$
ORD-LAX	$x_{5,1}$	$x_{5,2}$
BOS-LAX	$x_{6,1}$	$x_{6,2}$

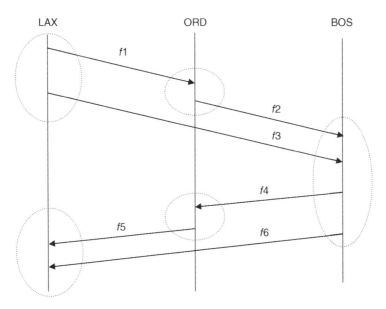

Figure 20.2 The interconnection nodes for the ISD-IFAM example.

Table 20.4 The decision variables for the ground arcs and the overnight arcs for the ISD-IFAM example.

$RON_{LAX\,e1}$	Number of aircraft of type $e1$ that remain overnight at LAX
$RON_{ORD\,e1}$	Number of aircraft of type $e1$ that remain overnight at ORD
$RON_{BOS\,e1}$	Number of aircraft of type $e1$ that remain overnight at BOS
$y_{(1-2)\,LAX\,e1}$	Number of aircraft of type $e1$ on the ground links and connects interconnection nodes 1 and 2 at LAX
$y_{(1-2)\,ORD\,e1}$	Number of aircraft of type $e1$ on the ground links and connects interconnection nodes 1 and 2 at ORD
$RON_{LAX\,e2}$	Number of aircraft of type $e2$ that remain overnight at LAX
$RON_{ORD\,e2}$	Number of aircraft of type $e2$ that remain overnight at ORD
$RON_{BOS\,e2}$	Number of aircraft of type $e2$ that remain overnight at BOS
$y_{(1\ 2)\,LAX\,e2}$	Number of aircraft of type $e2$ on the ground links and connects interconnection nodes 1 and 2 at LAX
$y_{(1-2)\,ORD\,e2}$	Number of aircraft of type $e2$ on the ground links and connects interconnection nodes 1 and 2 at ORD

Given the interconnection nodes, the list of decision variables that are related to the ground arcs and the overnight arcs are given in Table 20.4.

20.2 The Constraints of the Problem

The coverage constraints of problem entail that each flight is assigned to one fleet type only. The problem has six coverage constraints as follows.

For the mandatory flights

$$x_{1,1} + x_{1,2} = 1$$

$$x_{3,1} + x_{3,2} = 1$$

$$x_{5,1} + x_{5,2} = 1$$

$$x_{6,1} + x_{6,2} = 1$$

For the optional flights

$$x_{2,1} + x_{2,2} \leq 1$$

$$x_{4,1} + x_{4,2} \leq 1$$

The problem includes 10 balance constraints that are defined as follows:

$$\text{RON}_{\text{LAX } e1} = y_{(1-2)\text{LAX } e1} + x_{1,1} + x_{3,1}$$

$$y_{(1-2)\text{LAX } e1} + x_{5,1} + x_{6,1} = \text{RON}_{\text{LAX } e1}$$

$$\text{RON}_{\text{LAX } e2} = y_{(1-2)\text{LAX } e2} + x_{1,2} + x_{3,2}$$

$$y_{(1-2)\text{LAX } e2} + x_{5,2} + x_{6,2} = \text{RON}_{\text{LAX } e2}$$

$$\text{RON}_{\text{ORD } e1} + x_{1,1} = y_{(1-2)\text{ORD } e1} + x_{2,1}$$

$$y_{(1-2)\text{ORD } e1} + x_{4,1} = \text{RON}_{\text{ORD } e1} + x_{5,1}$$

$$\text{RON}_{\text{ORD } e2} + x_{1,2} = y_{(1-2)\text{ORD } e2} + x_{2,2}$$

$$y_{(1-2)\text{ORD } e2} + x_{4,2} = \text{RON}_{\text{ORD } e2} + x_{5,2}$$

$$\text{RON}_{\text{BOS } e1} + x_{2,1} + x_{3,1} = \text{RON}_{\text{BOS } e1} + x_{4,1} + x_{6,1}$$

$$\text{RON}_{\text{BOS } e2} + x_{2,2} + x_{3,2} = \text{RON}_{\text{BOS } e2} + x_{4,2} + x_{6,2}$$

For simplicity, it is assumed that there is no limitation on resources, and other constrains such as maintenance and crew constraints are ignored.

20.3 The Objective Function

The problem minimizes an objective function in the form of the sum of the operating cost, the spill cost minus the recapture revenue, and the change in unconstrained revenue due to schedule change $(S - M + \Delta R)$. To explain the different components of the ISD-IFAM, consider the Excel sheets given in Figures 20.3 through 20.35. These figures describe the solution of the ISD-IFAM for the hypothetical airline network presented above. The spreadsheets give the formulas used for the calculations as well as the resulting values.

In Figures 20.3 and 20.4, column B (rows 4–9) gives the list of mandatory and optional flights considered in the schedule. Each flight is defined by its origin–destination pair. Columns C and D give the operating cost associated with assigning each of the six flights to fleet types $e1$ and $e2$, respectively. For example, it costs the airline $15,500 and $45,000 to operate flight LAX-BOS using fleet type $e1$ and fleet type $e2$, respectively. Column B (rows 10–12) also gives the variables for the overnight arcs (RON) for the three stations LAX, ORD, and BOS. Finally, the variables of the ground arcs between the interconnection nodes are given in rows 13 and 14.

Columns C and D give the cost associated with an aircraft remaining overnight at each of the three stations for each fleet type, which are assumed to be zero in this example. In addition, the cost associated with each aircraft that remains idle between the interconnection nodes for each fleet type is given.

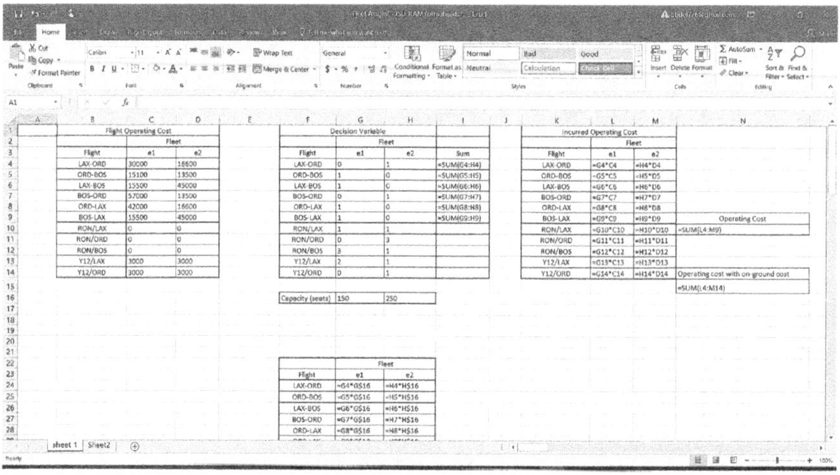

Figure 20.3 The decision variables and formulas for cost calculations for the ISD-IFAM example.

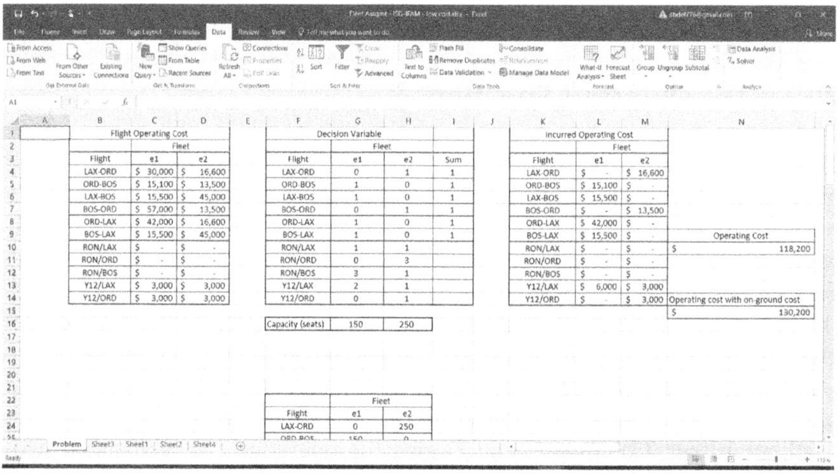

Figure 20.4 The decision variables and initial cost values for the ISD-IFAM example.

It costs the airline $3000 for each aircraft of both fleet types to remain on the ground between the defined interconnection nodes.

Columns G and H give the initial values of the assignment decision variables. For example, if the value of cell H4 is equal to 1, it implies that the LAX-ORD flight is assigned initially to fleet type $e2$. The value of cell G12 implies that there are three aircraft of type $e1$ to remain overnight at BOS. The value of cell H14 implies that there is one aircraft of fleet type $e2$ to remain on the ground during the time between the two interconnection nodes at ORD.

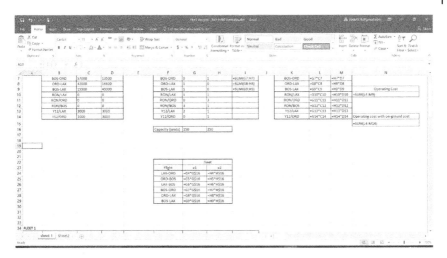

Figure 20.5 The formulas representing the seat capacity of each flight for the ISD-IFAM example.

The table in columns K, L, and M gives the incurred operating cost based on the initial values of the decision variables. The total operations cost is given in cell N15, which is the sum of the cost elements given in columns L and M (rows 4–14). For example, given the initial values of the decision variables, the operating cost is \$130,200. This cost corresponds to the value C in the objective function of the ISD-IFAM problem ($C + S - M + \Delta R$). Finally, column I (rows 4–9) gives the sum of the flight assignment decision variables for all fleet types. This value is used to set the coverage constraints of the fleet assignment problem for the mandatory and optional flights.

In Figures 20.5 and 20.6, cells G16 and H16 give the seat capacity of the aircraft of fleet types $e1$ and $e2$, respectively. An aircraft belonging to fleet type $e1$ has a seat capacity of 150 seats, while an aircraft of fleet type $e2$ has a seat capacity of 250 seats. Columns G and H (rows 24–29) give the seat capacity of each flight based on the initial values of the decision variables. The spreadsheet in Figure 20.5 gives the formula of the cells, while the spreadsheet in Figure 20.6 gives the corresponding values. For example, since the value in cell H4 is set to one, it implies that flight LAX-ORD is initially assigned to fleet type $e2$ and has a seat capacity of 250 seats.

Figures 20.7 and 20.8 show the balance constraints (starting at row 35) of the ISD-IFAM problem. The spreadsheet in Figure 20.7 gives the formula of the cells, while the spreadsheet in Figure 20.8 gives the corresponding values. As given above, there are five interconnection nodes for each fleet type, and hence ten balance constraints are defined for the problem. For each constraint, the values of the right-hand side (RHS) and the left-hand side (LHS) are given. At this stage, the balance constraints are not satisfied yet, and the RHS and the LHS of each constraint might not be equal.

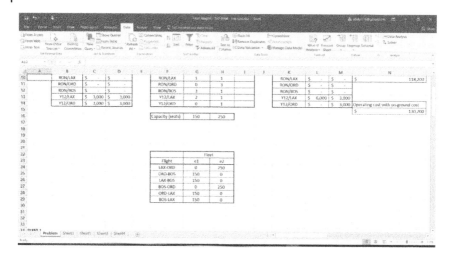

Figure 20.6 The initial values of the seat capacity of each flight for the ISD-IFAM example.

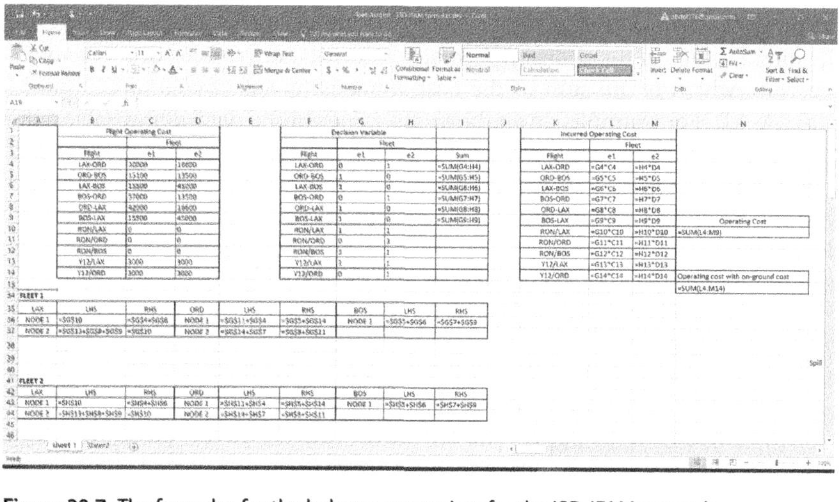

Figure 20.7 The formulas for the balance constraints for the ISD-IFAM example.

In Figures 20.9 and 20.10, the table in columns Q–X (rows 4–9) gives the itinerary–flight incidence matrix. The matrix defines the relationship between the flights and the itineraries in the network. In this matrix, the flights are given in the rows, while the itineraries are given in the columns. A cell in the matrix is equal to 1, if the itinerary represented by the cell includes the flight that corresponds to the same cell, and zero otherwise. For example, the value of cell W5 is equal to 1 because flight ORD-BOS is part of itinerary

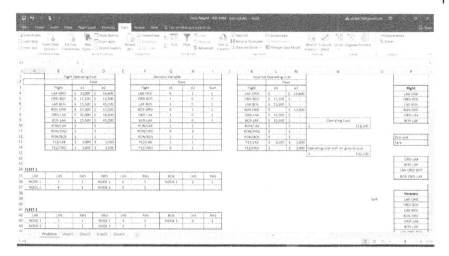

Figure 20.8 The initial values of the balance constraints for the ISD-IFAM example.

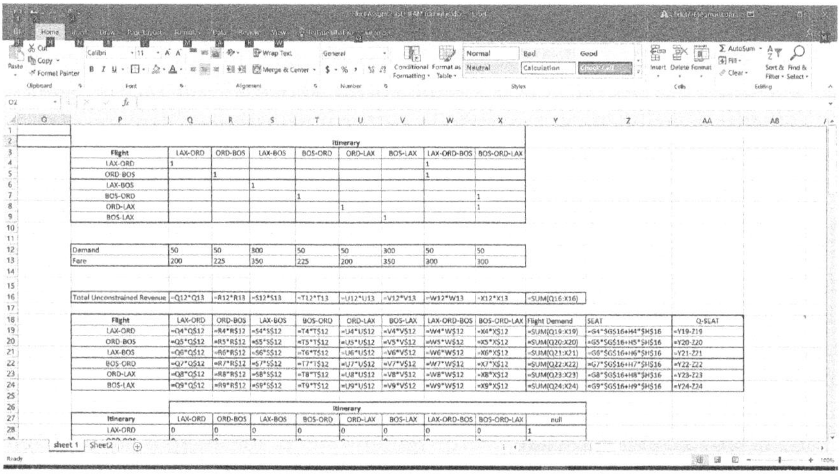

Figure 20.9 The formulas for the unconstrained demand for each flight for the ISD-IFAM example.

LAX-ORD-BOS. Similarly, itinerary BOS-ORD-LAX has two flights BOS-ORD and ORD-LAX. Thus, X7 and X8 are equal to 1.

Rows 12 and 13 (columns Q–X) give the unconstrained passenger demand and the fare for each of the eight itineraries. The unconstrained demand of an itinerary is defined as the number of passengers that choose this itinerary assuming unlimited seat capacity. The unconstrained revenue is calculated in row 16. The total unconstrained revenue is given in cell Y16.

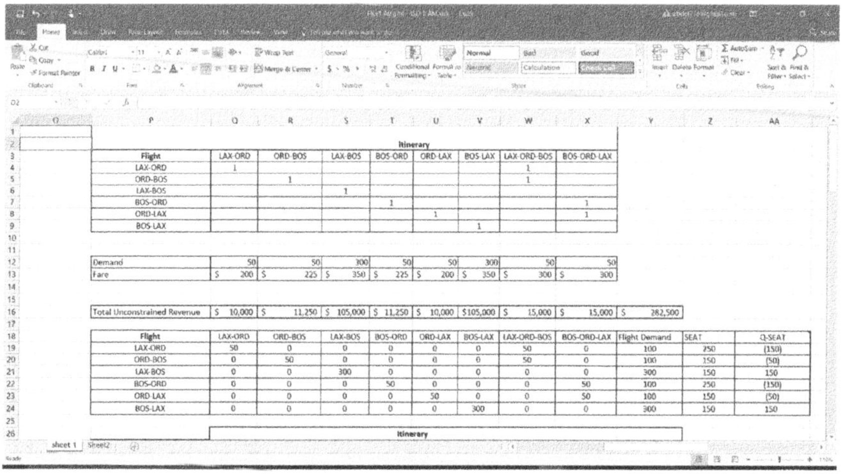

Flight	LAX-ORD	ORD-BOS	LAX-BOS	BOS-ORD	ORD-LAX	BOS-LAX	LAX-ORD-BOS	BOS-ORD-LAX
	Itinerary							
LAX-ORD	1						1	
ORD-BOS		1					1	
LAX-BOS			1					
BOS-ORD				1				1
ORD-LAX					1			1
BOS-LAX						1		

Demand	50	50	300	50	50	300	50	50	
Fare	$ 200	$ 225	$ 350	$ 225	$ 200	$ 350	$ 300	$ 300	
Total Unconstrained Revenue	$ 10,000	$ 11,250	$ 105,000	$ 11,250	$ 10,000	$105,000	$ 15,000	$ 15,000	$ 282,500

Flight	LAX-ORD	ORD-BOS	LAX-BOS	BOS-ORD	ORD-LAX	BOS-LAX	LAX-ORD-BOS	BOS-ORD-LAX	Flight Demand	SEAT	Q-SEAT
LAX-ORD	50	0	0	0	0	0	50	0	100	250	(150)
ORD-BOS	0	50	0	0	0	0	50	0	100	150	(50)
LAX-BOS	0	0	300	0	0	0	0	0	300	150	150
BOS-ORD	0	0	0	50	0	0	0	50	100	250	(150)
ORD-LAX	0	0	0	0	50	0	0	50	100	150	(50)
BOS-LAX	0	0	0	0	0	300	0	0	300	150	150

Itinerary

Figure 20.10 The values for the unconstrained demand for each flight for the ISD-IFAM example.

Figures 20.9 and 20.10 also give the calculations of the unconstrained demand at the flight level. First, the unconstrained demand of each itinerary is mapped to all flights included in the itinerary. Then, the unconstrained demand of each flight is calculated as the sum of the unconstrained demand of all itineraries that include this flight. For example, flight BOS-ORD is included in two itineraries, which are BOS-ORD and BOS-ORD-LAX. Each of these itineraries has 50 passengers (unconstrained demand). Thus, the unconstrained demand of this flight is equal to 100, which is the sum of the unconstrained demand of these two itineraries. The total unconstrained demand (Q) of each flight is given in column Y (rows 19–24). Column Z (rows 19–24) gives the seat capacity of each flight. The seat capacity of each flight depends on the fleet assignment solution of the flight. The fleet assignment solutions are given in columns G and H (rows 4–9). For example, the initial value of the assignment decision variable of flight LAX-BOS indicates that this flight is assigned to fleet type $e1$ (i.e. cell G6 = 1). Hence, the seat capacity of this flight is 150 seats. Finally, column AA (rows 19–24) gives the difference between the unconstrained demand of the flight and its seat capacity. For example, flight LAX-BOS (row 21) has a total demand of 300 passengers and a seat capacity of 150 seats, resulting in an excess demand of 150 passengers (cell AA21).

Figure 20.11 shows the itinerary–itinerary spill–recapture relationship matrix. This matrix is given in columns Q–Y and rows 28–35. The rows in the matrix represent itineraries that spill demand, while the columns represent itineraries that recapture demand. The cells in the matrix indicate whether there is a spill–recapture relationship between the two itineraries.

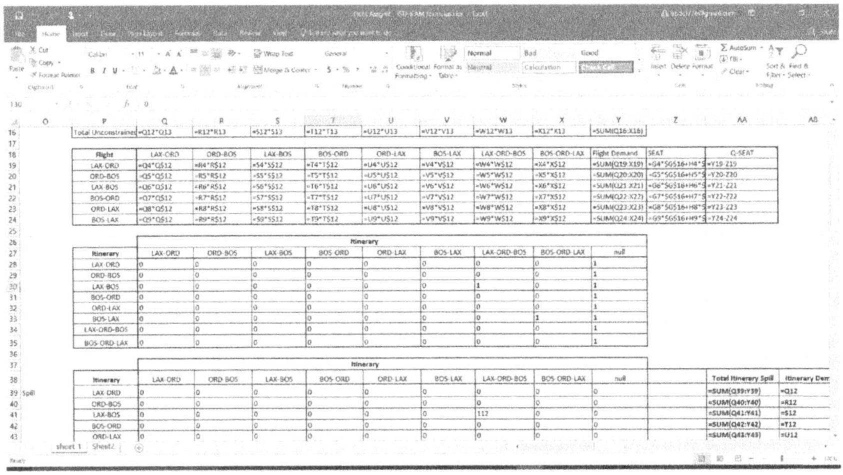

	LAX-ORD	ORD-BOS	LAX-BOS	BOS-ORD	ORD-LAX	BOS-LAX	LAX-ORD-BOS	BOS-ORD-LAX	
Total Unconstrained	=Q12*Q13	=R12*R13	=S12*S13	=T12*T13	=U12*U13	=V12*V13	=W12*W13	=X12*X13	=SUM(Q16:X16)

Flight	LAX-ORD	ORD-BOS	LAX-BOS	BOS-ORD	ORD-LAX	BOS-LAX	LAX-ORD-BOS	BOS-ORD-LAX	Flight Demand	SEAT	Q-SEAT
LAX-ORD	=Q4*Q$12	=R4*R$12	=S4*S$12	=T4*T$12	=U4*U$12	=V4*V$12	=W4*W$12	=X4*X$12	=SUM(Q19:X19)	=G4*$G$16+H4*$	=Y19-Z19
ORD-BOS	=Q5*Q$12	=R5*R$12	=S5*S$12	=T5*T$12	=U5*U$12	=V5*V$12	=W5*W$12	=X5*X$12	=SUM(Q20:X20)	=G5*$G$16+H5*$	=Y20-Z20
LAX-BOS	=Q6*Q$12	=R6*R$12	=S6*S$12	=T6*T$12	=U6*U$12	=V6*V$12	=W6*W$12	=X6*X$12	=SUM(Q21:X21)	=G6*$G$16+H6*$	=Y21-Z21
BOS-ORD	=Q7*Q$12	=R7*R$12	=S7*S$12	=T7*T$12	=U7*U$12	=V7*V$12	=W7*W$12	=X7*X$12	=SUM(Q22:X22)	=G7*$G$16+H7*$	=Y22-Z22
ORD-LAX	=Q8*Q$12	=R8*R$12	=S8*S$12	=T8*T$12	=U8*U$12	=V8*V$12	=W8*W$12	=X8*X$12	=SUM(Q23:X23)	=G8*$G$16+H8*$	=Y23-Z23
BOS-LAX	=Q9*Q$12	=R9*R$12	=S9*S$12	=T9*T$12	=U9*U$12	=V9*V$12	=W9*W$12	=X9*X$12	=SUM(Q24:X24)	=G9*$G$16+H9*$	=Y24-Z24

	Itinerary								
Itinerary	LAX-ORD	ORD-BOS	LAX-BOS	BOS-ORD	ORD-LAX	BOS-LAX	LAX-ORD-BOS	BOS-ORD-LAX	null
LAX-ORD	0	0	0	0	0	0	0	0	1
ORD-BOS	0	0	0	0	0	0	0	0	1
LAX-BOS	0	0	0	0	0	0	1	0	1
BOS-ORD	0	0	0	0	0	0	0	0	1
ORD-LAX	0	0	0	0	0	0	0	0	1
BOS-LAX	0	0	0	0	0	0	0	1	1
LAX-ORD-BOS	0	0	0	0	0	0	0	0	1
BOS-ORD-LAX	0	0	0	0	0	0	0	0	1

		Itinerary										
	Itinerary	LAX-ORD	ORD-BOS	LAX-BOS	BOS-ORD	ORD-LAX	BOS-LAX	LAX-ORD-BOS	BOS-ORD-LAX	null	Total Itinerary Spill	Itinerary Dem
Spill	LAX-ORD	0	0	0	0	0	0	0	0	0	=SUM(Q39:Y39)	=Q12
	ORD-BOS	0	0	0	0	0	0	0	0	0	=SUM(Q40:Y40)	=R12
	LAX-BOS	0	0	0	0	0	0	112	0	0	=SUM(Q41:Y41)	=S12
	BOS-ORD	0	0	0	0	0	0	0	0	0	=SUM(Q42:Y42)	=T12
	ORD-LAX	0	0	0	0	0	0	0	0	0	=SUM(Q43:Y43)	=U12

Figure 20.11 The itinerary–itinerary spill–recapture relationship matrix for the ISD-IFAM example.

For example, having cell W30 equal to 1 indicates that itinerary LAX-BOS can spill demand to itinerary LAX-ORD-BOS. Each itinerary can spill demand to a dummy itinerary, which represents the case when the airline cannot recapture the spilled demand on any of its itineraries.

Figure 20.12 shows the spilled demand decision variables. These variables are given in columns Q–Y and rows 39–46. It gives the number of spilled and recaptured passengers among itineraries. For example, it is initially assumed that 107 passengers are to be spilled from itinerary BOS-LAX to BOS-ORD-LAX (cell X44). These numbers are associated with the itinerary–itinerary spill–recapture relationship matrix defined above. In other words, the number of passengers to be spilled and recaptured between any two itineraries is valid, only if the corresponding value in the itinerary–itinerary spill–recapture relationship matrix is equal to 1. At this stage, preliminary numbers are given in the table. Solving the ISD-IFAM problem, the optimal values of the spilled demand are determined in conjunction with the optimal fleet assignment.

Figures 20.13 and 20.14 give the calculations of the total spilled demand of each itinerary. The total itinerary's spilled demand is the sum of the demand spilled from all recommended (substitute) itineraries including the dummy itinerary. Column AA gives the calculations of the total spilled demand of each itinerary (rows 39–46). Columns AB and AC (rows 39–46) give the itinerary demand and the itinerary fares. Column AD (rows 39–46) gives the spilled revenue of each itinerary. The spilled revenue is calculated by multiplying the spilled demand of the itinerary by its fare. Finally, cell AD47 gives the total spilled revenue, which is the sum of spill revenue of all itineraries. For example,

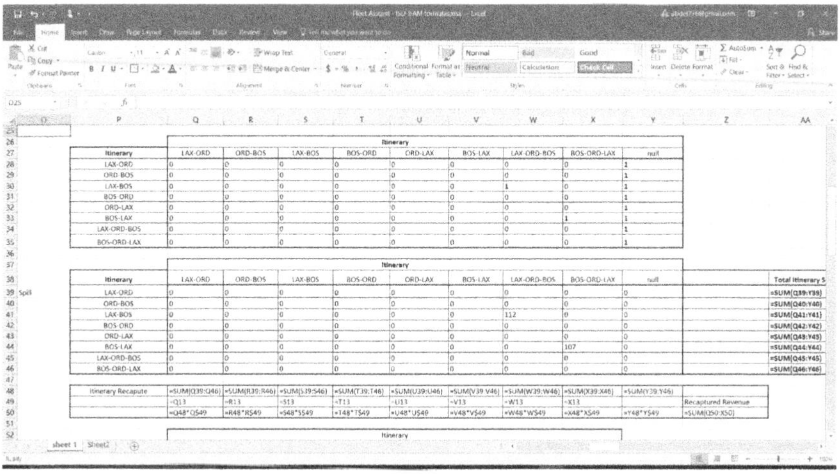

Figure 20.12 The initial values of the spilled demand decision variables for the ISD-IFAM example.

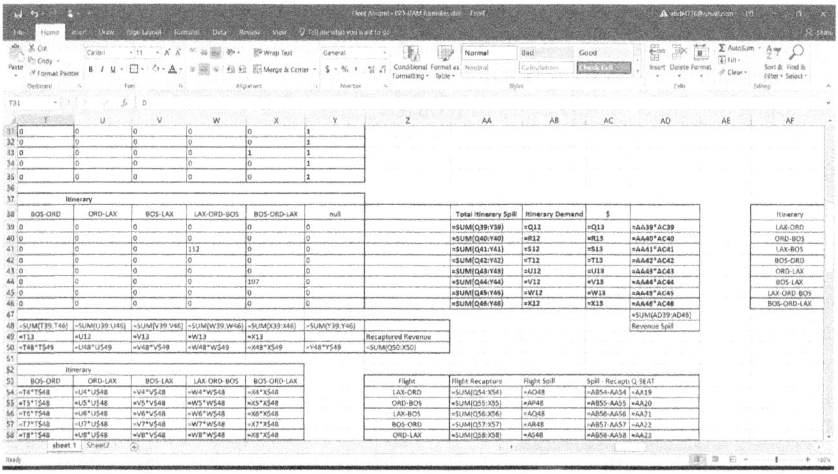

Figure 20.13 The formulas for the calculations of the total demand spill for each itinerary for the ISD-IFAM example.

given the initial values of the decision assignment and the spill variables, the spill revenue is \$76,650. The spilled revenue is the value of the variable S in the objective function $(C + S - M + \Delta R)$.

Figure 20.15 gives the itinerary–flight incidence matrix after transposing. As mentioned above, the incidence matrix defines the relation between the

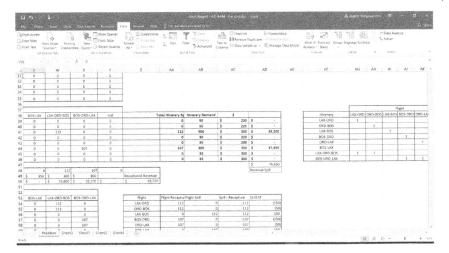

Figure 20.14 The initial values of the total demand spill for each itinerary for the ISD-IFAM example.

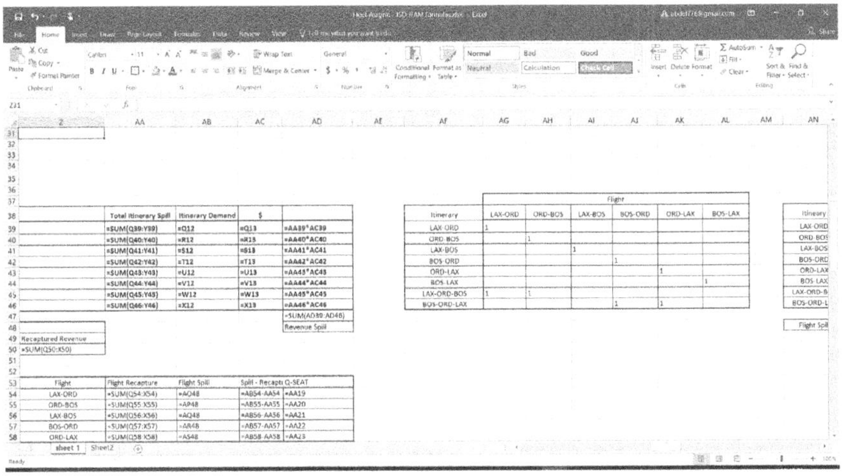

Figure 20.15 The transposed form of the itinerary–flight incidence matrix for the ISD-IFAM example.

flights and the itineraries in the network. The transposed matrix is given in columns AG–AL (rows 39–46), where the columns define the flights and the rows define the itineraries.

In Figures 20.16 and 20.17, the table given in columns AO–AT (rows 39–46) shows the calculations of the spilled demand at the flight level. The logic here is that in case a demand is spilled of an itinerary, it is by definition spilled from

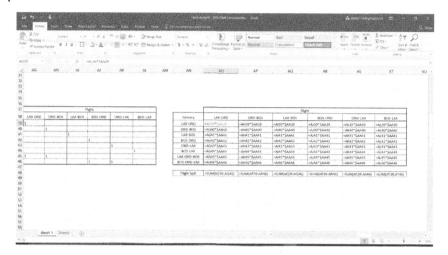

Figure 20.16 The formulas for the calculations of the demand spill at the flight level for the ISD-IFAM example.

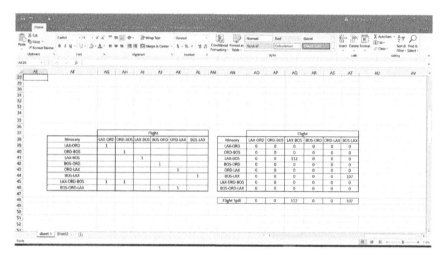

Figure 20.17 The initial values of the demand spill at the flight level for the ISD-IFAM example.

all flights of that itinerary. A flight could be part of more than one itinerary. Hence, a flight's spilled demand is calculated as the sum of the spilled demand of all itineraries that include the flight. The spilled demand of the flights is given in row 48 (columns AO–AT).

Figures 20.18 and 20.19 give the itineraries' recaptured demand. As described earlier, in case demand is spilled from an itinerary, the airline suggests another

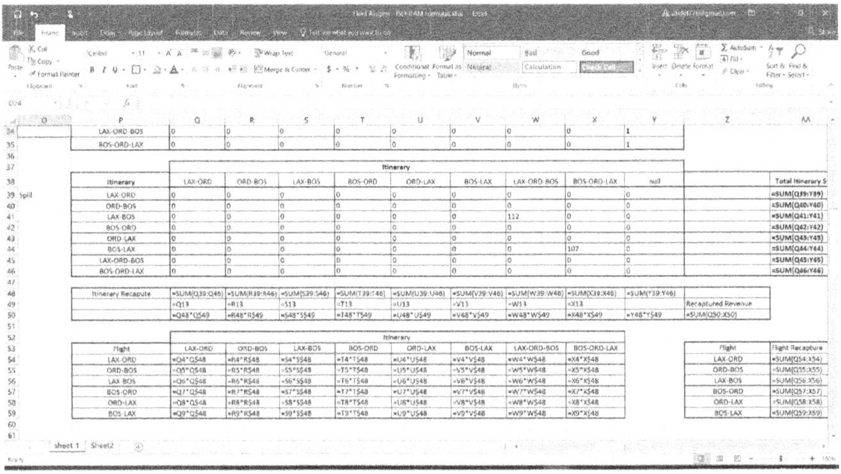

Figure 20.18 The formulas for the calculations of the recaptured demand and recaptured revenue for the ISD-IFAM example.

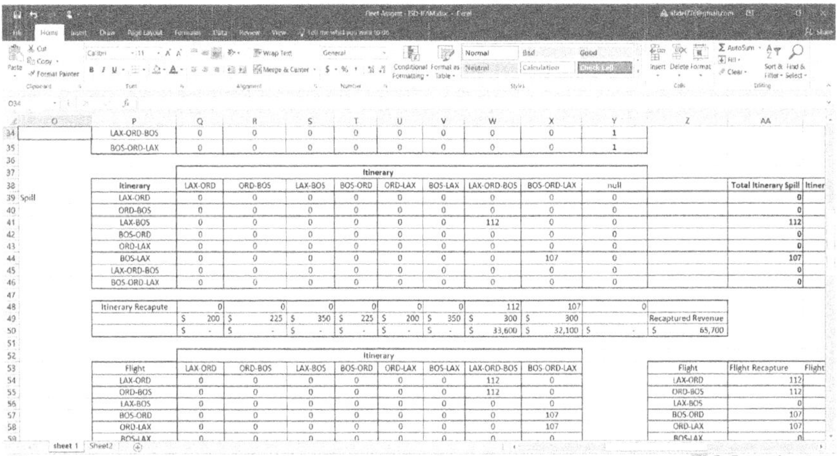

Figure 20.19 The initial values of the recaptured demand and recaptured revenue for the ISD-IFAM example.

itinerary (itineraries) to recapture this demand. An itinerary can recapture demand spilled from more than one itinerary. Thus, the demand recaptured on an itinerary is the sum of the demand spilled from other itineraries to this itinerary. Row 48 (columns Q–Y) gives the demand recaptured on each itinerary including the dummy itinerary.

At this stage, the values of the recaptured demand of each itinerary depend on the initial values of the decision variables (i.e. the fleet assignment variables and the demand spill variables). These initial values are not necessarily optimal or satisfying the problem's constraints. Accordingly, the recaptured demand of each itinerary is not necessarily satisfying the seat capacity constraints of the itineraries. Row 49 (columns Q–Y) gives the itinerary fare. Finally, the recaptured revenue of each itinerary is estimated by multiplying the recaptured demand of the itinerary by its fare. The total recaptured revenue of the airline is given in cell Z50, which is the sum of the recaptured revenue of all itineraries except for that of the dummy itinerary. For example, given the initial values of the fleet assignment and spill decision variables, the recaptured revenue is $65,700. The recaptured revenue corresponds to the value M in the objective function $(C + S - M + \Delta R)$.

Figures 20.18 and 20.19 also show the calculations of the recaptured demand at the flight level in the table given in columns Q–X (rows 54–59). The logic here is that if a demand is recaptured on an itinerary, it is recaptured on all flights that belong to the itinerary. A flight could be part of more than one itinerary. Thus, the demand recaptured on a flight is calculated as the sum of the demand recaptured on all itineraries that include the flight. The recaptured demand of each of the six flights is given in column AA (rows 54–59).

Figures 20.20 and 20.21 give the main statistics for the flights. The table in columns AA–AD (rows 54–59) gives the recaptured demand (column AA) and spilled demand (column AB) for each flight. The difference between

Figure 20.20 The calculations of the flight characteristics and the objective function of the ISD-IFAM example.

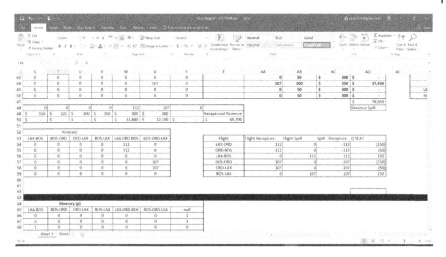

Figure 20.21 The initial values of the flight characteristics and the objective function of the ISD-IFAM example.

recaptured and spilled demand for each flight is given in column AC. Finally, the difference between the unconstrained demand and the seat capacity for each flight is given in column AD.

As mentioned above, the set of optional flights includes two flights, which are ORD-BOS and BOS-ORD. The decision to include these flights as part of the schedule is obtained based on the solution of this problem. The set of itineraries P^o that include these two optional flights includes four itineraries, which are ORD-BOS, BOS-ORD, LAX-ORD-BOS, and BOS-ORD-LAX. As explained above, if any of these itineraries is removed from the network, the unconstrained demand of the city-pair and the unconstrained demand of the alternative itineraries of the city-pair are expected to change. As mentioned above, the amount of change in the unconstrained demand of each itinerary depends on its attractiveness compared with other competing itineraries. The estimation of the share of each itinerary can be estimated using a methodology similar to the methodology presented in Chapter 8.

Figures 20.22 and 20.23 give the relationship between the itineraries in the set P^o, indexed by q, and alternative itineraries in the network. As mentioned above, the set P^o includes four itineraries that are included in the airline's schedule, only if the two optional flights ORD-BOS and BOS-ORD are scheduled. These itineraries are shown in rows 65–69. The table in columns Q–Y (rows 65–69) gives the incidence matrix for the alternative itineraries for each of the optional itineraries. In case an optional itinerary is removed from the schedule, the demand of this optional itinerary is shifted to alternative

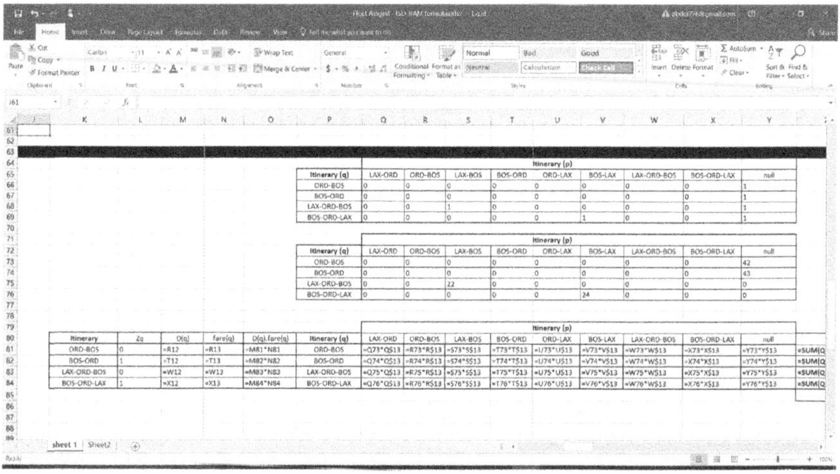

Figure 20.22 The formulas that give revenue change due to the suggested deletion of the optional itineraries for the ISD-IFAM example (the left-hand side).

itineraries. The alternative itineraries include a dummy itinerary, which represents the case when demand of the optional itinerary is spilled to competitors or other transportation modes. The value of a cell in this matrix is equal to 1 if the itinerary in the column is an alternative to the optional itinerary given in the row. For example, the optional itinerary LAX-ORD-BOS has the itinerary LAX-BOS and the dummy itinerary as alternative itineraries. The optional itinerary ORD-BOS has no alternative itineraries except the dummy itinerary. The table in column Q–Y (rows 73–76) gives the shifted demand from each optional itinerary to its alternative itineraries, in case the optional itinerary is removed from the schedule. The values in the cells of this table give the change in the demand of the alternative itinerary in case the optional itinerary is removed. For example, in case itinerary LAX-ORD-BOS is removed from the schedule, the demand of the alternative itinerary LAX-BOS is expected to increase by 22 passengers. The values in the cells of this table correspond to the values ΔD_q^p defined above, where q and p refer to the optional and alternative itineraries, respectively.

The table in columns K–AA (rows 81–84) gives the revenue change due to the deletion of the optional itineraries. The left side of this table is shown in Figures 20.22 and 20.23. The right side is shown in Figures 20.24 and 20.25. Column K (rows 81–84) gives the list of optional itineraries, as defined above. Column L (rows 81–84) gives the values of the decision variable Z_q for each optional itinerary q. As defined above, the decision variable z_q is equal to 1, if itinerary q is included in the proposed schedule, and zero otherwise. For example, the initial values suggest that the optional itineraries ORD-BOS and LAX-ORD-BOS are to be removed from the schedule. The values given at

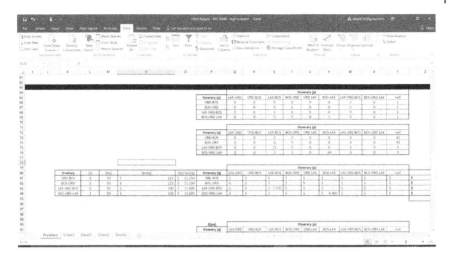

Figure 20.23 The revenue change due to the suggested deletion of the optional itineraries for the ISD-IFAM example (the left-hand side).

Figure 20.24 The formulas that give revenue change due to the suggested deletion of the optional itineraries for the ISD-IFAM example (the right-hand side).

this stage are initial values, which are not necessarily feasible nor optimum. The columns M and N (rows 81–84) give the demand and fare for each of the optional itineraries. These values are as defined in rows 12 and 13 (columns Q–X), respectively. The revenue of each of the optional itineraries is given in column O. This revenue is lost in case the optional itinerary is removed from the network. The revenues recaptured on alternative itineraries, if the

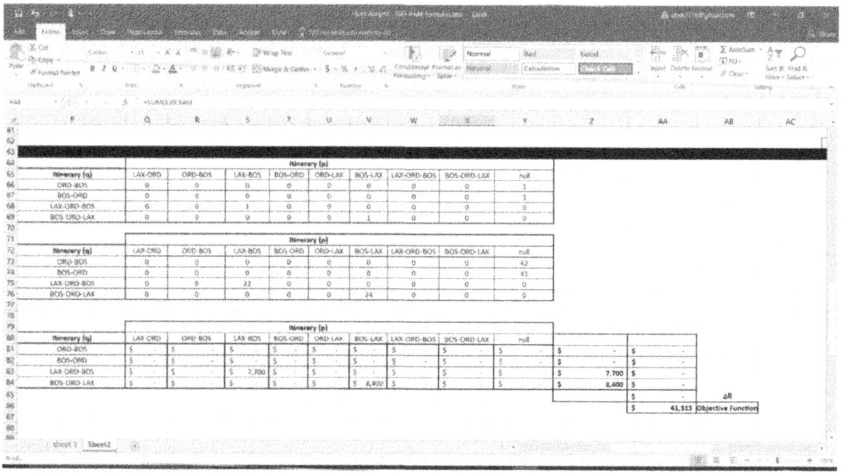

Figure 20.25 The revenue change due to the suggested deletion of the optional itineraries for the ISD-IFAM example (the right-hand side).

optional itineraries are removed, are given in column Z. The difference between the revenue loss and the recaptured revenue due to the removal of each of the optional itineraries is given in column AA (rows 81–84). Cell AA85 gives the term ΔR, which is the change in unconstrained revenue due to the elimination of the optional itineraries. Finally, cell AB86 gives the value of the objective function of the problem, which is defined as $C + S - M + \Delta R$. As shown in Figure 20.25, the initial value of the objective function is $41,313.

Figures 20.26 through 20.29 give additional calculations at the flight level. For example, the table in columns AG–AL (rows 81–84) gives the itinerary–flight incidence matrix for the optional itineraries. Columns AM and AN give the number of flights N_q and the value $1 - N_q$ for each optional itinerary q. Row 87 (columns AG–AL) gives the RHS of the coverage constraints of each flight. This value is equal to 1, if the flight is included in the schedule, and zero otherwise. Columns AP–AU (rows 81–84) redefine the itinerary–flight incidence matrix only if the flight is considered in the solution (i.e. when the RHS of the coverage constraint is not equal to 0). Columns AV and AW in Figures 20.28 and 20.29 give the value of the terms $\left(\sum_{f \in L(q)} \sum_{e \in E} x_{fe} \right)$ and $\left(Z_q - \sum_{f \in L(q)} \sum_{e \in E} x_{fe} \right)$ for each optional flight, respectively.

In Figures 20.30 and 20.31, row 96 (columns Q–Y) gives the change in the demand for each of the alternative itineraries resulting from removing the optional itineraries. Row 97 (columns Q–Y) gives the total demand for each alternative itinerary. This total demand is calculated as the unconstrained demand of the itinerary, which is given in row 12 (columns Q–Y) plus the

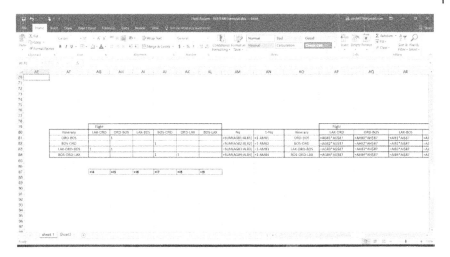

Figure 20.26 The formulas that show additional calculations for the optional flights for the ISD-IFAM example (part 1).

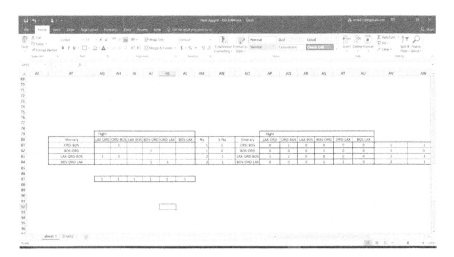

Figure 20.27 Additional calculations for the optional flights for the ISD-IFAM example (part 1).

additional demand calculated in row 96 (columns Q–Y). The table in rows 103–107 (columns Q–Y) shows the calculations of the additional demand at the flight level, which is shifted to the alternative itineraries, if the optional itineraries are eliminated. For example, given the initial values of the decision variable Z_q, there are 22 additional passengers on the flight LAX-BOS (cell S104).

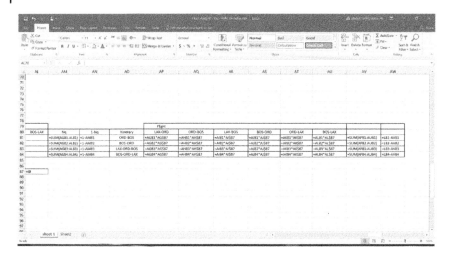

Figure 20.28 The formulas that show additional calculations for the optional flights for the ISD-IFAM example (part 2).

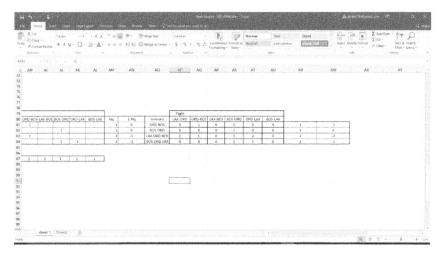

Figure 20.29 Additional calculations for the optional flights for the ISD-IFAM example (part 2).

In Figures 20.32 and 20.33, column AF (rows 102–107) gives the total additional demand shifted to each flight. Columns AB (rows 102–107) and AC (rows 102–107) in this table give the recaptured and spilled demand at the flight level. The difference between the recaptured and spilled demand at the flight level is given in column AE (rows 102–107). Finally, column AG (rows 102–107) gives the sum of columns AE and AF.

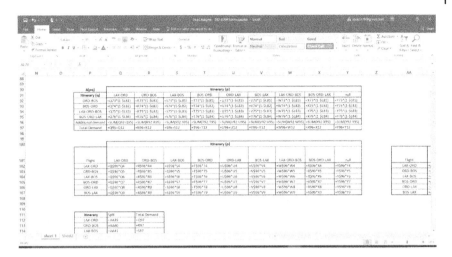

Figure 20.30 The formulas for the calculations of the demand change for each of the alternative itineraries, because of removing the optional itineraries for the ISD-IFAM example.

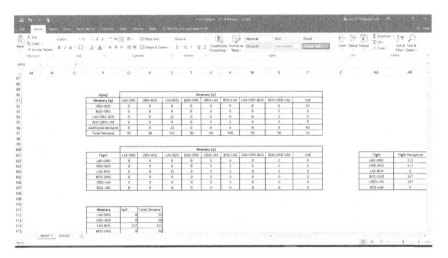

Figure 20.31 The demand change for each of the alternative itineraries, because of removing the optional itineraries for the ISD-IFAM example.

In Figures 20.34 and 20.35, the total demand spilled of each itinerary and the total demand of each itinerary are given in columns Q (rows 112–119) and R (rows 112–119), respectively. These two variables are used in the demand constraints, as illustrated in the next section.

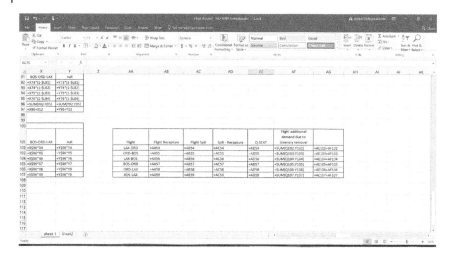

Figure 20.32 Formulas for calculating main flight characteristics for ISD-IFAM example.

Figure 20.33 Main flight characteristics for ISD-IFAM example.

20.4 Problem Solving

The Excel Solver is used to solve the problem as shown in Figures 20.36 through 20.38. A minimization objective function is defined in cell AA86. The decision variables of the problem are also defined. The decision variables include the flight-fleet assignment variables, which are given in columns G and H (rows 4–9). They also include the number of aircraft to stay on ground between the

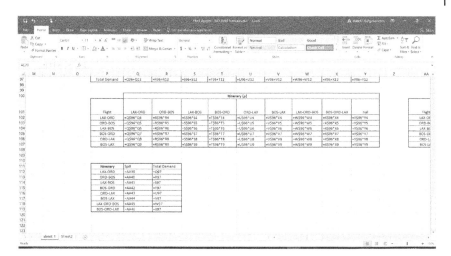

Figure 20.34 The formulas for the total demand spilled for each itinerary and the total demand of each itinerary for the ISD-IFAM example.

Figure 20.35 The total demand spilled for each itinerary and the total demand of each itinerary for the ISD-IFAM example.

interconnection nodes and overnight at the different stations for each fleet type, which are given in columns G and H (rows 10–14). The amount of demand spilled among itineraries is also included in the decision variables, which are cell W41, cell X44, and cells Y39–Y46. Finally, the decision variable Z_q for each optional itinerary q is given in cells L81–L84.

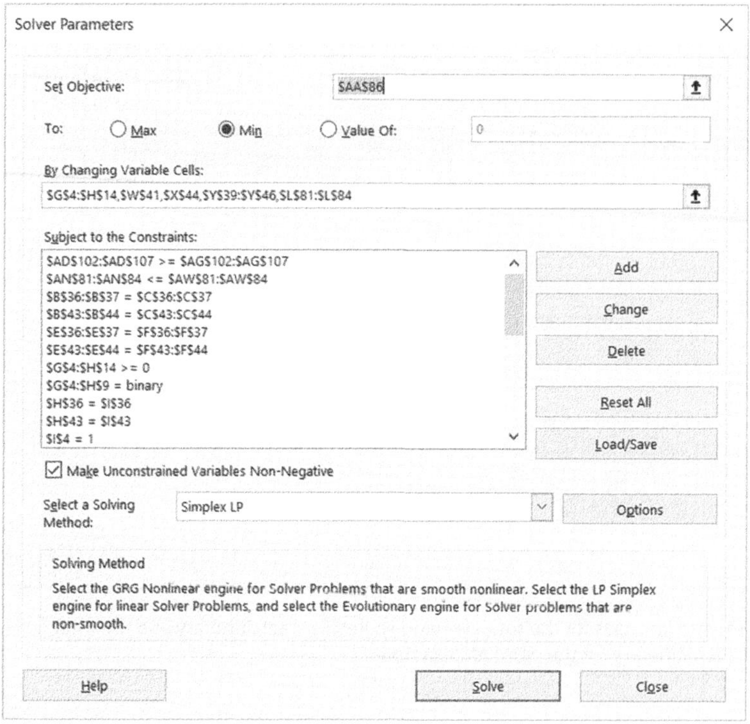

Figure 20.36 The Solver Parameters window for the ISD-IFAM example (part 1).

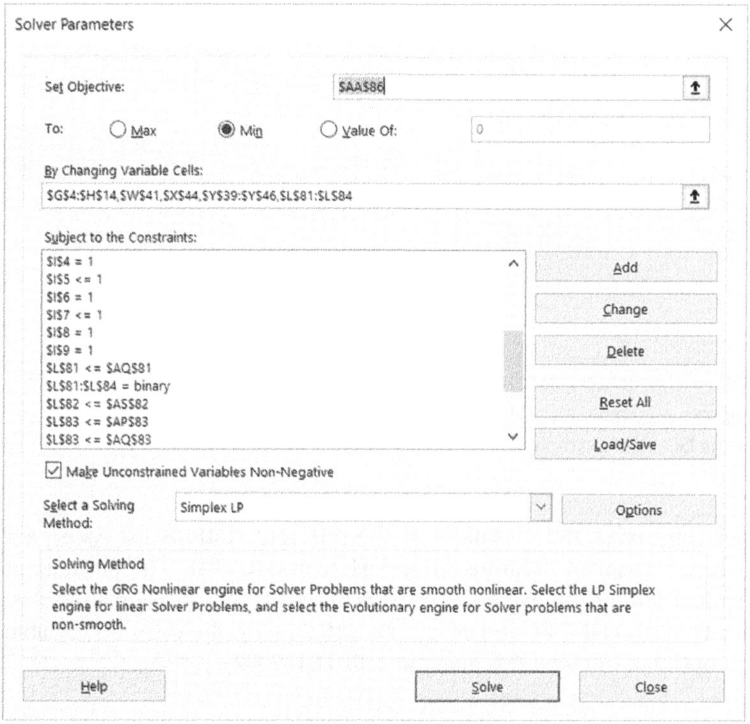

Figure 20.37 The Solver Parameters window for the ISD-IFAM example (part 2).

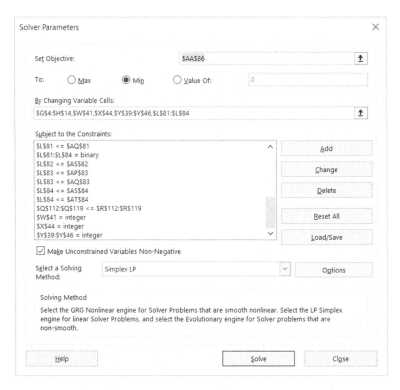

Figure 20.38 The Solver Parameters window for the ISD-IFAM example (part 3).

Next, the different constraints of the problem are defined. These constraints include the coverage constraints for the mandatory flights and the optional flights, the balance constraints, the flight capacity constraints, the itinerary demand constraints, and the constraints that link the optional flights to their corresponding itineraries. This final set of constraints ensure that if any optional flight is removed from the schedule, all itineraries that include this flight are removed from the schedule, and vice versa. Considering such setting for the problem, it is solved as indicated in Figure 20.39.

20.5 Solution Interpretation

The optimal solution of the problem is given in Figures 20.40 through 20.50. As shown in Figure 20.40, the solution recommends that the two optional flights are included in the schedule. The two flights LAX-BOS and BOS-LAX are assigned to fleet type e_1, while the remaining four flights are assigned to fleet type e_2.

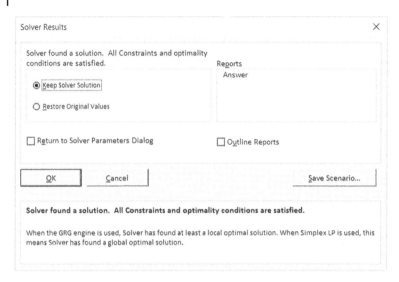

Figure 20.39 The Solver Results window for the ISD-IFAM example.

Figure 20.40 The values of the decision variables at the optimal solution for the ISD-IFAM example.

The solution also requires that one aircraft of each fleet type is to remain overnight at LAX. Figure 20.41 shows that all balance constrains are satisfied.

Figure 20.42 shows the results of the spill variables. As shown in the figure, the model suggests that 150 passengers are spilled from itinerary LAS-BOS to itinerary LAX-ORD-BOS. Another 150 passengers are spilled from itinerary

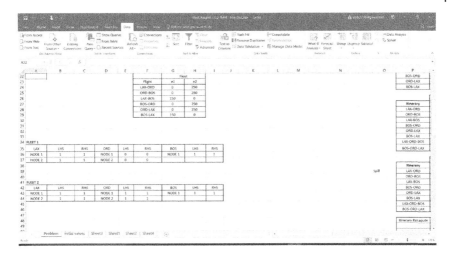

Figure 20.41 The values of the RHS and LHS for the balance constraints at the optimal solution for the ISD-IFAM example.

Figure 20.42 The values of the spill variables at the optimal solution for the ISD-IFAM example.

BOS-LAX to itinerary BOS-ORD-LAX. As shown in Figure 20.43, the recaptured revenue is \$90,000, and the spilled revenue is \$105,000. The recaptured and spilled demand at the flight level is given in Figure 20.44.

Figure 20.45 shows the values of the decision variable Z_q for each optional itinerary. As shown in column L (rows 81–84), $Z_q = 1 \ \forall q$. In other words, all optional itineraries are included in the schedule. The value of the objective

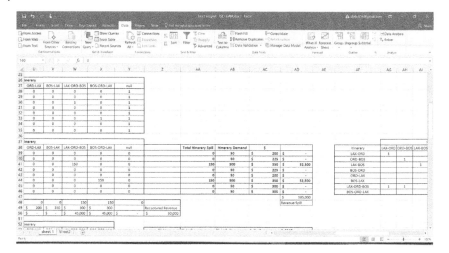

Figure 20.43 The value of the total revenue spill and the total revenue recaptured at the optimal solution for the ISD-IFAM example.

Figure 20.44 The value of demand spill and recapture at the flight level at the optimal solution for the ISD-IFAM example.

function is given in cell AA86, as shown in Figure 20.46, which is equal to $106,200. Figures 20.47 through 20.50 give the different values of the remaining elements of the problem. As shown in these figures, all constraints are satisfied at the optimal values.

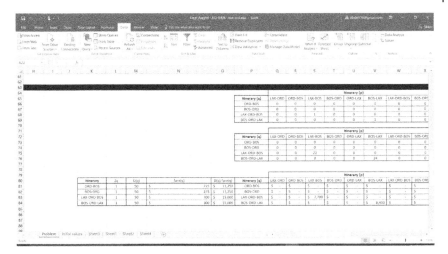

Figure 20.45 The values of the decision variable Z_q for each optional itinerary at the optimal solution for the ISD-IFAM example.

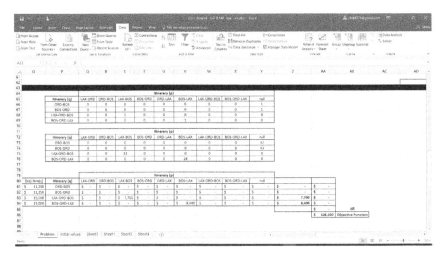

Figure 20.46 The value of the objective function at optimal solution for the ISD-IFAM example.

20.6 Changing the Operations Cost

This example is solved again after significantly increasing the operations cost of the two optional flights ORD-BOS and BOS-ORD for the two fleet types considered in the problem. The new cost values are given in Table 20.5.

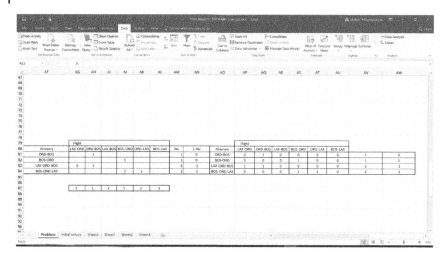

Figure 20.47 The values of the problem elements at optimal solution for the ISD-IFAM example (part 1).

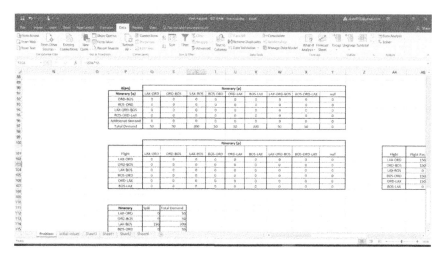

Figure 20.48 The values of the problem elements at optimal solution for the ISD-IFAM example (part 2).

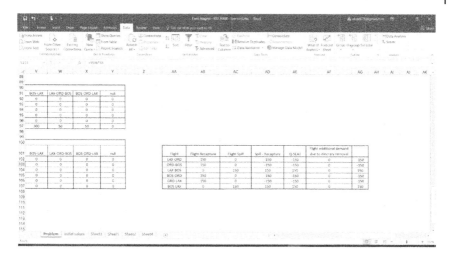

Figure 20.49 The values of the problem elements at optimal solution for the ISD-IFAM example (part 3).

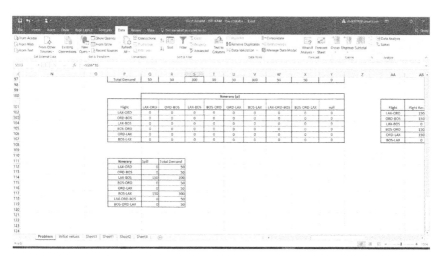

Figure 20.50 The values of the problem elements at optimal solution for the ISD-IFAM example (part 4).

Table 20.5 The new operating cost of each flight for each fleet type for the ISD-IFAM problem.

Flight		Fleet	
		e1 ($)	e2 ($)
LAX-ORD	f1	30,000	16,600
ORD-BOS	f2	95,000	95,000
LAX-BOS	f3	15,500	45,000
BOS-ORD	f4	95,000	95,000
ORD-LAX	f5	42,000	16,600
BOS-LAX	f6	15,500	45,000

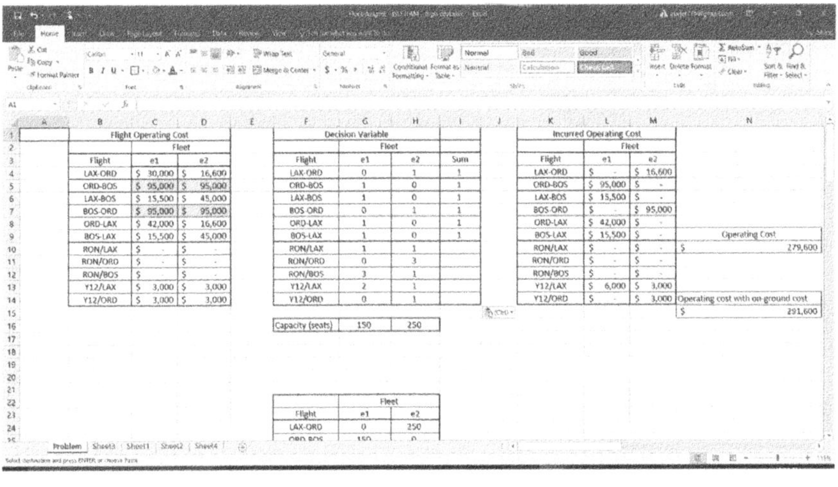

Figure 20.51 The decision variables and formulas for cost calculations for the modified ISD-IFAM example.

As given in the table, the costs of the two flights ORD-BOS and BOS-ORD are increased to $95,000, when assigned to fleet type e_1 or fleet type e_2.

Similar to the previous example, Figures 20.51 through 20.54 show several snapshots for the modified ISD-IFAM problem. Except the assignment costs of the two flights ORD-BOS and BOS-ORD, all initial values are similar to the ones given in the previous example. The problem is again solved using the Excel Solver as shown in Figures 20.55 and 20.56.

Figures 20.57 through 20.68 give the values of the different elements of the problem at optimality. As shown in Figure 20.57, the model suggests that the

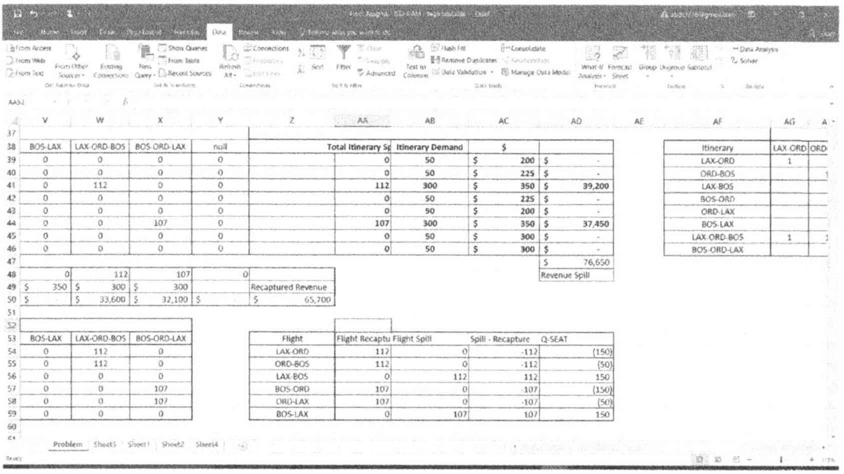

Figure 20.52 The revenue spill and the revenue recaptured for the modified ISD-IFAM example.

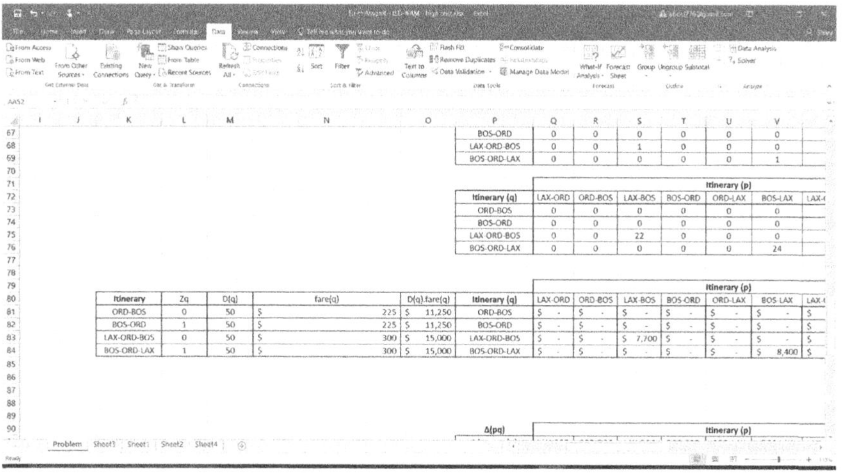

Figure 20.53 The revenue change due to the suggested deletion of the optional itineraries for the modified ISD-IFAM example.

flights ORD-BOS and BOS-ORD are to be removed from the schedule as they are expensive to operate. All four mandatory flights are assigned to fleet type e_2. The solution requires two aircraft of type e_2 to remain overnight at LAX and one aircraft of type e_2 to remain on the ground between the two interconnection nodes at ORD. Figure 20.58 shows that all balance constraints are satisfied at the optimal solution. Figure 20.59 shows the spilled demand among

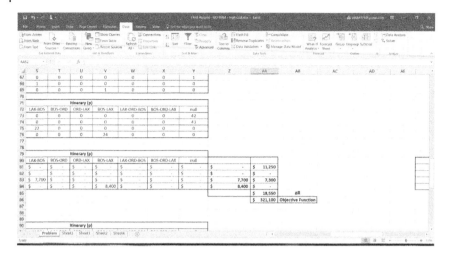

Figure 20.54 The initial value of the objective function for the modified ISD-IFAM example.

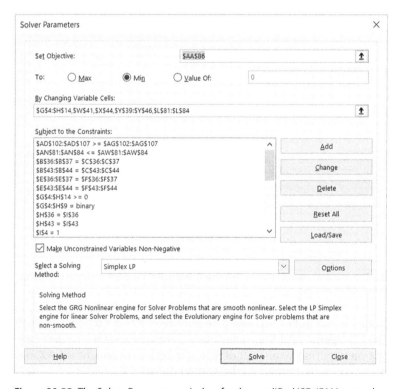

Figure 20.55 The Solver Parameters window for the modified ISD-IFAM example.

Solver Results   ×

Solver found a solution. All Constraints and optimality
conditions are satisfied.

    Reports

    Answer

◉ Keep Solver Solution

○ Restore Original Values

☐ Return to Solver Parameters Dialog   ☐ Outline Reports

  OK   Cancel   Save Scenario...

Solver found a solution. All Constraints and optimality conditions are satisfied.

When the GRG engine is used, Solver has found at least a local optimal solution. When Simplex LP is used, this means Solver has found a global optimal solution.

Figure 20.56 The Solver Results window for the modified ISD-IFAM example.

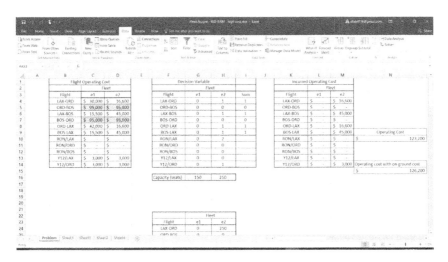

Figure 20.57 The values of the decision variables at the optimal solution for the modified ISD-IFAM example.

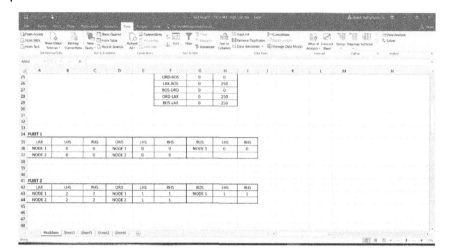

Figure 20.58 The values of the RHS and LHS for the balance constraints at the optimal solution for the modified ISD-IFAM example.

Figure 20.59 The demand spilled between itineraries at the optimal solution for the modified ISD-IFAM example.

itineraries. As shown in the figure, all demand is spilled to the dummy itinerary. The total recaptured revenue is $0, since all demand is spilled to the dummy itinerary. The total spilled revenue is $103,600. The calculation of the spilled revenue is given in Figure 20.60.

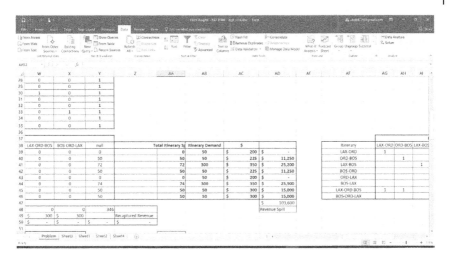

Figure 20.60 The spilled and recaptured revenue at the optimal solution for the modified ISD-IFAM example.

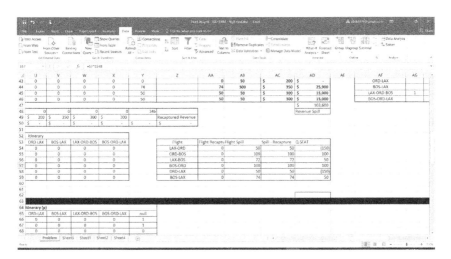

Figure 20.61 The demand spill and recapture at the flight level at the optimal solution for the modified ISD-IFAM example.

Figure 20.61 shows the recaptured and spilled demand at the flight level. Figure 20.62 shows the values of the decision variable Z_q for each optional itinerary. As shown in column L (rows 81–84), $Z_q = 0$, $\forall q$. In other words, all optional itineraries are removed from the schedule. The value of the objective

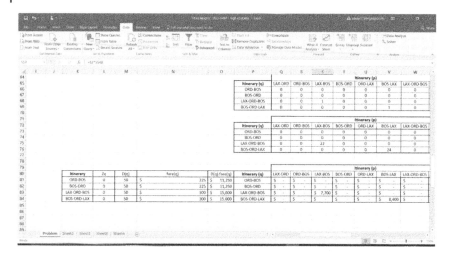

Figure 20.62 The values of the decision variable Z_q for each optional itinerary at the optimal solution for the modified ISD-IFAM example.

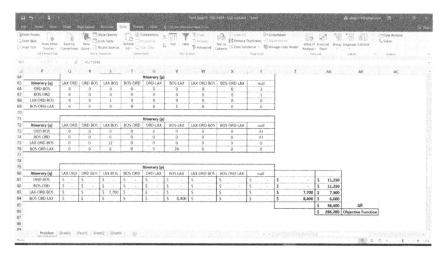

Figure 20.63 The value of the objective function at the optimal solution for the modified ISD-IFAM example.

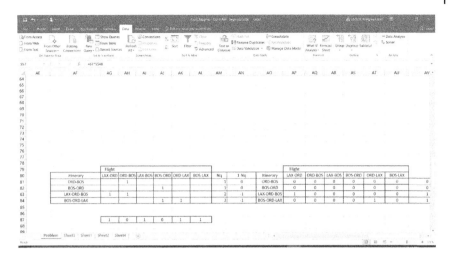

Figure 20.64 The values of the problem elements at optimal solution for the modified ISD-IFAM example (part 1).

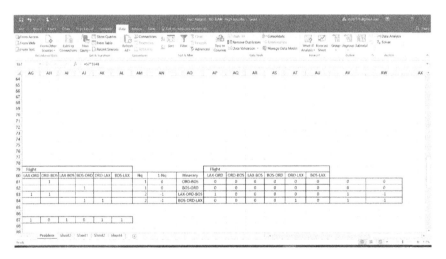

Figure 20.65 The values of the problem elements at optimal solution for the modified ISD-IFAM example (part 2).

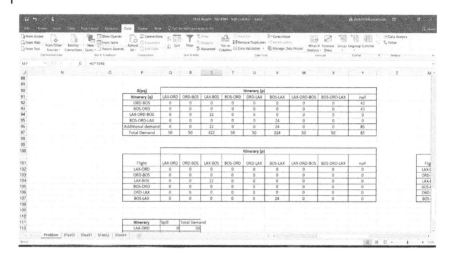

Figure 20.66 The values of the problem elements at optimal solution for the modified ISD-IFAM example (part 3).

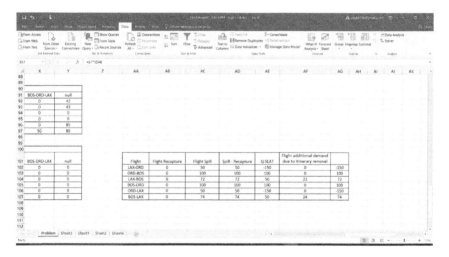

Figure 20.67 The values of the problem elements at optimal solution for the modified ISD-IFAM example (part 4).

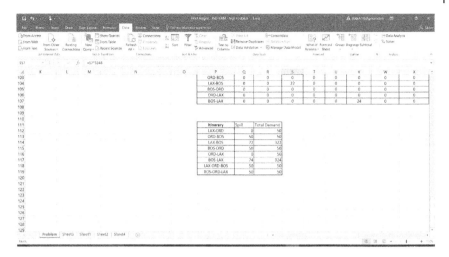

Figure 20.68 The values of the problem elements at optimal solution for the modified ISD-IFAM example (part 5).

function is given in cell AA86, as shown Figure 20.63, which is equal to $266,200. Finally, Figures 20.64 through 20.68 give the different values of the problem elements at optimality. As shown in these figures, all constraints are satisfied at the optimal solution.

Section 7

21

Schedule Robustness

21.1 Introduction

The main objective of the airline planning process is to develop the future flight schedule as well as the schedule of all resources (aircraft, gates, crew, ground staff, etc.) used to operate all scheduled flights. Airlines spend significant effort to develop an optimal schedule for flights and resources in order to maximize its net revenue. However, during the operation phase, the scheduled flights and resources could be subject to several sources of unexpected disruptions. These disruption sources include, for example, adverse weather conditions, aircraft breakdown, unexpected air traffic congestion, and sometimes computer system glitches. These disruptions could result in a reduced system capacity or purging resources needed to operate the flights. Thus, they lead to undesired flight delays and cancellations that jeopardize the optimality of the originally planned schedules.

Disruption scenarios vary by source, location, severity, impact, and recoverability. Large-scale airline operating hundreds of daily flights is encountering these disruptions on a daily basis. In the United States, for example, the average on-time performance rate for scheduled flights is in the lower 70%. The on-time performance rate measures the percentage of flights that arrive at its destination within 15 minutes from its scheduled arrival time. This measure indicates that only three flights out of each four flights operate as planned. Such low on-time arrival performance motivates the airlines to understand the behavior and recoverability of their proposed schedules considering different disruption scenarios.

To deal with irregular operations, one might suggest to reuse the methods adopted to develop the original schedules and develop new schedules considering the encountered disruption scenario. However, this approach is impractical for two reasons. First, the flight scheduling tools used in the planning phase require considerable computational time. Such time is not available during

Airline Network Planning and Scheduling, First Edition. Ahmed Abdelghany and Khaled Abdelghany.
© 2019 John Wiley & Sons, Inc. Published 2019 by John Wiley & Sons, Inc.

operations as schedule recovery decisions need to be made in near real time. Second, rerunning the tools used for schedule planning might result in new schedules for the resources that are completely different from the initially published ones, which is undesired by the airlines.

Major airlines deal with irregular operations following either a reactive or a proactive approach. The reactive approach allows the airlines to respond to operation irregularity by adopting specialized tools specifically designed for that purpose. These tools are used to develop schedules for disruption recovery with minimum deviation from the original schedules and can be generated in near real time. On the other hand, the proactive approach allows the airlines to develop schedules that are robust against disruptions. Two common methods are used to develop the proactive approach. The first method is to add slack times in the schedule to make the schedule less prone to disruptions. The second method is to align flights and resources in a way that makes the schedule easy to recover in case irregular operations occur. This chapter discusses these two proactive methods.

21.2 Less-prone-to-disruptions Schedules: The Concept of Adding Slack Times

21.2.1 Slack in Flight Block Time

As discussed in Chapter 3, the scheduled departure and arrival times of a flight are linked together by the block time planned for this flight. A flight block time is defined as the period between the aircraft pushback time at the departure gate and the arrival time at the destination gate. Defining the scheduled departure time of a flight, the flight's scheduled arrival time is determined by adding the block time of the flight to its scheduled departure time. Alternatively, if the scheduled arrival time of the flight is defined, its scheduled departure time is determined by subtracting the block time from the flight's scheduled arrival time. To link the scheduled departure time and arrival times of the flight, the block time of each flight needs to be accurately estimated.

The flight block time consists of three main components, which are the taxi-out time, the flight (airborne) time, and the taxi-in time. The taxi-out time is the period between the aircraft pushback time at the departure gate and the aircraft wheels-off time at the departure runway. The flight time is the period between the aircraft wheels-off time at the departure runway and the aircraft wheels-on time at the destination runway. The taxi-in time is the period between the aircraft wheels-on time at the arrival runway and the aircraft arrival time at the destination gate. Block times for all flights in the schedule are typically estimated using historical data of similar flights.

It is critical for the airlines to precisely estimate the block time for each flight in the schedule. In case the flight block time is underestimated, the flight is expected to arrive later than its scheduled arrival time, which results in poor on-time performance. On the other hand, if the flight block time is overestimated, the flight arrives ahead of its scheduled arrival time. Overestimating the flight block time leads to inefficiency in aircraft/crew utilization, since the aircraft remains idle for the period between the actual and the scheduled arrival times. Also, this time unnecessarily counts as part of the crew duties. The inefficient utilization of the aircraft and crew contributes to an additional cost to the airlines. It also contributes to loss of potential revenue, as these idle resources could have been allocated to other flights. Figure 21.1 shows an example of a flight with underestimated (*case a*) and overestimated (*case b*) block times, respectively.

Block time is uncertain as it is affected by the congestion level at the airports and by the weather conditions (e.g. wind speed). When estimating the block time, the airline should select a balanced value that minimizes the chance of having a late flight while simultaneously minimizing the chance of having idle resources. Increasing the accuracy of the block-time estimation is crucial for the airlines to improve on-time performance and resources utilization.

In case the airline is critical on improving its on-time performance, it tends to set longer flight block times. Longer flight block time is equivalent to incorporating a slack time in the schedule for the flight. It absorbs any unexpected delay as long as this delay is shorter than the slack time. However, as mentioned above, extending the flight block time has negative impact on the resources utilization. Any unused slack time represents undesirable idle time for the aircraft, crew, and gates.

21.2.2 Slack Time of a Connecting Resource

Another form of the schedule slack is the slack time of the connecting resource (aircraft and crew). The connection time of a resource is the period between the arrival time of its inbound flight and the departure time of the outbound flight. This connection time has to be equal to or greater than the minimum time required for the resource to make the connection. For example, for an aircraft connecting at a station, there is a minimum time required for the inbound aircraft to get ready for the next outbound flight. This time is needed for deplaning passengers, aircraft cleaning, baggage and cargo handling, catering, refueling, and boarding passengers of the outbound flight. This time is determined based on the aircraft type, airport, ground operations, etc. Similarly, connecting crew members are provided a minimum time to connect between two flights. This minimum time is set according to the crew legality rules. If the connection time at a station is scheduled to be greater than the

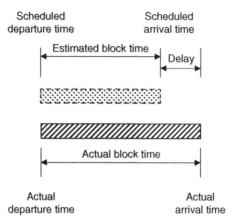

Case a: Underestimating the flight block time

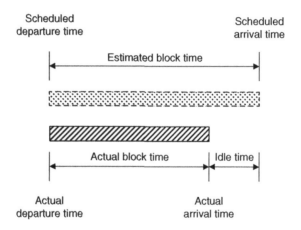

Case b: Overestimating the flight block time

Figure 21.1 The relationship between the actual and the estimated flight block time.

minimum required connection time for a resource, the difference between the scheduled connection time and the minimum required connection (service) time of the resource represents the slack time of that resource (Abdelghany et al. 2004b).

Figure 21.2 shows an example of the slack time of an aircraft connecting between two flights. As shown in the figure, the slack time of the aircraft (connecting resource) is calculated as the difference between the scheduled

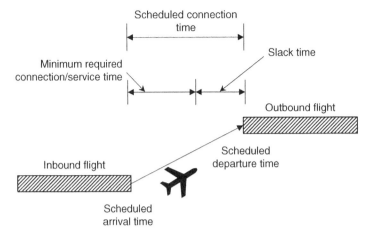

Figure 21.2 Example of a slack time of a connecting aircraft.

connection time and the aircraft minimum required connection (service) time. The connection time is the difference between the departure time of the outbound flight and the arrival time of the inbound flight. The minimum required connection time is a predefined known number for each connecting resource. The slack time of the resource can absorb any delay for the inbound flight of the resource, as long as this delay is less than or equal to the slack time. Thus, the slack time of a resource can be defined as the maximum time this resource can be delayed without affecting its outbound flight.

Similar to the flight block time, when the airline is scheduling its flights and resources, it can select to have longer connections for the resources between their connecting flights. Having longer connections is equivalent to incorporating slack time for these resources. The advantage of having this slack time is that it can absorb any unexpected delay of the resources. On the other hand, in case no delay occurs, this slack time represents an idle time for the resource. This idle time is undesirable as it reduces the utilization of the resources and increases the overall cost.

21.2.3 Slack Time of an Inbound Flight

The slack time of an inbound flight is the maximum time this flight can be delayed without impacting any of the outbound flights that share its resources (i.e. aircraft and crew) (Abdelghany et al. 2004b; Wu 2005; Wong and Tsai 2012). Figure 21.3 shows an example of an inbound flight F0, where its aircraft is connecting to the outbound flight F1 and its crew is connecting to the outbound flight F2. Each connecting resource has a slack time as explained above. As shown in the figure, the aircraft and the crew have slack times

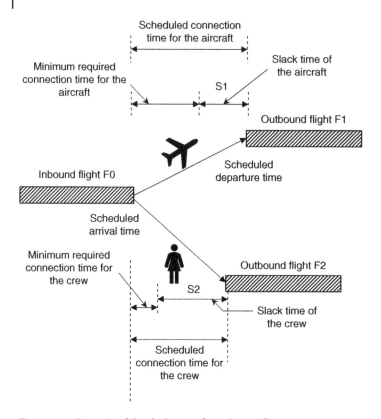

Figure 21.3 Example of the slack time of an inbound flight.

of S1 and S2, respectively. The slack time of the flight is the minimum of the slack times among all its connecting resources. In the example in Figure 21.3, the slack time of flight F0 is the minimum of S1 and S2. This slack time is the maximum amount the flight F0 can be delayed without affecting any downline flight (i.e. F1 and F2). The slack time of the flight is valuable during irregular operations, as it prevents any unexpected delay to propagate through the network. However, during normal operations, a slack time is undesirable as it represents idle time for the connecting resources.

In summary, there are two views on how to consider the slack time, when developing the schedule of flights and resources. The first is in favor of incorporating some slack times for selected flights in the schedule. These slack times absorb any delay that might occur and prevent its downline propagation. The second view is against incorporating slack times, as they deteriorate the resources utilization and increase cost. Overall, it is difficult at the schedule planning stage to identify the appropriate trade-off between enhancing the level of schedule robustness and the associated possible increase in cost.

However, it is argued that the benefits of having robust schedule with minimal delays and cancellations exceed the expected increase in costs. Nonetheless, since such benefits are difficult to quantify at the planning stage, airlines are often reluctant to increase the cost to achieve higher levels of schedule robustness.

21.3 Recoverable Flight Schedules

21.3.1 Background

Several techniques have been suggested over the past decade to make the airline schedule easy to recover in case of irregular operations (Kohl et al. 2007; Barnhart 2009; Clausen et al. 2010). Similar to the previous method, adopting these techniques might reduce resources utilization and increase cost. During irregular operations, airlines try to quickly develop a recovery plan that minimizes the impact of the disruption in terms of flight delays and cancellations. These recovery plans are typically developed and implemented by the airlines operations center (AOC). There are four main recovery actions considered by AOC including (i) flight delays, (ii) flight cancellations, (iii) resources replacement and swapping, and (iv) passenger rebooking.

Delaying an outbound flight to wait for its delayed inbound resources (aircraft and crew) is a common recovery action. Flight delays also happen when delay programs are issued for airports that are under or anticipate adverse weather conditions. For example, in the United States, ground delay programs (GDP) are used to control flow to airports with adverse weather conditions. During GDP, inbound flights are mandated to remain on the ground (i.e. delayed) at their origins, until capacity permits at the destination. Typically, the airline is allocated a new set of landing slots at the destination. The arrival times associated with these slots are later than the originally assigned landing slots. In case an airline has more than one flight impacted by the GDP, it can strategically reassign its flights to the given slots in a way that minimizes the impact of the GDP on its schedule (known as flight substitution). In other words, the airline prioritizes the landing of flights that is expected to result in significant downline disruption in case these flights are delayed. Prioritizing flights to manage a GDP situation requires managing the delay decision of each impacted flight to minimize the overall impact of the GDP on the schedule (Luo and Yu 1997; Chang et al. 2001; Abdelghany et al. 2004a; Ball and Lulli 2004; Ball et al. 2010).

Flight cancellations are decided when there is a reduction in airport capacity due to adverse conditions. Cancellations also occur when one or more resources needed to operate the flight are unexpectedly not available (e.g. aircraft breakdown). During GDP management, less important flights are canceled to free landing slots for more important flights (i.e. higher revenue flights or flights with resources required for other downline flights). Flight cancellation results in stranded resources at the origin of the flight. In addition, if there is an additional downline flight(s) scheduled for these resources, this downline flight misses its resources. Thus, if cancellation occurs, the airline should find a way that enables the stranded resources to catch their downline flights (Jarrah et al. 1993; Cao and Kanafani 1997; Xiong and Hansen 2013).

In case a scheduled resource is unavailable to operate a flight on time (delayed or totally unavailable), this resource can be replaced by or swapped with another similar resource. Aircraft swapping is a common practice to avoid delays. Ideally, aircraft swapping is preferred among aircraft of the same fleet type or similar configuration (seat capacity, range, etc.). Also, some airlines might have a few spare aircrafts deployed at their major hubs to replace any aircraft with unexpected breakdown. Similarly, in case of a crew member delay or no show, airline can swap a duty between crew members. Swapped crew members should have the same qualifications. Airlines usually deploy standby crew members at the hub airports to cover any crew openings. Also, on-call reserve crew can be called from home, if needed. Furthermore, crew members can be flown as passengers (also known as deadhead) to recover crew problems at other destinations. In all cases, if a crew member is assigned a new duty, the new assignment must follow all regulations related to crew workloads. Swapping usually happens at one of the hubs where the assignments of the resources frequently intersect (Abdelghany et al. 2004c, 2008).

Consider the example given in Figure 21.4, which shows the connection of two aircraft at a hub (e.g. Chicago O'Hare International Airport ORD). As shown in *case a*, the first aircraft is scheduled to fly flights F1 and F2, while the second aircraft is scheduled to fly flights F3 and F4. In *case b*, it is assumed that the inbound flights F1 and F3 are delayed. Due to these delays, the aircraft that is scheduled to fly the two flights F3 and F4 cannot fly flight F4 on time. A recovery action is suggested in *case c*, where the schedule of the two aircraft is swapped. In this case, the first aircraft is assigned to the delayed inbound flight F1 followed by flight F4, and the late inbound aircraft of flight F3 is assigned to the outbound flight F2. In this case, the schedule of the outbound flights is not impacted by the delay of the inbound flights.

Figure 21.5 shows the impact of this swapping on the routes of both aircraft. *Case a* shows the routes of the two aircraft before swapping. The first aircraft has a three-day route, which is starting and ending at Denver International Airport (DEN). The second aircraft has a four-day route, which is starting and ending at San Francisco International Airport (SFO). *Case b* shows the routes of the two aircraft after swapping at ORD. The main problem of swapping is the displacement of the resources after swapping. The aircraft that started at DEN ends up at SFO after one day of its scheduled flying hours. The aircraft

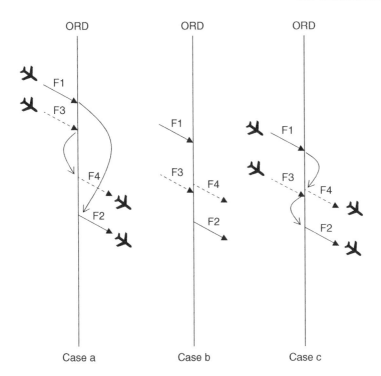

Figure 21.4 Example on swapping of airline resources to recover flights.

that started at SFO ends up at DEN one day before its scheduled flying hours. The same problem is encountered after swapping crew members with different domiciles (home base). A swapped crew member could be ending her/his trip-pair (a terminology used for the crew schedule) at a station that is different from her/his domicile. Thus, other complementary actions have to be considered to make sure that each crew member is returning to her/his domicile.

21.3.2 Station Purity

As mentioned earlier, for airlines that operate more than one fleet type, the objective of the fleet assignment problem is to assign the scheduled flights to the most suitable fleet type. If a station is served by more than one inbound flight, these flights can be assigned to either same or different fleet types. For example, in Figure 21.4, flights F1 and F3 could be assigned to the same fleet type or to two different fleet types. Typically, when these two flights are assigned to the same fleet type, it is easier to swap their resources (e.g. aircraft and crew), if needed. Limiting fleet types serving a station creates opportunities for aircraft and crew swapping to recover disruptions.

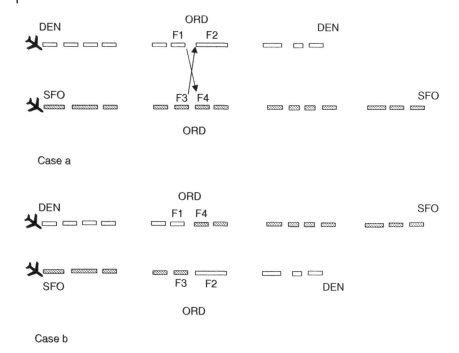

Figure 21.5 The impact of the swapping of the two aircraft on their routes.

To maximize recovery and swapping options at every station, as the airline develops its schedule, it limits fleet types used to serve each station. The number of different fleet types serving a station defines the station's purity level. Station purity ensures that there are opportunities to swap aircraft or crews. However, forcing a certain level of station purity in the fleet assignment solution might deteriorate the quality of the fleet assignment solution, where some flights might not be allocated to the most suitable fleet type in terms of seat capacity (Smith and Johnson 2006).

21.3.3   Short Cancellation Cycles

As mentioned above, flight cancellations are unavoidable during severe irregular operations. The problem with flight cancellation is that the resources of the canceled flight are stranded at the origin, resulting in resource shortage at the destination of the flight. For example, Figure 21.6 shows a hypothetical aircraft route. In *case a*, if flight F1 from ORD to IAD is canceled, the aircraft is stranded at ORD. In addition, there is an aircraft shortage at IAD. In other words, another aircraft is needed to complete the remaining flights in the route. To overcome this problem, when a flight is canceled, the cancellation should include a closed cycle of flights. For example, for the aircraft route given in

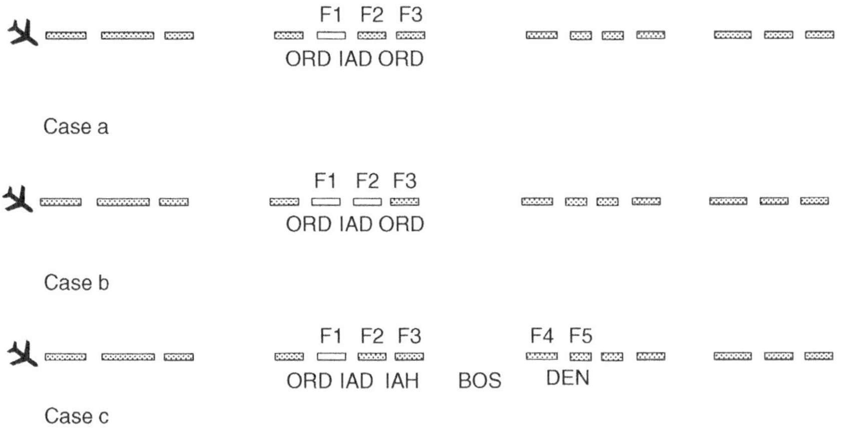

Figure 21.6 The impact of cancellation on aircraft rotation.

Figure 21.6 (*case b*), if flight F1 (ORD-IAD) is canceled, it is highly recommended that flight F2 (IAD-ORD) is also canceled. The cancellation of the cycle ORD-IAD-ORD allows the stranded aircraft at ORD to resume its original route later in the day starting from flight F3.

The problem of recovering from cancellation is more complex in case the schedule of the aircraft lacks these short cycles, as shown in Figure 21.6 (*case c*). Short cycles are proposed for a robust fleet assignment and aircraft routing model that produces a large number of short cycles with a hub connectivity (Rosenberger et al. 2004). Including short cycles in the schedule of the aircraft assists in recovering the schedule after a single cancellation.

21.3.4 Maximizing Swapping Possibility

As explained above, aircraft and crew swapping is an important strategy used to recover schedule disruptions during irregular operations. Swapping is typically beneficial in case one resource is late and the other resource's connection incorporates a large slack. The main problem with resources swapping is the displacement of the resources after swapping. To overcome this problem, Ageeva (2000) proposes an aircraft routing model that maximizes the number of times different aircraft routes intersect. When aircraft intersect more than once, this provides an opportunity for resources to swap routes and return to their original route at some point in the future. For example, consider the two aircraft routes given in Figure 21.7 (*case a*). These two routes meet at ORD (on the second day) and LAX (on the third day). If the two aircraft are swapped at ORD, they can be swapped again at LAX. As such, each aircraft returns to its original route. Figure 21.7 (*case b*) shows the new routes of the two aircraft after the two-step swapping is completed at ORD and LAX. As shown in the figure, both aircraft continue their original routes and return to their designated stations.

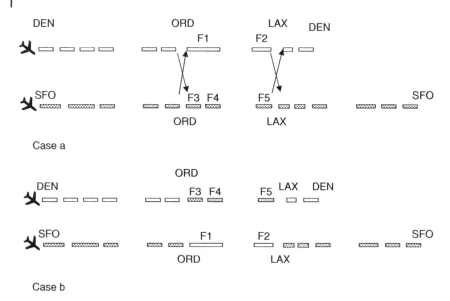

Figure 21.7 Multiple swapping opportunities for the aircraft.

21.3.5 Allocating Standby and Reserve Crew

The schedule recovery problem occasionally relies on standby and reserve crew to substitute for any crew shortage at the hubs of the airline. Standby crew are ready at the airport (i.e. hub) for any assignment. Standby crew are typically assigned a four-hour duty period at the airport. They might receive any new assignment at any time during this period; otherwise they return home when this standby period ends. Reserve crew, on the other hand, are on call at home for future assignments. If called, reserve crews would need more time to get ready for transportation to the airport. Thus, they are called at least four to five hours before their assigned duties. It is important for the airline to optimize the number of standby and reserve crew at the different stations. In case the airline operates more than one fleet type, the number of standby and reserve crews for each fleet type has to be optimized. In addition, the standby duty periods for the standby crew at the different stations have to be linked to the types of fleet arriving/departing within this duty period. For example, if the airline schedule includes departures/arrivals of flights that are assigned to a certain fleet type in the morning at a given airport, the airline has to assign morning duties for standby crew that are qualified to fly this fleet type. The number of standby and reserve crew should be adjusted for seasons with high disruptions due to adverse weather. For example, if more schedule disruptions are expected during the winter season at any hub, the airline should schedule more standby and reserve crew to recover these seasonal expected disruptions.

References

Abdelghany, A. and Abdelghany, K. (2007a). Evaluating airlines ticket distribution strategies: a simulation-based approach. *International Journal of Revenue Management* 1(3): 231–246.

Abdelghany, A. and Abdelghany, K. (2007b). Modelling air-carrier's portfolio of business travel. *Journal of Revenue and Pricing Management* 6(1): 51–63.

Abdelghany, A. and Abdelghany, K. (2008). A micro-simulation approach for Airline Competition Analysis and Demand Modelling. *International Journal of Revenue Management* 2(3): 287–306.

Abdelghany, A. and Abdelghany, K. (2012). *Modeling Applications in the Airline Industry*. Ashgate Publishing, Ltd.

Abdelghany, A., Abdelghany, K., and Ekollu, G. (2004a). A genetic algorithm approach for ground delay program management: the airlines' side of the problem. *Air Traffic Control Quarterly* 12(1): 53–74.

Abdelghany, K.F., Shah, S.S., Raina, S., and Abdelghany, A.F. (2004b). A model for projecting flight delays during irregular operation conditions. *Journal of Air Transport Management* 10(6): 385–394.

Abdelghany, A., Ekollu, G., Narasimhan, R., and Abdelghany, K. (2004c). A proactive crew recovery decision support tool for commercial airlines during irregular operations. *Annals of Operations Research* 127(1): 309–331.

Abdelghany, A., Abdelghany, K., and Narasimhan, R. (2006). Scheduling baggage-handling facilities in congested airports. *Journal of Air Transport Management* 12(2): 76–81.

Abdelghany, K.F., Abdelghany, A.F., and Ekollu, G. (2008). An integrated decision support tool for airlines schedule recovery during irregular operations. *European Journal of Operational Research* 185(2): 825–848.

Abdelghany, A., Sattayalekha, W., and Abdelghany, K. (2009). On airlines code-share optimisation: a modelling framework and analysis. *International Journal of Revenue Management* 3(3): 307–330.

Airline Network Planning and Scheduling, First Edition. Ahmed Abdelghany and Khaled Abdelghany.
© 2019 John Wiley & Sons, Inc. Published 2019 by John Wiley & Sons, Inc.

Abdelghany, A., Abdelghany, K., and Azadian, F. (2017). Airline flight schedule planning under competition. *Computers & Operations Research* 87: 20–39.

Ageeva, Y. (2000). Approaches to incorporating robustness into airline scheduling. Doctoral dissertation, Massachusetts Institute of Technology, Cambridge, MA.

Agresti, A. (2003). Logit models for multinomial responses. In: *Categorical Data Analysis*, 2e, 267–313. Wiley.

Alamdari, F. and Mason, K. (2006). The future of airline distribution. *Journal of Air Transport Management* 12(3): 122–134.

Alderighi, M., Cento, A., Nijkamp, P., and Rietveld, P. (2005). Network competition – the coexistence of hub-and-spoke and point-to-point systems. *Journal of Air Transport Management* 11(5): 328–334.

Alderighi, M., Cento, A., Nijkamp, P., and Rietveld, P. (2007). Assessment of new hub-and-spoke and point-to-point airline network configurations. *Transport Reviews* 27(5): 529–549.

Armstrong, J.S. (ed.) (2001). *Principles of Forecasting: A Handbook for Researchers and Practitioners*, vol. 30. Springer Science & Business Media.

Backx, M., Carney, M., and Gedajlovic, E. (2002). Public, private and mixed ownership and the performance of international airlines. *Journal of Air Transport Management* 8(4): 213–220.

Baker, K.R. (2012). *Optimization Modeling with Spreadsheets*. Wiley.

Ball, M.O. and Lulli, G. (2004). Ground delay programs: optimizing over the included flight set based on distance. *Air Traffic Control Quarterly* 12(1): 1–25.

Ball, M.O., Hoffman, R., and Mukherjee, A. (2010). Ground delay program planning under uncertainty based on the ration-by-distance principle. *Transportation Science* 44(1): 1–14.

Barnhart, C. (2009). Irregular operations: schedule recovery and robustness. In: *The Global Airline Industry*, vol. 23 (ed. P. Belobaba), 253. Wiley.

Barnhart, C., Boland, N.L., Clarke, L.W. et al. (1998). Flight string models for aircraft fleeting and routing. *Transportation Science* 32(3): 208–220.

Barnhart, C., Kniker, T.S., and Lohatepanont, M. (2002). Itinerary-based airline fleet assignment. *Transportation Science* 36(2): 199–217.

Barnhart, C., Belobaba, P., and Odoni, A.R. (2003a). Applications of operations research in the air transport industry. *Transportation Science* 37(4): 368–391.

Barnhart, C., Cohn, A.M., Johnson, E.L. et al. (2003b). Airline crew scheduling. In: *Handbook of Transportation Science* (ed. R.W. Hall), 517–560. Springer US.

Barrett, S.D. (2001). Market entry to the full-service airline market – a case study from the deregulated European aviation sector. *Journal of Air Transport Management* 7(3): 189–193.

Bazargan, M. (2016). *Airline Operations and Scheduling*. Routledge.

Belobaba, P., Odoni, A., and Barnhart, C. (eds.) (2015). *The Global Airline Industry*. Wiley.

Ben-Akiva, M.E. and Lerman, S.R. (1985). *Discrete Choice Analysis: Theory and Application to Travel Demand*, vol. 9. MIT Press.

Bihr, R.A. (1990). A conceptual solution to the aircraft gate assignment problem using 0, 1 linear programming. *Computers & Industrial Engineering* 19(1–4): 280–284.

de Boer, S.V., Freling, R., and Piersma, N. (2002). Mathematical programming for network revenue management revisited. *European Journal of Operational Research* 137(1): 72–92.

Borenstein, S. (1989). Hubs and high fares: dominance and market power in the US airline industry. *The RAND Journal of Economics* 20: 344–365.

Borenstein, S. and Netz, J. (1999). Why do all the flights leave at 8 am?: Competition and departure-time differentiation in airline markets. *International Journal of Industrial Organization* 17(5): 611–640.

Bowen, J.T. (2010). *The Economic Geography of Air Transportation: Space, Time, and the Freedom of the Sky*. Routledge.

Brueckner, J.K. (2001). The economics of international codesharing: an analysis of airline alliances. *International Journal of Industrial Organization* 19(10): 1475–1498.

Brueckner, J.K. and Luo, D. (2014). Measuring strategic firm interaction in product-quality choices: the case of airline flight frequency. *Economics of Transportation* 3(1): 102–115.

Brueckner, J.K. and Zhang, Y. (2001). A model of scheduling in airline networks: how a hub-and-spoke system affects flight frequency, fares and welfare. *Journal of Transport Economics and Policy (JTEP)* 35(2): 195–222.

Brueckner, J.K., Dyer, N.J., and Spiller, P.T. (1992). Fare determination in airline hub-and-spoke networks. *The Rand Journal of Economics* 23: 309–333.

Bryan, D.L. and O'kelly, M.E. (1999). Hub-and-spoke networks in air transportation: an analytical review. *Journal of Regional Science* 39(2): 275–295.

Buck, S. and Lei, Z. (2004). Charter airlines: have they a future? *Tourism and Hospitality Research* 5(1): 72–78.

Burghouwt, G. (2007). *Airline Network Development in Europe and Its Implications for Airport Planning*. Ashgate Publishing, Ltd.

Burghouwt, G., Hakfoort, J., and van Eck, J.R. (2003). The spatial configuration of airline networks in Europe. *Journal of Air Transport Management* 9(5): 309–323.

Button, K. (2009). The impact of US–EU "Open Skies" agreement on airline market structures and airline networks. *Journal of Air Transport Management* 15(2): 59–71.

Button, K. and Taylor, S. (2000). International air transportation and economic development. *Journal of Air Transport Management* 6(4): 209–222.

Cao, J.M. and Kanafani, A. (1997). Real-time decision support for integration of airline flight cancellations and delays part I: mathematical formulation. *Transportation Planning and Technology* 20(3): 183–199.

Caves, D.W., Christensen, L.R., and Tretheway, M.W. (1984). Economies of density versus economies of scale: why trunk and local service airline costs differ. *The RAND Journal of Economics* 15(4): 471–489.

Chang, K., Howard, K., Oiesen, R. et al. (2001). Enhancements to the FAA ground-delay program under collaborative decision making. *Interfaces* 31(1): 57–76.

Chang, Y.C., Williams, G., and Hsu, C.J. (2004). The evolution of airline ownership and control provisions. *Journal of Air Transport Management* 10(3): 161–172.

Clarke, L.W., Hane, C.A., Johnson, E.L., and Nemhauser, G.L. (1996). Maintenance and crew considerations in fleet assignment. *Transportation Science* 30(3): 249–260.

Clarke, L., Johnson, E., Nemhauser, G., and Zhu, Z. (1997). The aircraft rotation problem. *Annals of Operations Research* 69: 33–46.

Clausen, T. and Pisinger, D. (2010). *Airport Ground Staff Scheduling*. Technical University of Denmark (DTU).

Clausen, J., Larsen, A., Larsen, J., and Rezanova, N.J. (2010). Disruption management in the airline industry – concepts, models and methods. *Computers & Operations Research* 37(5): 809–821.

Coldren, G.M., Koppelman, F.S., Kasturirangan, K., and Mukherjee, A. (2003). Modeling aggregate air-travel itinerary shares: logit model development at a major US airline. *Journal of Air Transport Management* 9(6): 361–369.

Condorelli, D. (2007). Efficient and equitable airport slot allocation. *Rivista di politica economica* 1: 81–104.

Daniel, J.I. and Harback, K.T. (2008). (When) Do hub airlines internalize their self-imposed congestion delays? *Journal of Urban Economics* 63(2): 583–612.

De Boe, V. (2005). Allocation of slots to airlines: what role for competition law? *Competition and Regulation in Network Industries* 6(4): 293–332.

De Neufville, R. (1994). The baggage system at Denver: prospects and lessons. *Journal of Air Transport Management* 1(4): 229–236.

De Wit, J. and Burghouwt, G. (2008). Slot allocation and use at hub airports, perspectives for secondary trading. *European Journal of Transport and Infrastructure Research* 8(2): 147–164.

Derudder, B., Devriendt, L., and Witlox, F. (2007). Flying where you don't want to go: an empirical analysis of hubs in the global airline network. *Tijdschrift voor economische en sociale geografie* 98(3): 307–324.

Desaulniers, G., Desrosiers, J., Dumas, Y. et al. (1997). Daily aircraft routing and scheduling. *Management Science* 43(6): 841–855.

Devriendt, L., Burghouwt, G., Derudder, B. et al. (2009). Calculating load factors for the transatlantic airline market using supply and demand data – a note on the identification of gaps in the available airline statistics. *Journal of Air Transport Management* 15(6): 337–343.

Ding, H., Lim, A., Rodrigues, B., and Zhu, Y. (2005). The over-constrained airport gate assignment problem. *Computers & Operations Research* 32(7): 1867–1880.

Dobruszkes, F. (2006). An analysis of European low-cost airlines and their networks. *Journal of Transport Geography* 14(4): 249–264.

Dobson, G. and Lederer, P.J. (1993). Airline scheduling and routing in a hub-and-spoke system. *Transportation Science* 27(3): 281–297.

Doganis, R. (2002). *Flying off Course: The Economics of International Airlines*. Psychology Press.

Dolnicar, S., Grabler, K., Grün, B., and Kulnig, A. (2011). Key drivers of airline loyalty. *Tourism Management* 32(5): 1020–1026.

Dorndorf, U., Drexl, A., Nikulin, Y., and Pesch, E. (2007). Flight gate scheduling: state-of-the-art and recent developments. *Omega* 35(3): 326–334.

Elhedhli, S. and Hu, F.X. (2005). Hub-and-spoke network design with congestion. *Computers & Operations Research* 32(6): 1615–1632.

Fageda, X. and Flores-Fillol, R. (2012). Air services on thin routes: regional versus low-cost airlines. *Regional Science and Urban Economics* 42(4): 702–714.

Fageda, X., Suau-Sanchez, P., and Mason, K.J. (2015). The evolving low-cost business model: network implications of fare bundling and connecting flights in Europe. *Journal of Air Transport Management* 42: 289–296.

Fan, T., Vigeant-Langlois, L., Geissler, C. et al. (2001). Evolution of global airline strategic alliance and consolidation in the twenty-first century. *Journal of Air Transport Management* 7(6): 349–360.

Farkas, A. (1996). The influence of network effects and yield management on airline fleet assignment decisions. Ph.D. dissertation, Massachusetts Institute of Technology, Cambridge, MA.

Forbes, S.J. and Lederman, M. (2007). The role of regional airlines in the US airline industry. *Advances in Airline Economics* 2: 193–208.

Fox, J. (2015). *Applied Regression Analysis and Generalized Linear Models*. Sage.

French, T. (1995). Transport: charter airlines in Europe. *Travel & Tourism Analyst* 4: 4–19.

French, T. (1998). Europe's regional airlines. *Travel & Tourism Analyst* 5: 1–18.

Fu, X., Oum, T.H., and Zhang, A. (2010). Air transport liberalization and its impacts on airline competition and air passenger traffic. *Transportation Journal* 49: 24–41.

Glover, F., Glover, R., Lorenzo, J., and McMillan, C. (1982). The passenger-mix problem in the scheduled airlines. *Interfaces* 12(3): 73–80.

Goetz, A.R. and Vowles, T.M. (2009). The good, the bad, and the ugly: 30 years of US airline deregulation. *Journal of Transport Geography* 17(4): 251–263.

Graham, B. (1997). Regional airline services in the liberalized European Union single aviation market. *Journal of Air Transport Management* 3(4): 227–238.

Gu, Z., Johnson, E.L., Nemhauser, G.L., and Yinhua, W. (1994). Some properties of the fleet assignment problem. *Operations Research Letters* 15(2): 59–71.

Gudmundsson, S.V. and Rhoades, D.L. (2001). Airline alliance survival analysis: typology, strategy and duration. *Transport Policy* 8(3): 209–218.

Hane, C.A., Barnhart, C., Johnson, E.L. et al. (1995). The fleet assignment problem: solving a large-scale integer program. *Mathematical Programming* 70(1): 211–232.

Hardy, M.A. and Bryman, A. (eds.) (2004). *Handbook of Data Analysis*. Sage.

Hassan, A., Abdelghany, K., and Abdelghany, A. (2009). Modeling Framework for Airline Competition in the US Domestic Network. *Transportation Research Record: Journal of the Transportation Research Board* 2106: 47–56.

Heshmati, A. and Kim, J. (2016). Introduction to efficiency and competitiveness of International Airlines. In: *Efficiency and Competitiveness of International Airlines*, 1–14. Singapore: Springer.

Holloran, T.J. and Byrn, J.E. (1986). United Airlines station manpower planning system. *Interfaces* 16(1): 39–50.

Horner, M.W. and O'Kelly, M.E. (2001). Embedding economies of scale concepts for hub network design. *Journal of Transport Geography* 9(4): 255–265.

IATA (2017). *Worldwide Slot Guidelines*, 8e. International Air Transport Association – Joint Advisory Group.

Ito, H. and Lee, D. (2007). Domestic code sharing, alliances, and airfares in the US airline industry. *The Journal of Law and Economics* 50(2): 355–380.

Jarrah, A.I., Yu, G., Krishnamurthy, N., and Rakshit, A. (1993). A decision support framework for airline flight cancellations and delays. *Transportation Science* 27(3): 266–280.

Jiang, H. and Barnhart, C. (2013). Robust airline schedule design in a dynamic scheduling environment. *Computers & Operations Research* 40(3): 831–840.

Karlaftis, M.G. and Papastavrou, J.D. (1998). Demand characteristics for charter air-travel. *International Journal of Transport Economics. Rivista internazionale di economia dei trasporti* XXV(3): 19–35.

Kleymann, B. and Seristö, H. (2001). Levels of airline alliance membership: balancing risks and benefits. *Journal of Air Transport Management* 7(5): 303–310.

Kohl, N. and Karisch, S.E. (2004). Airline crew rostering: problem types, modeling, and optimization. *Annals of Operations Research* 127(1): 223–257.

Kohl, N., Larsen, A., Larsen, J. et al. (2007). Airline disruption management—perspectives, experiences and outlook. *Journal of Air Transport Management* 13(3): 149–162.

Lee, D. and Luengo-Prado, M.J. (2005). The impact of passenger mix on reported "hub premiums" in the US airline industry. *Southern Economic Journal* 72: 372–394.

Levine, M.E. (2011). Regulation and the nature of the firm: the case of US regional airlines. *The Journal of Law and Economics* 54(S4): S229–S248.

Lohatepanont, M. and Barnhart, C. (2004). Airline schedule planning: integrated models and algorithms for schedule design and fleet assignment. *Transportation Science* 38(1): 19–32.

Long, J.S. and Freese, J. (2006). *Regression Models for Categorical Dependent Variables Using Stata*. Stata Press.

Lordan, O., Sallan, J.M., and Simo, P. (2014). Study of the topology and robustness of airline route networks from the complex network approach: a survey and research agenda. *Journal of Transport Geography* 37: 112–120.

Lordan, O., Sallan, J.M., Simo, P., and Gonzalez-Prieto, D. (2015). Robustness of airline alliance route networks. *Communications in Nonlinear Science and Numerical Simulation* 22(1): 587–595.

Luo, S. and Yu, G. (1997). On the airline schedule perturbation problem caused by the ground delay program. *Transportation Science* 31(4): 298–311.

Madas, M.A. and Zografos, K.G. (2008). Airport capacity vs. demand: mismatch or mismanagement? *Transportation Research Part A: Policy and Practice* 42(1): 203–226.

Marti, L., Puertas, R., and Calafat, C. (2015). Efficiency of airlines: hub and spoke versus point-to-point. *Journal of economic studies* 42(1): 157–166.

Maruyama, G. (1997). *Basics of Structural Equation Modeling.* Sage.

Mimouni-Chaabane, A. and Volle, P. (2010). Perceived benefits of loyalty programs: scale development and implications for relational strategies. *Journal of Business Research* 63(1): 32–37.

Montgomery, D.C., Peck, E.A., and Vining, G.G. (2012). *Introduction to Linear Regression Analysis*, vol. 821. Wiley.

Morandi, V., Malighetti, P., Paleari, S., and Redondi, R. (2015). Codesharing agreements by low-cost carriers: an explorative analysis. *Journal of Air Transport Management* 42: 184–191.

Nero, G. (1999). A note on the competitive advantage of large hub-and-spoke networks. *Transportation Research Part E: Logistics and Transportation Review* 35(4): 225–239.

O'Connell, J.F. and Williams, G. (2005). Passengers' perceptions of low cost airlines and full service carriers: a case study involving Ryanair, Aer Lingus, Air Asia and Malaysia Airlines. *Journal of Air Transport Management* 11(4): 259–272.

Osborne, J.W. (ed.) (2008). *Best Practices in Quantitative Methods.* Sage.

Oum, T.H. and Tretheway, M.W. (1990). Airline hub and spoke systems. *Journal of the Transportation Research Forum* 30(2): 380–393.

Pai, V. (2010). On the factors that affect airline flight frequency and aircraft size. *Journal of Air Transport Management* 16(4): 169–177.

Park, J.H. and Zhang, A. (2000). An empirical analysis of global airline alliances: cases in North Atlantic markets. *Review of industrial organization* 16(4): 367–384.

Pitfield, D.E., Caves, R.E., and Quddus, M.A. (2010). Airline strategies for aircraft size and airline frequency with changing demand and competition: a simultaneous-equations approach for traffic on the north Atlantic. *Journal of Air Transport Management* 16(3): 151–158.

Proussaloglou, K. and Koppelman, F. (1995). Air carrier demand. *Transportation* 22(4): 371–388.

Reiss, P.C. and Spiller, P.T. (1989). Competition and entry in small airline markets. *The Journal of Law and Economics* 32(2, Part 2): S179–S202.

Rexing, B., Barnhart, C., Kniker, T. et al. (2000). Airline fleet assignment with time windows. *Transportation Science* 34(1): 1–20.

Reynolds-Feighan, A.J. (1998). The impact of US airline deregulation on airport traffic patterns. *Geographical Analysis* 30(3): 234–253.

Rosenberger, J.M., Schaefer, A.J., Goldsman, D. et al. (2002). A stochastic model of airline operations. *Transportation Science* 36(4): 357–377.

Rosenberger, J.M., Johnson, E.L., and Nemhauser, G.L. (2004). A robust fleet-assignment model with hub isolation and short cycles. *Transportation Science* 38(3): 357–368.

Rushmeier, R.A. and Kontogiorgis, S.A. (1997). Advances in the optimization of airline fleet assignment. *Transportation Science* 31(2): 159–169.

Santos, G. and Robin, M. (2010). Determinants of delays at European airports. *Transportation Research Part B: Methodological* 44(3): 392–403.

Seredyński, A., Rothlauf, F., and Grosche, T. (2014). An airline connection builder using maximum connection lag with greedy parameter selection. *Journal of Air Transport Management* 36: 120–128.

Shaw, S.L. (1993). Hub structures of major US passenger airlines. *Journal of Transport Geography* 1(1): 47–58.

Shaw, S. (2016). *Airline Marketing and Management*. Routledge.

Sherali, H.D., Bish, E.K., and Zhu, X. (2006). Airline fleet assignment concepts, models, and algorithms. *European Journal of Operational Research* 172(1): 1–30.

Slack, B. (1999). Satellite terminals: a local solution to hub congestion? *Journal of Transport Geography* 7(4): 241–246.

Smith, B.C. and Johnson, E.L. (2006). Robust airline fleet assignment: imposing station purity using station decomposition. *Transportation Science* 40(4): 497–516.

Sohoni, M., Lee, Y.C., and Klabjan, D. (2011). Robust airline scheduling under block-time uncertainty. *Transportation Science* 45(4): 451–464.

Stavins, J. (2001). Price discrimination in the airline market: the effect of market concentration. *The Review of Economics and Statistics* 83(1): 200–202.

Stolletz, R. (2010). Operational workforce planning for check-in counters at airports. *Transportation Research Part E: Logistics and Transportation Review* 46(3): 414–425.

Subramanian, R., Scheff, R.P., Quillinan, J.D. et al. (1994). Coldstart: fleet assignment at delta air lines. *Interfaces* 24(1): 104–120.

Teodorovic, D. (2017). *Airline Operations Research*, vol. 3. Routledge.

Tiernan, S., Rhoades, D., and Waguespack, B. (2008). Airline alliance service quality performance—an analysis of US and EU member airlines. *Journal of Air Transport Management* 14(2): 99–102.

Toh, R.S. and Higgins, R.G. (1985). The impact of hub and spoke network centralization and route monopoly on domestic airline profitability. *Transportation Journal* 24: 16–27.

Train, K.E. (2009). *Discrete Choice Methods with Simulation*. Cambridge University Press.

Truitt, L.J. and Haynes, R. (1994). Evaluating service quality and productivity in the regional airline industry. *Transportation Journal* 33: 21–32.

Tsoukalas, G., Belobaba, P., and Swelbar, W. (2008). Cost convergence in the US airline industry: an analysis of unit costs 1995–2006. *Journal of Air Transport Management* 14(4): 179–187.

Vidović, A., Štimac, I., and Vince, D. (2013). Development of business models of low-cost airlines. *The International Journal for Traffic and Transport Engineering* 3(1): 69–81.

Vowles, T.M. and Lück, M. (2016). Low cost carriers in the USA and Canada. In: *The Low Cost Carrier Worldwide* (ed. S. Gross and M. Lück), 61. Taylor & Francis.

Wang, S.W. (2014). Do global airline alliances influence the passenger's purchase decision? *Journal of Air Transport Management* 37: 53–59.

Weber, M. and Dinwoodie, J. (2000). Fifth freedoms and airline alliances. The role of fifth freedom traffic in an understanding of airline alliances. *Journal of Air Transport Management* 6(1): 51–60.

Wei, W. and Hansen, M. (2005). Impact of aircraft size and seat availability on airlines' demand and market share in duopoly markets. *Transportation Research Part E: Logistics and Transportation Review* 41(4): 315–327.

Williams, G. (2001). Will Europe's charter carriers be replaced by "no-frills" scheduled airlines? *Journal of Air Transport Management* 7(5): 277–286.

Williams, G. (2017). *The Airline Industry and the Impact of Deregulation*. Routledge.

Witt, S.F. and Witt, C.A. (1995). Forecasting tourism demand: a review of empirical research. *International Journal of Forecasting* 11(3): 447–475.

Wong, J.T. and Tsai, S.C. (2012). A survival model for flight delay propagation. *Journal of Air Transport Management* 23: 5–11.

Wu, C.L. (2005). Inherent delays and operational reliability of airline schedules. *Journal of Air Transport Management* 11(4): 273–282.

Wu, C.L. and Caves, R.E. (2000). Aircraft operational costs and turnaround efficiency at airports. *Journal of Air Transport Management* 6(4): 201–208.

Wu, C.L. and Caves, R.E. (2004). Modelling and optimization of aircraft turnaround time at an airport. *Transportation Planning and Technology* 27(1): 47–66.

Xiong, J. and Hansen, M. (2013). Modelling airline flight cancellation decisions. *Transportation Research Part E: Logistics and Transportation Review* 56: 64–80.

Zhang, A. and Zhang, Y. (2002). Issues on liberalization of air cargo services in international aviation. *Journal of Air Transport Management* 8(5): 275–287.

Zou, L. and Chen, X. (2017). The effect of code-sharing alliances on airline profitability. *Journal of Air Transport Management* 58: 50–57.

Zou, B. and Hansen, M. (2012). Impact of operational performance on air carrier cost structure: evidence from US airlines. *Transportation Research Part E: Logistics and Transportation Review* 48(5): 1032–1048.

Index

Airline Network Planning and Scheduling, First Edition. Ahmed Abdelghany
and Khaled Abdelghany.
© 2019 John Wiley & Sons, Inc. Published 2019 by John Wiley & Sons, Inc.

Printed and bound by CPI Group (UK) Ltd, Croydon, CR0 4YY

16/04/2025

14658520-0001